DISEASES OF HORTICULTURAL CROPS

Diagnosis and Management

Volume 1: Fruit Crops

Diseases of Horticultural Crops: Diagnosis and Management, 4 Volume Set:

Volume 1: Fruit Crops

Volume 2: Vegetable Crops

Volume 3: Ornamental Plants and Spice Crops

Volume 4: Important Plantation Crops, Medicinal Crops, and Mushrooms

Innovations in Horticultural Science

DISEASES OF HORTICULTURAL CROPS

Diagnosis and Management

Volume 1: Fruit Crops

Edited by
J. N. Srivastava, PhD
A. K. Singh, PhD

AAP | APPLE
ACADEMIC
PRESS

First edition published 2022

Apple Academic Press Inc.
1265 Goldenrod Circle, NE,
Palm Bay, FL 32905 USA

4164 Lakeshore Road, Burlington,
ON, L7L 1A4 Canada

CRC Press
6000 Broken Sound Parkway NW,
Suite 300, Boca Raton, FL 33487-2742 USA

2 Park Square, Milton Park,
Abingdon, Oxon, OX14 4RN UK

Library and Archives Canada Cataloguing in Publication

Title: Diseases of horticultural crops : diagnosis and management / edited by J.N. Srivastava, PhD, A.K. Singh, PhD.

Names: Srivastava, J. N. (Plant pathologist), editor. | Singh, A. K. (Plant pathologist), editor.

Series: Innovations in horticultural science.

Description: First edition. | Series statement: Innovations in horticultural science | Includes bibliographical references and indexes. | Content: Volume 1: Fruit Crops.

Identifiers: Canadiana (print) 20210324392 | Canadiana (ebook) 2021032452X | ISBN 9781771889896 (v. 1 ; hardcover) ISBN 9781771889889 (set) | ISBN 9781774639429 (v. 1 ; softcover) | ISBN 9781003160397 (v. 1 ; ebook).

Subjects: LCSH: Plant diseases—Diagnosis. | LCSH: Phytopathogenic microorganisms—Control.

Classification: LCC SB731 .D57 2022 | DDC 632—dc23

Library of Congress Cataloging-in-Publication Data

Names: Srivastava, J. N. (Plant pathologist), editor. | Singh, A. K. (Plant pathologist), editor.

Title: Diseases of horticultural crops: diagnosis and management / edited by J. N. Srivastava, A. K. Singh.

Other titles: Innovations in horticultural science.

Description: First edition. | Palm Bay, FL, USA : Apple Academic Press, [2022] | Series: Innovations in horticultural science | Includes bibliographical references and index. | Contents: V. 1. -- Fruit crops -- v.2. -- Vegetable crops -- v. 3. -- Ornamental plants and spice crops -- v. 4. -- Important plantation crops, medicinal crops, and mushrooms. | Summary: "Diseases of Horticultural Crops: Diagnosis and Management: Volume 1: Fruit Crops is the first of the 4-volume set that provides valuable information on major diseases of a selection of fruit crops along with effective management techniques. Each chapter in the volume includes an introduction, disease symptoms, causal organisms, disease cycles, epidemiology, and management of economically important plants. With contributions from national scientists who are engaged in teaching, research, and extension services who share their experiences here, the chapters explore apples, amla (or Indian gooseberry), avocado, Indian bael, banana, Indian jujube, citrus, grapes, guava, hazelnut, and more. Other volumes in this four-volume set focus on vegetable crops, ornamental plants and spice crops, and a selection of plantation and medicinal crops and mushrooms. The information provided here will be valuable to researchers, scientists, growers, faculty and students, extension services, and many others engaged in the discipline of not only plant pathology but also in the other fields of agriculture"-- Provided by publisher.

Identifiers: LCCN 2021045619 (print) | LCCN 2021045620 (ebook) | ISBN 9781771889896 (v. 1 ; hardback) | ISBN 9781774639429 (v. 1 ; paperback) | ISBN 9781003160397 (v. 1 ; ebook)

Subjects: LCSH: Plant diseases. | Plant diseases--Diagnosis. | Horticultural crops.

Classification: LCC SB731 .D574 2022 (print) | LCC SB731 (ebook) | DDC 632/.3--dc23

LC record available at https://lccn.loc.gov/2021045619

LC ebook record available at https://lccn.loc.gov/2021045620

ISBN: 978-1-77188-989-6 (hbk)
ISBN: 978-1-77463-942-9 (pbk)
ISBN: 978-1-00316-039-7 (ebk)

Dedicated to

My beloved parents for blessing me
My Uncle (Mama)
Dr. H. K. Srivastava
For my successes in every sphere of life

INNOVATIONS IN HORTICULTURAL SCIENCE

Editor-in-Chief:

Dr. Mohammed Wasim Siddiqui, Assistant Professor-cum- Scientist
Bihar Agricultural University | www.bausabour.ac.in
Department of Food Science and Post-Harvest Technology
Sabour | Bhagalpur | Bihar | P. O. Box 813210 | INDIA
Contacts: (91) 9835502897
Email: wasim_serene@yahoo.com | wasim@appleacademicpress.com

The horticulture sector is considered as the most dynamic and sustainable segment of agriculture all over the world. It covers pre- and postharvest management of a wide spectrum of crops, including fruits and nuts, vegetables (including potatoes), flowering and aromatic plants, tuber crops, mushrooms, spices, plantation crops, edible bamboos etc. Shifting food pattern in wake of increasing income and health awareness of the populace has transformed horticulture into a vibrant commercial venture for the farming community all over the world.

It is a well-established fact that horticulture is one of the best options for improving the productivity of land, ensuring nutritional security for mankind and for sustaining the livelihood of the farming community worldwide. The world's populace is projected to be 9 billion by the year 2030, and the largest increase will be confined to the developing countries, where chronic food shortages and malnutrition already persist. This projected increase of population will certainly reduce the per capita availability of natural resources and may hinder the equilibrium and sustainability of agricultural systems due to overexploitation of natural resources, which will ultimately lead to more poverty, starvation, malnutrition, and higher food prices. The judicious utilization of natural resources is thus needed and must be addressed immediately.

Climate change is emerging as a major threat to the agriculture throughout the world as well. Surface temperatures of the earth have risen significantly over the past century, and the impact is most significant on agriculture. The rise in temperature enhances the rate of respiration, reduces cropping periods, advances ripening, and hastens crop maturity, which adversely affects crop productivity. Several climatic extremes such as droughts, floods, tropical cyclones, heavy precipitation events, hot extremes, and heat waves cause a negative impact on agriculture and are mainly caused and triggered by climate change.

In order to optimize the use of resources, hi-tech interventions like precision farming, which comprises temporal and spatial management of resources in horticulture, is essentially required. Infusion of technology for an efficient utilization of resources is intended for deriving higher crop productivity per unit of inputs. This would be possible only through deployment of modern hi-tech applications and precision farming methods. For improvement in crop production and returns to farmers, these technologies have to be widely spread and adopted. Considering the above-mentioned challenges of horticulturist and their expected role in ensuring food and nutritional security to mankind, a compilation of hi-tech cultivation techniques and postharvest management of horticultural crops is needed.

This book series, Innovations in Horticultural Science, is designed to address the need for advance knowledge for horticulture researchers and students. Moreover, the major advancements and developments in this subject area to be covered in this series would be beneficial to mankind.

Topics of interest include:

1. Importance of horticultural crops for livelihood
2. Dynamics in sustainable horticulture production
3. Precision horticulture for sustainability
4. Protected horticulture for sustainability
5. Classification of fruit, vegetables, flowers, and other horticultural crops
6. Nursery and orchard management
7. Propagation of horticultural crops
8. Rootstocks in fruit and vegetable production
9. Growth and development of horticultural crops
10. Horticultural plant physiology
11. Role of plant growth regulator in horticultural production
12. Nutrient and irrigation management
13. Fertigation in fruit and vegetables crops
14. High-density planting of fruit crops
15. Training and pruning of plants
16. Pollination management in horticultural crops
17. Organic crop production
18. Pest management dynamics for sustainable horticulture
19. Physiological disorders and their management
20. Biotic and abiotic stress management of fruit crops
21. Postharvest management of horticultural crops
22. Marketing strategies for horticultural crops
23. Climate change and sustainable horticulture
24. Molecular markers in horticultural science
25. Conventional and modern breeding approaches for quality improvement
26. Mushroom, bamboo, spices, medicinal, and plantation crop production

BOOKS IN THE SERIES

- **Spices: Agrotechniques for Quality Produce**
  Amit Baran Sharangi, PhD, S. Datta, PhD, and Prahlad Deb, PhD

- **Sustainable Horticulture, Volume 1: Diversity, Production, and Crop Improvement**
  Editors: Debashis Mandal, PhD, Amritesh C. Shukla, PhD, and Mohammed Wasim Siddiqui, PhD

- **Sustainable Horticulture, Volume 2: Food, Health, and Nutrition**
  Editors: Debashis Mandal, PhD, Amritesh C. Shukla, PhD, and Mohammed Wasim Siddiqui, PhD

- **Underexploited Spice Crops: Present Status, Agrotechnology, and Future Research Directions**
  Amit Baran Sharangi, PhD, Pemba H. Bhutia, Akkabathula Chandini Raj, and Majjiga Sreenivas

- **The Vegetable Pathosystem: Ecology, Disease Mechanism, and Management**
  Editors: Mohammad Ansar, PhD, and Abhijeet Ghatak, PhD

- **Advances in Pest Management in Commercial Flowers**
  Editors: Suprakash Pal, PhD, and Akshay Kumar Chakravarthy, PhD

- **Diseases of Fruits and Vegetable Crops: Recent Management Approaches**
  Editors: Gireesh Chand, PhD, Md. Nadeem Akhtar, and Santosh Kumar

- **Management of Insect Pests in Vegetable Crops: Concepts and Approaches**
  Editors: Ramanuj Vishwakarma, PhD, and Ranjeet Kumar, PhD

- **Temperate Fruits: Production, Processing, and Marketing**
  Editors: Debashis Mandal, PhD, Ursula Wermund, PhD, Lop Phavaphutanon, PhD, and Regina Cronje

- **Diseases of Horticultural Crops: Diagnosis and Management, Volume 1: Fruit Crops**

  Editors: J. N. Srivastava, PhD, and A. K. Singh, PhD

- **Diseases of Horticultural Crops: Diagnosis and Management, Volume 2: Vegetable Crops**

Editors: J. N. Srivastava, PhD, and A. K. Singh, PhD

- **Diseases of of Horticultural Crops: Diagnosis and Management, Volume 3: Ornamental Plants and Spice Crops**

  Editors: J. N. Srivastava, PhD, and A. K. Singh, PhD

- **Diseases of Horticultural Crops: Diagnosis and Management, Volume 4: Important Plantation Crops, Medicinal Crops, and Mushrooms**

  Editors: J. N. Srivastava, PhD, and A. K. Singh, PhD

- **Biotic Stress Management in Tomato**
  Editors: Shashank Shekhar Solankey, PhD, and Md. Shamim, PhD

- **Medicinal Plants: Bioprospecting and Pharmacognosy**
  Editors: Amit Baran Sharangi, PhD, and K. V. Peter, PhD

- **Tropical and Subtropical Fruit Crops: Production, Processing, and Marketing**
  Editors: Debashis Mandal, PhD, Ursula Wermund, PhD, Lop Phavaphutanon, PhD, and Regina Cronje

ABOUT THE EDITORS

J. N. Srivastava, PhD, has about 20 years of experience of teaching undergraduate, postgraduate, and PhD classes in the courses of plant pathology in the capacity of Assistant Professor/ Junior Scientist (Plant Pathology) at Sher-e-Kashmir University of Agriculture Sciences and Technology, Jammu (J&K), and as Associate Professor cum Senior Scientist (Plant Pathology) at Bihar Agricultural University, Sabour, Bhagalpur, Bihar. His vast experience in agriculture also includes experience in extension programs, consultancy, development, administration, etc.

Dr. Srivastava has extensively pursued research work on different aspects of plant diseases. Dr. Srivastava has made significant research contributions in biological control of diseases in crop plants. He has identified bio-control agents (*Trichodema viride*) to manage guava wilt disease in guava, and he has also done research on integrated disease management of vegetable crops. He is working on biological control/integrated disease management as an eminent scientist and has handled 10 projects as principal investigator and co-principal investigator. In addition to teaching and research, Dr. Srivastava is credited with many publications, including two books, over 45 research papers, over 47 book chapters, over 63 extension articles, four practical bulletins, five technical bulletins, and many leaflets also.

Dr. Srivastava was the recipient of many awards, including Outstanding Achievement Award (2015), Kunwar Sexena Bahadur–SRDA Award (2015), Excellence in Teaching Award (2016, 2017), Dr. R. S. Paroda Medal (2016), Outstanding Scientist Award (2016), Excellence in Science Communication Award (2017), Distinguished Faculty Award (2017, 2018), Eminent Scientist Award (2018), Best Faculty Award (2019), Distinguished Teacher Award (2020). He is also the recipient of several fellowships, including F.H.A.S.-2010, F.B.R.S.-2013, F.S.S.D.A.T.-2014, F.P.P.S.-2015, F.S.E.R.S.-2016, F.S.B.S.R.D.-2016, F.S.A.I.D.-2017, F.P.S.I.-2018, F.N.S.F.-2018,

F.I.N.S.O.P.P.,-2019, from various national and international academic/ scientific societies.

Dr. Srivastava is a life member of 16 academic/scientific societies, namely, Indian Phyto-pathological Society, Society of Mycology and Plant Pathology, Indian Society of Plant Pathologist, Association of Plant Pathologists of India, Society of Plant Protection Sciences, Society for Recent Development in Agriculture, Society for Scientific Development in Agriculture and Technology, National Academy of Biological Sciences (NABS), Association for Advancement in Plant Protection, Society for Plant Research, Indian Botanical Society, and Society of Human Resource and Innovation. He is also associated with many international/ national/provincial scientific/ cultural/academic/educational bodies.

Dr. Srivastava has served as a member of editorial boards of the *International Journal of Plant Protection*. He is reviewer of various international journals and has also visited Kasetsart University, Thailand. Dr. Srivastava has also attended many national and international workshops. He is also engaged in various agriculture extension activity, namely, farm advisory services, Kisan Chaupal Programme, On Farm Trial (OFT) and Front Line Demonstration (FLD), etc. He has also been involved with exhibitions, campaigns, farmers' fairs, Kishan Gosthi, etc., video conferencing, radio talks, TV talks and vocational training for rural youth (men and women), skill/entrepreneurship development programs, farmer training, in-service training, and field days on maize, sunflowers, rajmash, mash, moong, and soybeans. He is also engaged in Mera Gaon Mera Gaurav.

Dr. Srivastava received his MSc in Agriculture and PhD in Plant Pathology.

A. K. Singh, PhD, is Assistant Professor in the Division of Plant Pathology, Sher-e-Kashmir University of Agricultural Sciences and Technology, Jammu, J&K, India. He has been engaged for more than 12 years in teaching (both undergraduate and postgraduate levels) and research and also involved in the transfer of technology through different extension activities.

He has published more than 30 research papers in national and international journals of repute, one practical manual, and more than 10 book chapters published in books along with eight popular articles. Dr. Singh has also presented many research papers at international, national, and regional symposiums, and seminars. He is the recipient of several prestigious awards for research contributions. He has also served as a member of editorial board of *Krishi Vikas Patrika*, published by Directorate of Extension, SKUAST of Jammu. He is a life member of more than four societies of plant pathologists in India.

He earned his DPhil from the University of Allahabad, Allahabad, UP, India.

CONTENTS

CONTRIBUTORS

Amar Bahadur
College of Agriculture, Tripura, Lembucherra, Agartala 799210, West Tripura, India

Prashant Bakshi
Division of Fruit Science, Sher-e-Kashmir University of Agriculture Sciences and Technology Jammu, Jammu and Kashmir, India

Saba Banday
Division of Plant Pathology, (FOA), Sher-E-Kashmir University of Agricultural Sciences and Technology (SKUAST-K) Srinagar, Kashmir, Jammu and Kashmir, India

Susanta Banik
Department of Plant Pathology, School of Agriculture Science and Rural Development (SASRD), Nagaland University, Medziphema 797106, Nagaland, India

Sharad Bisen
Department of Horticulture, College of Agriculture, Jawahar Lal Nehru Krishi Viswavidyalay (JNKVV), Jabalpur, Madhya Pradesh, India

Pezangulie Chakruno
Department of Plant Pathology, School of Agriculture Science and Rural Development (SASRD), Nagaland University, Medziphema 797106, Nagaland, India

B. R. Choudhary
ICAR-Central Institute for Arid Horticulture, Beechwal, Bikaner 334 006, Rajasthan, India

Upma Dutta
Division of Microbiology, Sher-e-Kashmir University of Agricultural Sciences & Technology of Jammu, Jammu and Kashmir, India

Ramyashree Devi
ICAR-Central Institute for Arid Horticulture, Beechwal, Bikaner 334 006, Rajasthan, India

Nasreen Fatima
Krishi Vigyan Kendra, Ladakh, Sher-E-Kashmir University of Agricultural Sciences and Technology (SKUAST-K) Srinagar, Kashmir, Jammu and Kashmir, India

N. M. Gohel
Department of Plant Pathology, B.A. College of Agriculture, Anand Agricultural University, Anand 388 110, Gujarat, India

S. T. Ingle
Department of Plant Pathology, Dr Panjabrao Deshmukh Krishi Vidyapeeth, Akola, Maharashtra, India

Anamika Jamwal
Krishi Vigyan Kendra—Kathua, Sher-e-Kashmir University of Agricultural Sciences & Technology of Jammu, Jammu and Kashmir, India

Sonika Jamwal
Advanced Centre for Rainfed Agriculture-Dhiansar, Sher-e-Kashmir University of Agricultural Sciences and Technology-SKUAST, Jammu, Jammu and Kashmir, India

K. Jayalakshmi
National Bureau of Agriculturally Important Microorganisms (NBAIM), Kushmaur, Maunath Bhanjan, Uttar Pradesh 275103, India

Manoj Kumar Kalita
Department of Plant Pathology, Biswanath College of Agriculture, Assam Agricultural University, Biswanath Chariali 784176, Assam, India

Shakti Khajuria
ICAR-Central Institute for Arid Horticulture, Beechwal, Bikaner 334 006, Rajasthan, India

B. R. Khatri
ICAR-Central Institute for Arid Horticulture, Beechwal, Bikaner 334 006, Rajasthan, India

Hare Krishna
ICAR-Central Institute for Arid Horticulture, Beechwal, Bikaner 334 006, Rajasthan, India

Ashish Kumar
Department of Plant Pathology, College of Agriculture, Jawahar Lal Nehru Krishi Viswavidyalay (JNKVV), Jabalpur, Madhya Pradesh, India

J. Kumar
Department of Plant Pathology, College of Agriculture, G B Pant University of Agriculture & Technology, Pantnagar, Udham Singh Nagar, Uttarakhand, India

Lokesh Kumar
ICAR-Central Institute for Arid Horticulture, Beechwal, Bikaner-334 006, Rajasthan, India

S. K. Maheshwari
ICAR-Central Institute for Arid Horticulture, Beechwal, Bikaner 334 006, Rajasthan, India

Roman Nissar
Division of Agronomy, (FOA), Sher-E-Kashmir University of Agricultural Sciences and Technology (SKUAST-K) Srinagar, Kashmir, Jammu and Kashmir, India

B. K. Prajapati
Agricultural Research Station, Sardarkrushinagar Dantiwada Agricultural University, District Banaskantha 385 535, Gujarat, India

S. Raghu
Crop Protection Division, National Rice Research Institute, Cuttack, Odisha 753006, India

J. Raju
Plant Quarantine Station, Ministry of Agriculture and Farmers Welfare, Government of India, Mangalore, Karnataka 575011, India

K. P. Singh
Department of Plant Pathology, College of Agriculture, G B Pant University of Agriculture & Technology, Pantnagar, Udham Singh Nagar, Uttarakhand, India

Priti S. Sonavane
Indian Institute of Horticultural Research, Hesaraghatta, Bangalore, Karnataka, India

Rakesh Kumar Sharma
Division of Fruit Science, Sher-e-Kashmir University of Agriculture Sciences and Technology, Jammu, Jammu and Kashmir, India

Efath Shahnaz
Division of Plant Pathology, (FOA), Sher-E-Kashmir University of Agricultural Sciences and
Technology (SKUAST-K) Srinagar, Kashmir, Jammu and Kashmir, India

K. K. Sharma
Regional Research Station (PAU), Ballowal Saunkhri, Balachaur 144521, SBS Nagar, Punjab, India

R. S. Shete
Department of Plant Pathology, Bihar Agricultural University, Sabour, Bhagalpur, Bihar, India

A. K. Singh
Division of Plant Pathology, Sher-e-Kashmir University of Agriculture Sciences and Technology
Jammu (J & K) India

V.B. Singh
Division of Plant Pathology, Sher-e-Kashmir University of Agriculture Sciences and Technology
Jammu, Jammu and Kashmir, India

P. P. Singh
ICAR-Central Institute for Arid Horticulture, Beechwal, Bikaner 334 006, Rajasthan, India

J. N. Srivastava
Department of Plant Pathology, Bihar Agricultural University, Sabour 813210, Bhagalpur, Bihar, India

Kavi Sumi
Department of Plant Pathology, School of Agriculture Science and Rural Development (SASRD),
Nagaland University, Medziphema 797106, Nagaland, India

V. Venkataravanappa
Central Horticultural Experiment Station, ICAR-IIHR, Chettalli 571248, Kodagu, Karnataka

A. K. Verma
ICAR-Central Institute for Arid Horticulture, Beechwal, Bikaner 334 006, Rajasthan, India

ABBREVIATIONS

ASP	apple scab predictor
BBTD	banana bunchy top disease
CEV	citrus exocortis viroid
CMBV	citrus yellow mosaic badnavirus
CMV	cucumber mosaic virus
CP	coat protein
CTV	citrus tristeza virus
EC	emulsifiable concentrates
EBI	ergosterol-biosynthesis-inhibiting (fungicides)
BFB	Bacterial Fruit Blotch
HLB	Huanglongbing
HPLC	high-performance liquid chromatography
PAD	potential ascospore dose
PCR	polymerase chain reaction
RH	relative humidity
RP	root pits
SBI	sterol-biosynthesis-inhibiting (fungicides)
SIF	sterol-inhibiting fungicides
S.O.	sweet orange
SP	stem pitting
STG	shoot tip grafting
SY	seedling yellow

FOREWORD 1

Fruits are very important for human beings, and they are not only delicious but also have many nutrients that are necessary for human health. Fruit crops play an important role in the human diet. Because of their ingredients fruits are very healthy and protective against a number of diseases. They are known to contain several health-promoting components, such as vitamins, essential minerals, antioxidants, and prebiotics (fibers).

India is the second largest producer of fruits in the world after China. India has diverse climates and soil for cultivation of horticulture crops, providing ample opportunities for the development of the fruit industry, but the greatest challenge at present is to produce sufficient fruit to feed the ever-increasing human population. This demands infusion of technology for an efficient utilization of resources for deriving higher output per unit of input while rataining excellent quality of the produce within short span of time.

In the present era of open economy, it has become increasingly necessary that our produce is competitive, both in the domestic as well as in international markets. This would be possible only through deployment of high-tech applications and precision farming methods, but diseases are one of the important constraints for productivity of fruits all over the world, including India.

Hence, understanding the fundamental components of disease, that is, symptoms, causal organisms, disease cycles, and epidemiology, is essential for the effective management of diseases and to avert significant losses in the yield of fruits.

It is in this context this book, *Diseases of Horticultural Crops: Diagnosis and Management, Vol. 1: Fruit Crops*, contains elaborate descriptions of all important diseases of fruit crops and is a valuable addition to the literature on fruit pathology in the country.

A commendable effort has be made by Dr. J. N. Srivastava, Bihar Agricultural University, Bhagalpur, Bihar, and Dr. A. K. Singh, Sher-e-Kashmir

University of Agricultural Sciences and Technology, Jammu, (J&K), in planning, compiling, and editing this publication.

I am sure the book will be found very useful not only by the students studying at undergraduate and postgraduate levels, researchers, and teachers of Plant Pathology but also by the planners, administrators and growers, and various organizations dealing with the export of fruits from the country.

I give compliment to the authors, Dr. J. N. Srivastava and Dr. A. K. Singh, for their effort on bringing out this publication and wish them success in this and future endeavors.

— Panjab Singh, DSc

President, National Academy of Agricultural Sciences;
Chancellor, RLB Central Agricultural University, Jhansi (U.P.);
President, FAARD Foundation, Delhi, India

FOREWORD 2

Fruit growing is one of the important and age old practices, practiced in India since ancient times. Cultivation of fruit crops plays an important role in overall status of the mankind and the nation. The standard of living of the people of a country depends upon the production and per capita consumption of fruits. Fruit crops also play an important role in the human diet. Because of their ingredients fruits are very healthy and protective against a number of diseases. They are known to contain several health-promoting components, such as vitamins, essential minerals, antioxidants, and prebiotics (fibers). There is a great potential of export of fruits from India, which can be achieved not only by improving the quality of fruits but also through judicious management of diseases. Diseases are one of the important constraints for productivity of fruits all over the world including India.

Hence, understanding the fundamental components of disease i.e. symptoms, causal organism, disease cycle and epidemiology is essential for effective management of diseases and to avert significant losses in the yield of fruits. It is in this context the book "Diseases of Horticultural Crops: Diagnosis and Management, Vol. I: Fruit Crops" contains elaborate descriptions of all important disease of fruit crops and is valuable addition to the literature on fruit pathology in the country. A commendable effort has be made by Dr. J. N. Srivastava, Bihar Agricultural University, Bhagalpur, Bihar and Dr. A. K. Singh, Sher-e-Kashmir University of Agricultural Sciences and Technology, Jammu, (J&K) in planning, compiling, and editing this publication.

I am sure the book will be found to be very useful not only by the students studying at undergraduate and postgraduate levels, researchers and teachers of plant pathology but also by the planners, administrators and growers, and various organizations dealing with the export of fruits.

I congratulate Dr. J. N. Srivastava and Dr. A. K. Singh for their efforts in updating and including information about fruits in the new edition of his book.

— **C. D. Mayee, PhD, DSc**
Adjunct Professor, Indian Agricultural Research Institute, New Delhi, India;
President, ISCI, Mumbai, India

FOREWORD 3

Fruits are very important in human diet, as they are the major source of vitamins and micronutrients. India provides ideal climatic conditions for growing a wide variety of fruits. The varieties of fruits, which are grown in India, are unmatched with any other part of the world.

In India, fruits production provides nutritional security but major constraints in improving the productivity of fruits in India are the pests and diseases. These constraints not only adversely affect the production but also impede export of fruits from India. There is a great potential of export of fruits from India, which can be achieved not only by improving the quality of fruits but also through judicious management of diseases.

It is in this context the book "Diseases of Horticultural Crops: Diagnosis and Management, Vol. I: Fruit Crops" contains elaborate descriptions of all important disease of fruit crops and is valuable addition to the literature on fruit pathology in the country.

The book deals with fruits crop diseases caused by fungi, bacteria, and viruses. The management practices included in the book are easy to follow and will certainly be useful in minimizing the losses caused by diseases of fruits in the country.

It is a commendable effort has be made by Dr. J. N. Srivastava, Bihar Agricultural University, Bhagalpur, Bihar, and Dr. A. K. Singh, Sher-e-Kashmir University of Agricultural Sciences and Technology, Jammu, (J&K), in planning, compiling, and editing this publication.

I am sure the book will be found very useful not only by students studying at undergraduate and postgraduate levels, researchers and teachers of plant pathology but also by the planners, administrators and growers and various organizations dealing with the export of fruits from the country.

I heartily congratulate the editors/authors, Dr. J. N. Srivastava and Dr. A. K. Singh, for bringing out this useful publication that provides comprehensive information about fruits crop diseases.

— Amar Nath Mukhopahyay, PhD, DSc

Ex-Vice Chancellor, Assam Agricultural University, Jorhat, Assam, India;
Ex-Director General, Tea Research Association India;
Ex-Director, Biopesticide Division, Super Agro India Pvt. Ltd., Kolkata, India

PREFACE

Fruit crops cultivation is one of the important and age old practices, practiced in India since ancient times. Fruit crops cultivation plays an important role in overall status of the mankind and the nation. The standard of living of the people of a country is depending upon the production and per capita consumption of fruits. Fruit growing has more economic advantages. Fruits crops are capable of giving higher tonnage of yield per unit area than other field crops.

According to recent research findings, many phytochemicals are found in fruits and act as powerful antioxidants protecting cells and organs from damage caused by free radicals, neutralizing their damaging effects. They are the biologically active substances in plants that give them color, flavor, odor, and protection against not only diseases affecting the plants but also provide health benefits to human beings. Consequently hundreds of such plant substances are being investigated now for their role in preventing cancer and other degenerative diseases. Some of the promising phytochemicals that act as antioxidants are bioflavanoids (vitamin P), phenolics, lycopene, carotenoids, antioxidant vitamins (C and E), and glucosinolates. Oranges, lemons, limes, and grape fruits in addition to being principal sources of vitamin C and folate are rich in a class of phytochemicals called limonoids. This antioxidant has been found to be very effective against cancer.

The potential of fruit crops in the growth of national economy is noteworthy. The prosperity of the country lies in building-up its foreign exchange reserves. Being a country having varied climatic conditions ranging from tropical to subtropical and to temperate, India has very immense potential for the production of different fruits and their export. Recent policies of the Government of India to encourage export of fruits and their products by announcing concessions to the fruit industry, such as reduced air freight charges and exemption for storage charges for refrigerated air cargo at international airports, have encouraged a number of private entrepreneurs and corporate bodies and NRIs to go in for planting larger areas under fruit crops with an aim to export fruits and fruit products.

The foregoing account is given with a view to emphasize the growing importance of the fruit industry. As could be seen from the details given therein, the fruit culture is vital to the health and economy of the nation, from the standpoint of increased food production, nutrition, trade, and fruit-based industries.

In spite of all above facts, the production of fruits in the country is less as compared to advanced countries because of different reasons; the most important being the prevalence of diseases in fruit crops. The management of any disease successfully involves its detailed study regarding symptoms, casual agent, disease cycle, and epidemiology.

This book, *Diseases of Horticultural Crops: Diagnosis and Management Volume-1 Fruit Crops*, presents with 23 chapters. Each chapter has detailed account of introduction, symptoms, causal organisms, disease cycles, epidemiology, and management of economically important diseases. The book chapters have been contributed by the authors who are engaged in teaching, research, and extension services and are well-known national scientists in their respective field. The authors, while writing the chapters, have incorporated their experience and knowledge with recent developments in the field of plant diseases.

It is hoped that the book will cater to the needs of students studying at undergraduate and postgraduate levels, researchers, teachers, planners, administrators, and growers, not only in the discipline of plant pathology but also in other fields of agriculture.

We sincerely acknowledge our thanks and gratitude to the esteemed scientists who have spared their time and contributed valuable chapters for this book. We are also heartily thankful to Apple Academic Press for publishing this book.

— **Dr. J. N. Srivastava**
Dr. A. K. Singh

CHAPTER 1

CURRENT STATUS OF APPLE SCAB DISEASE AND MANAGEMENT STRATEGIES IN UTTARANCHAL HIMALAYAS

K. P. SINGH* and J. KUMAR

Department of Plant Pathology, College of Agriculture, Govind Ballabh Pant University of Agriculture & Technology, Pantnagar, Udham Singh Nagar, Uttarakhand, India

Corresponding author. E-mail: kpsingh.gbpuat@gmail.com

ABSTRACT

Loss from apple scab over a period of years is greater than that from any other disease of apple. Scab reduces leaf areas, defoliation, and poor spur development, fruit size, shelf life, and quality of the fruit. Yield losses during epidemic years can go up to 70% or even more. In Uttarakhand Himalayas, control of apple scab is achieved primarily through a protectant fungicides spray program. In a protectant program for primary scab control, fungicides are generally applied after every 10 days of new growth. Applications are made regardless of whether infection periods have occurred. In wet growing seasons, Gangotri fruit belt growers make up to 10–12 fungicides applications for scab control. An alternative to a protectant program is the after-infection or curative spray program. In a curative program, the fungicide is applied after the initiation of an infection period but before symptom development.

1.1 INTRODUCTION

All the diseases collectively did not discourage the fruit growers to that level as the one scavenger disease called "SCAB" did in several years. Scab is not a newly discovered disease since it was already recorded in 1819 in Sweden by Fries. It is more severe during spring, summer, and rainy season and almost absent in the desiccate and warmer areas. It was first observed in 1935 in Ambri variety in Kashmir, 1977 in Himachal Pradesh at Mauhli village, and 1987 in Uttarakhand at Gangotri valley on about 35,000 apple plants, and soon after in 1989 Purola–Naugaon fruit belt was also affected. It occurred in scourge form in 1996, 2008, and again in 2013 in the Harsil fruit belt of Uttarakhand (Prasad et al., 2018; Singh and Kumar 1999a, b, and c, 2004, 2007, 2008, 2009; Singh et al., 2010, 2015, 2016). In Uttarakhand, management of scab is observed primarily through a protectant fungicides spray program. In a protectant program for primary scab management, fungicides are normally applied after each week of new growth. Applications are made regardless of whether infection periods have occurred. In wet growing seasons, Harsil fruit belt growers make up to 10–14 sprays for apple scab management (Singh and Kumar, 2005, Singh et al., 1995, 1997, 2008). An alternative to a protectant program is the after-infection or curative spray program. In a curative program, the fungicide is applied after the commencement of an infection period but before symptom development. Since Mills and his coworkers published their findings on the environmental parameters necessary for apple scab infection (Mills 1944; Mills and LaPlante, 1951), we have the facility to observe infection periods. Although apple scab spray advisory programs have been formulated with reference to Mills system earlier, they have not been broadly established and are not used by orchardists in Harsil fruit belt (Singh and Kumar 1999a, b, and c; Singh et al., 2010; 2015; 2016). The most important factors are causative to be deficient of fungicides with dependable curative activity able to 3 or 4 days behind the commencement of an infection period. The preamble of the ergosterol-biosynthesis-inhibiting (EBI) fungicides, which have excellent curative activity (Singh and Kumar, 1999a, b, and c; Singh et al., 1997, 2001), could make scab prediction systems more attractive to growers because they would have the capability to manage the scab after infection periods were identified. A computer program was generated as a device that combined electronic environmental monitoring sensors with a microcomputer designed to provide simple and rapid on-site identification of scab infection periods. This unit has evolved into the apple scab predictor (ASP) presently being manufactured and marketed by Reuter-Stokes, METOS, Cleveland, and other manufacturers. In Garhwal Himalayas, we developed scab warning service in the majority of key apple-growing districts, based on electronic scab predictors (Singh and Kumar 1999a, b, and c, 2004, Singh et al., 2005, 2006, 2010, 2015; 2016). To get better spraying effectiveness, with reduced fungicide use, more consistent scab warnings are needed. GBPUAT established the usefulness of the scab predictor for scheduling several EBI fungicides that have postinfection control activity against apple scab. We have started experiments to learn a few of the relevant epidemiologic questions related to apple scab management.

1.2 ECONOMIC IMPORTANCE

Loss from apple scab over a time of years is superior than that from any other disease of apple. Scab reduces leaf areas, defoliation and poor spur development, fruit size, shelf life, and quality of the fruit. Yield losses throughout the epidemic years can go up to 70% or even more. It occurred in epidemic proportion in 1996, 2008, and 2013 in the Harsil fruit belts of Uttarakhand resulting in a loss of up to Rs. 1.25 crores, 74 lakhs, and 1 crores, respectively (Figure 1.1). In 1996, the government had taken decision to purchased scab infected apples @ Rs. 2.00/kg and destroyed simultaneously. In 2008, the scab has made nearly 23% of the apple crop unhealthy for either market use or processing (Prasad et al., 2018; Singh and Kumar, 2009a, b, Kumar et al., 1998, 1999). The losses due to disease result from the reduction in fruit set due to attack on flower stalks, pedicel infection and early fruit infection cause premature fruit drops, spoil the shape and appearance of fruits, impairs the keeping quality of fruits, scabbed fruits encourage more storage rot, less CO_2 assimilation, restricted leaf growth, premature defoliation, stunted or reduced growth of young infected plants, high expenditure on pesticides, sprayers and spray operations and ancillary industries like sawmills and transport badly affected.

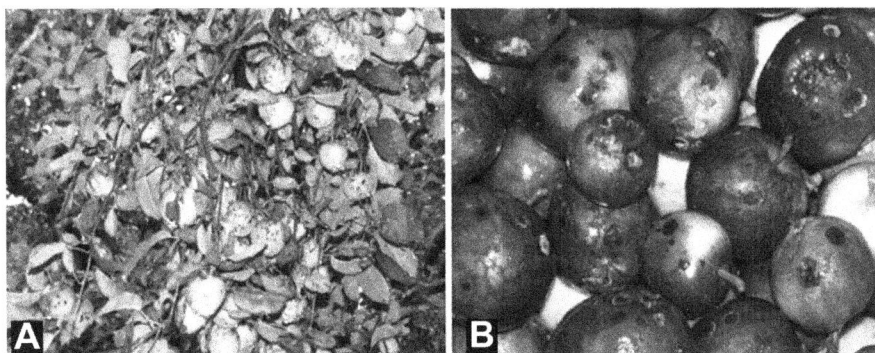

FIGURE 1.1 (A) and (B) Scab infected apple fruits at Harsil, Uttarkashi.

1.3   DISEASE IDENTIFICATION

The symptoms occur both on foliage and fruit are generally common but less distinct on the twigs. The first symptoms of scab appear on the new leaves as olive-green spots that become gradually brown and like to velvet appearance. Later, the velvety surface disappears, the lesions appear metallic black in color and slightly raised. The affected leaves may become dwarfed and curled and may later fall off prematurely. Individual infection emerge as rough, spherical, brown to black olive green spot that often appear slightly fuzzy or like velvet in texture. Scab lesions along the margins or vein often cause affected areas to deform or wrinkle (Prasad et al., 2018; Singh and Kumar, 2005, 2008). Primary infections are usually restricted to one or two individual spot per leave, whereas secondary infections are often greatly more abundant and cover the entire surface of the leaf areas. Twig and blossom infection emerge as minute scab spots, but at most places they are unusual and of slight importance. Lesions also develop as olivaceous spots on the pedicel (Figure 1.2).

FIGURE 1.2   Symptoms and signs of apple scab on leaves.

The scab spots are generally more positively defined on the fruit. The infections may take place on the fruit of any epoch or size. The lesions on immature fruits primarily very small and usually enlarge more slowly but turn brownish black and almost black with the passage of time. The mature lesions are converted into bare, brown, and corky in the center as the fungus dies. Several lesions may join together and the fruit becomes indistinct and may crack through and around the lesion. Cracking occurs because diseased skin cannot maintain swiftness with the growth of the underlying flesh. Early scab spot on fruit occur near calyx end, and stalk end lesions occur later. It has been observed that the majority of the lesion is at the stalk end as this part holds water that is favorable for germination of the spores (Singh and Kumar, 2005, 2008). Severely affected prematurely fruit may drop to the ground (Figure 1.3).

FIGURE 1.3 Several young and old (raised, corky) scab lesions (center) on apple fruit.

1.4 THE PATHOGEN

The apple scab pathogen attacks simply a member of the genus *Malus* and belong to the class Ascomycetes, subclass Pyrenomycete, order Sphaeriales, family Pleosporacae. The pathogen has two distinct stages, one the conidial state that occurs on the leaves, fruits, floral peduncles, and rarely on bud scales and the further ascigerous state that is found on deceased fallen leaves.

The teleomorph state (perithecia) of this pathogen was initially noted by Cooke in 1866, who named it as *Sphaerella inaequalis* to indicate that the bicelled ascospores have unequal cells. In 1875 winter studied the perithecial state as occurring on deceased leaves of apple and renamed it as *Venturia inaequalis* (Cke.) Wint. Aderhold in 1897 observed the connection between the conidial and the ascigerous stage and named it as *V. inaequalis* (Cke.) Aderh. Being a later homonym, the correct name of this fungus is *V. inaequalis* (Cke.) Winter. Both stages the imperfect or asexual (*Spilocea pomi* Fr.) and perfect or sexual (*V. inaequalis* (Cke.) Wint.) of the fungus have been encountered in the natural world. The perithecial stage (*Venturia*) is saprophytic, whereas asexual conidial stage is parasitic.

1.5 ETIOLOGY

The mycelium of the pathogen at first is light in color (hyaline) but later turns brownish or olivaceous and septate in the host tissues. In young leaf lesions, the mycelium develops radially in branched ribbons of hyphae, but in older leaves and on fruits the mycelia strands are compact, thick, and in several superimposed layers. In living tissue, the mycelium is located only between the cuticle and the epidermal cells and produces short, erect, and brownish conidiophores that successively furnish rise to several, one-or two-celled, reddish-brown *Fusicladium* type conidia of variable shape. The conidiophore is brown, arises from ribbon-like hyphal strands or more compact stroma. The stromatic cushion maybe 10 or more cells in thickness. The conidiophore bears a single conidium at its tip. Conidia are continuous (one-celled), ovate to lanceolate, with truncate base, and somewhat pointed above that is, flame-shaped. With age, some conidia may become two-celled and very rarely three-celled. They measure 10–20 × 6–9 µm. Lesions having conidia in the center become visible within 9–17 days of infection. Conidia are then the principal source concerned in the buildup of the disease during the summer and early autumn. The imperfect stage is now called *Spilocaea pomi* Fr. In dead leaves the mycelium grows through the leaf tissue and extends into mesophyll where it gives rise to primordial that becomes the perithecia on the leaves. Perithecia measure 90–150 µm in diameter. The mature perithecia ensuing from the fertilization of ascogonia and antherdia are dark brown to black with slight beak and ostiolate openings. The wall of the perithecium is composed of brownish cells 2–5 layers in thickness. Inside the perithecium 60–136 asci have been counted but it will be differ in number and every one containing eight ascospores. The asci occur from the cushion of ascogenous cells at the base of the perithecium. Since all the asci do not mature simultaneously, both mature and immature asci may be found together. Asci are spatulate, bitunicate, 55–75 × 6–12 µm with short stipe and thin wall. Ascospores are eight in each ascus, one row above and two rows below, oval, olive-brown in color and measure 11–15 × 5–7 µm. Each ascospore consists of two cells of unequal size, which are hyaline at first, but on maturity these turn brown (Figure 1.4).

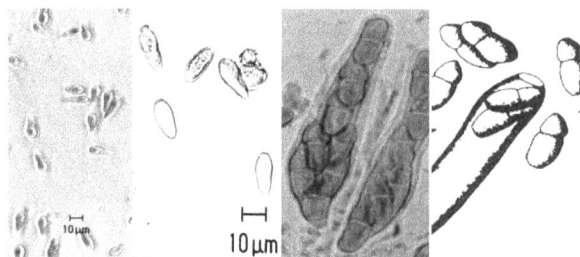

FIGURE 1.4 Imperfect and perfect stages of scab.

1.6 DISEASE CYCLE

Scab has two distinctive phases, that is, saprophytic (perfect stage) and parasitic (imperfect stage). In wintry weather, the fungus overwinters in infected dead apple leaves that lying on the orchard ground. During autumn, the fungus begins to form minute fruiting bodies, which initiate to develop within the pseudothecia by delayed winter and probably continue to grow during warm period in the early spring. The ascospores go on to develop and mature in spring when weather conditions become favorable. A few ascospores in the pseudothecia are generally full-grown at the time of bud break (green tip), and maturity progresses gradually until about the tight cluster stage of blossom development. The majority of the ascospores have matured by the ending of bloom and forcibly discharge the ascospores into the atmosphere. All the pseudothecia and asci do not mature simultaneously. Some may shed the ascospores before the apple buds start to open in the spring; however, the majority of the ascospores in the perithecia mature in the period during which the fruit buds open. When dead leaves containing perithecia become thoroughly covered with water in the spring, the asci elongate, push through the ostiole, and discharge the ascospores forcibly into the atmosphere and are being passed through the air currents on to the susceptible host tissues. Ascospore discharge may continue for 3–5 weeks later than petal fall. The development of ascospores is slow at 4–7 °C but faster at 12–16 °C, with most favorable at 20 °C but beyond 24 °C their formation ceases. Ascospores are blown to close by trees by wind currents. The ascospores can grow in a layer of water on the surface of leaves and cause infection only when temperature ranging from 10 to 18 °C. Thus for infection the spores must be continuously wet for 28 h at 6 °C, for 14 h at 10°C, for 9 h at 18–24 °C, and for 12 h at 26 °C. Upon germination on an apple leaves or fruit, the ascospore produces a disk-like appressorium from which a slender mycelia tube pierces the cuticle, and after developing into a hypha of normal diameter, it grows in the middle of the cuticle and the outer cell wall of the epidermal cells. For a few days after infection, the epidermal cells show no injury at all. These cells showed a gradual depletion of their contents and eventually collapse and die by the time the lesion appears. Soon the palisade and later the mesophyll cells show the same reactions, whereas the fungus still remains largely in the subcuticular position. The parasitic stage invades the present year growth, which later reproduces the countless number of secondary or summer spores. With the establishment of mycelium in the host, it produces an enormous number of conidia that are pushed outward by rupturing the cuticle and form olive-green, velvety scab lesions within 8–15 days. Conidia get attached to the conidiophores in dry weather, but upon wetting during the rains these are easily detached and may be washed down or blown away to other leaf or fruit on which they grow and cause infection in the same way ascospores do. Additional infections by conidia continue all through the growing season following a rain of adequate duration. Infection, however, is more profuse in the cool, wet periods of spring and early summer and yet again in the fall and infrequent or nearly absent in the dry hot summer weather (Prasad et al., 2018; Singh and Kumar, 2005, 2008). This stage is more often known as the secondary infection that is responsible for epidemics.

1.7   CURRENT STATUS OF SCAB

In Garhwal Himalayas, the scab was found to be more common and prevalent. Scab outbreak was observed in Harsil fruit belt of district Uttarkashi during 1996 recording severe infection on foliage (30%–80%) and fruits (70%–95%). It occurred as an outbreak in 1996, 2008, and 2013 in the Harsil fruit valley of Uttarakhand. In all the valley, cent percent orchards and trees were found to be infected, the maximum percentage of infected trees was in Harsil fruit belt (Figure 1.5). Scab epidemics are initiated in most apple-growing regions by ascospores discharged during wet periods, appropriate for infection, with the severity of epidemic governed by favorable weather for infection and maximum amount of ascospores discharged. However, in general, the disease severity remained low (Prasad et al., 2018; Singh and Kumar 1999a, b, and c; Singh, 2005; 2006; 2016; 2017) in Garhwal Himalayas. Most of the places of Uttarakhand Himalayas fruit belts were apparently free of scab.

1.8   EPIDEMIOLOGY

The four major factors governing the development of an epidemic are the potential pathogen inoculum, host susceptibility, favorable environmental conditions, and human interventions. Apple scab, caused by *V. inaequalis*, is an ideal example of the disease that develops into epidemic proportions wherever the management options are not adopted. The pathogen completes the infection cycle in two phases. The sexual or the saprophytic phase occurs in the fallen leaves that are infected and the asexual phase occurs on the tree during growing season on all above-ground plant parts namely leaves, shoots, and fruits the extend of the occurrence of scab epidemic is determined by both the environmental and biological factors. The pathogen survives on the fallen infected leaves as pseudothecia. The initiation of pseudothecia formation occurs once the leaves fall off from the tree and is dependent on apple cultivar, leaf moisture, atmospheric temperature, and infection time. Our experimental results showed that in Uttarakhand Himalayas, the greater risk of increasing primary inoculum is posed by the leaves that fall late in the season as compared to those falling in early stages, that is, August–first week of September. Low to medium temperature during February to April was more related to pseudothecial development and time of ascospore discharge as compared to the mean rainfall and temperature during the subsequent months. Depending upon the temperature and moisture conditions, the pseudothecia develop within 4 weeks after leaf fall. In spring, with the initiation of the emergence of new growth, minute sacs of spores are formed on the undersurface of the dead leaves. The maturation of pseudothecia occurs by the end of March in Gwaldam, Talwari, Purola–Naugaon, Nainbagh, Tuni, Chakrata, Koti-Kanasar, and Tal regions; whereas it occurs by the last week of April in Munshiyari and Joshimath areas. In Gangotri valley the pseudothecia mature late (second week of May). A single pseudothecium releases about 1000–1200 ascospores (bicelled with one cell greater than the other). The release and maturation of ascospores continue till the end of spring and is dependent on the atmospheric temperature and leaf wetting period that is highly influenced by intermittent rain or heavy dew. Mature ascospores are discharged into the air during periods of rain. However, prolonged wetting and direct sunlight are harmful for the ascospore maturation and release. We observed that the maximum discharge and percentage maturation of spores occur during the end of bloom to petal fall stage. In Garhwal Himalayas of Uttarakhand, the peak period of ascospore release is 3–8 weeks (Table 1.1). The ascospores release increases till May and then decreases gradually thereafter. During day time, the ascospore discharge begins soon after initiation of rain (within 30 min) and is largely completed within 3–6 h. If rain occurs during the night, the discharge is delayed till dawn. Nearly 2000 pseudothecia (with about two million ascospores) may occur on a single leaf and ascospore discharge will continue for 3–5 weeks after petal fall (June end). Ascospores mature at the same rate in all the cultivar. But the ascospore productivity is influenced by the host cultivar (Singh et al., 2016).

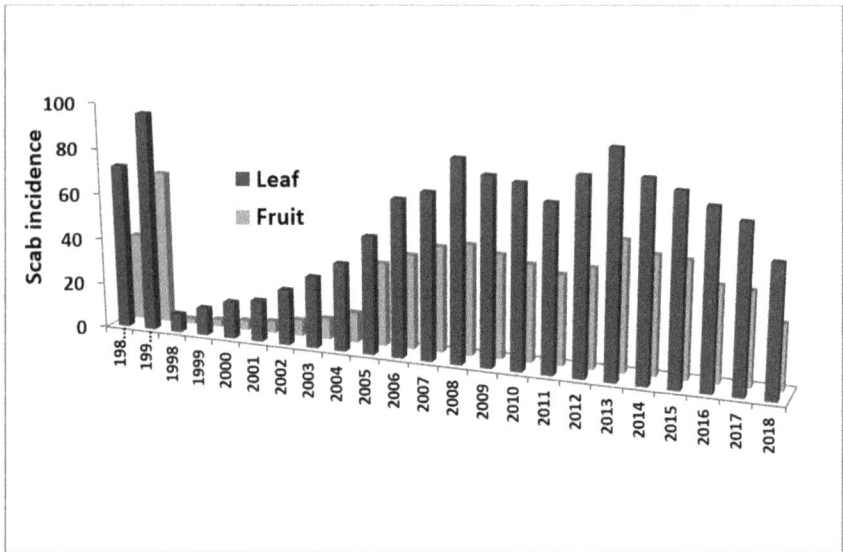

FIGURE 1.5   Scab incidence in Harsil valley of Garhwal Himalayas.

TABLE 1.1   Ascospore Infection on Apple Leaves at Different Place of Garhwal Himalayas

Apple Orchards	Growth Stage of Tree and Actual Time for Primary Infection		
	Tree stage	**Time**	**Primary infection**
Harsil Valley, Uttarkashi (>2500 m asl) Harsil, Dharali, Jhalla, Sukhi, etc.	Bloom stage to fruit stage (pea size)	First week to last week of May	Last week of May to second week of June
Joshimath–Auli, Chamoli (2200–2500 m asl)	Pink bud stage to bloom and petal fall stage	First week to third week of May	Last week of May to first week of June
Naugaon–Syori, Uttarkashi (1950–2200 m asl) (Syori, Koti, Talwadi Gwaldam)	Pink bud stage to petal fall stage	Last week of March to third week of April	Third week of April to second week of May
Tal–Talwari, Gwaldam, Chamoli (1750–1950 m asl)	Tight cluster to petal fall stage	First week to last week of April	Third week of April to second week of May
Tuni, Koti-Kanasar, Dehradun (1750–1950 m asl)	Pink bud stage to petal fall stage	Last week of March to second week of April	Second week of April to second week of May

1.8.1 SECONDARY INFECTION

The secondary infection through conidia takes place in an overlapping period of primary infection during spring. The pathogen now enters from its vegetative to reproductive phase. Once the fungus is established in the host, it breaks through the cuticle and begins to produce enormous numbers of conidiophores and conidia. Conidia are released only upon wetting by rain and soon they germinate to cause secondary infections on the host. The conidial development can occur in a wide range of temperature (0 °C–30 °C) and the optimum is about 16 °C–20 °C. Relative humidity of around 90% is congenial for its development. Free water and/or high humidity promote the rapid production of conidia. These are the principal cause (up to 100,000 conidia in each leaf lesion) for scab development during summer and early autumn. The latent period rapidly decreases with decrease in temperature. It is 8 days at 19°C and only 3 weeks at 5°C. Conidia production and dispersal in late summer depend upon temperature and occur both under rainy and dry conditions depending upon the wind velocity. During the day time, the release is more in the afternoon hours. Young leaves, 1–5 days after unfolding, and 3–4 weeks old fruits are more susceptible to infection. Management strategies must be adopted if there are more than two primary scab lesions per tree.

The dispersal of spores largely depends on the air movement in the orchard. Singh and Kumar (2009) studied the wind dissemination of conidia in Gangotri valley and have reported that it highly influences the ascospores discharge. They trapped conidia of *V. inaequalis* at 2500 m asl height and postulated that airborne spores were important epidemiologically in establishing disease in scab free orchards. He also suggested that compared to early morning the spore discharge is high during the afternoon hours. The ascospore release is higher when the pseudothecia are wetted during day time and more ascospore release occurs during warm and humid conditions (Singh and Kumar, 2009a, b). The spread of the summer spore occurs mostly via washing action of rain as compared to the air current. The development of the scab epidemic is influenced by the meteorological parameters either by their direct influence on the pathogen or by alteration of host physiology. A great number of studies have been undertaken in the G. B. Pant University of Agriculture and Technology through continues trials at Ranichauri station. A better understanding of disease cycle and epidemiology can be obtained through detailed studies on the impact weather parameters on the two distinct phases of the pathogen life cycle (Singh, 2005; Singh et al., 2015; 2016).

1.8.2 POTENTIAL ASCOSPORE DOSE

The potential of using potential ascospore dose (PAD) values for forecasting scab epidemic was suggested by Gadoury and MacHardy (1986) and has been successfully employed to reduce the number of fungicide applications in apple (MacHardy 1996). Accordingly, we have studied different PAD levels under Indian conditions. The potential ascospore dose (PAD=LD×LLD×PD×AD×n) for each orchard was the product of the lesion density (number of lesions on leaves per square meter of orchard floor at leaf fall), leaf litter density (proportion of the orchard floor covered by leaf litter at bud break), pseudothecial density (number of mature pseudothecia/ visible lesion multiplied by a lesion fertility factor), ascus density (number of asci/pseudothecium), and number of ascospores/ascus. To evaluate PAD, trials were conducted from 1994 to 2011 at Harsil of district Uttarkashi. During these years, recommended fungicide sprays during several infection periods, both before and after petal fall, had been left out. The trials gave promising results as there was no scab development in orchards with low initial inoculum levels. The PAD value in various orchards varied during the years under study and value during the spring had the highest correlation with the scab development in the following season.

1.8.3 DEGREE-DAYS

Gadoury and MacHardy (1982) made a model for estimation of the apple scab maturity. According to them, 95% of the spores germinated after 477 degree-days. The ascospore discharge of 50% and 95% were reported after 498 and 900 degree-days in South Africa (Schwabe et al., 1989). In Finland, the end of the primary inoculum period varied between 328 and 690 degree-days in between 10 years of observation. Stensvand et al. (1992), in Norway, found 50% and 95% ascospore discharge after approximately 250, 400, 350, and 600 degree-days in 1989 and 1990.

At the GB Pant University of Agriculture, Ranichauri, India, the ascospore discharge data accumulated over 15 years was plotted against the accumulated number of degree-days from the date of first ascospore discharge. The relation between ascospore maturation and degree-days was analyzed using standard regression analysis. When accumulated degree-days were plotted against cumulative percent ascospore discharge, a sigmoid curve (high–low values) was obtained. The curve represented three distinct phases, namely, lag, accelerated, and final phase, and was used for ASP and management program in the state. The lag phase is characterized by the slow build-up of inoculum and is comparatively easy to control with the fungicides. During the accelerated phase, the crop is at high risk and if adequate management options are not adopted, it can lead to a severe epidemic. The end of the primary season is marked by the final phase and emphasis should on the management of the other apple diseases. A protectant strategy is preferred over postinfection strategy especially in orchards where scab is endemic and sufficient management strategy was not adopted during the previous year. Cumulative degree-days were computed from daily maximum and minimum temperature values. The temperature of 0 °C was used as the base temperature for all degree-day calculations. The accumulation of degree-days was started with the discharge of the first ascospore. The mean number of cumulative degree-days for 50% and 75% ascospore discharge were following degree-days (Table 1.2).

TABLE 1.2 Degree-Day for Primary Infection of Scab in Different Places of Garhwal Himalayas

Apple Fruit Belts	Degree-Day Required for Ascospores Maturation (%)	
	50%	75%
Uttarkashi (Harsil fruit valley)	899.5	1080
Uttarkashi (Sukhi fruit valley)	557.2	842
Uttarkashi (Syouri–Navgaon fruit valley)	286.7	456.35
Dehradun (Koti–Kanasar fruit valley)	278.45	412.25
Chamoli (Tal–Talwari fruit valley)	480.15	587.75
Chamoli (Auli fruit valley)	603.45	983
Chamoli (Joshimath fruit valley)	338.6	493.7

Based on the observation during 2002–2012, two linear lines were developed, one for the use when the cumulative degree-days from 1 February to 15 May were < 618 and another for use when the cumulative degree-days for those dates were >618. Our observation showed for the orchards situated at 1900–2200 m asl, 50% and 95% ascospore maturation occurred after 338 and 859 cumulative degree-days (Syori, Talwadi, Joshimath, Gwaldam and Koti); whereas for orchards situated above 2200 m asl (Dharali, Sukhi, Harsil, Auli, Jhalla) degree-days were > 1167. We observed a longer duration of ascospore release, which was dependent on weather, under field conditions and it varied from one place to another. The seasonal variation of ascospore maturation was evident. The winter remains for a longer duration in the hills (Auli and Gangotri valley) and hence apple remains in the green tip stage till the end of March. The same phonological stage appears much earlier at lower altitude regions. A compression of two linear lines showed that in orchards at lower altitude with the increase in temperature the maturity period of ascospores increased proportionally. It was found 88.98% of accuracy for ascospore maturity under low altitude areas (1900–2200 m asl) by degree-days, contrary more higher accuracy was obtained for ascospore maturity (93.87%) at high altitude situated orchards.

1.8.4   MILLS INFORMATION

The first attempt to discourse the correlation between the scab infection severity and prevailing temperature and leaf wetting duration was done by Mills (1944). The Mills Table was subsequently used for the prediction of the scab epidemic throughout the world. Mills reported that a minimum 9 h and 5.9 leaf wetness is required for the ascospore infection and conidial infection, respectively. The symptoms would appear by 9 days at an optimum atmospheric temperature (18.2 °C–23.8 °C). Sixteen days were required for the same at 10 °C. Our observations indicated that 5–8 light infection periods during April and May would initiate primary infection and 9–14 days were required for symptom expression (Table 1.3). During 1990–2012, six to eight moderate infection periods were recorded under orchard conditions in each month and almost all indicated delay by a day in symptom expression (1–3 days). According to Mills' third criteria, that is, severe infection period, 2–5 infection periods were observed almost all the months that had an average temperature and leaf wetness period of 11.4 °C–15.2 °C and 3.4–27.2 h, respectively. Under these conditions, there was a delay of 1–2 days for symptom expression. However, in our studies, we observed that for light, moderate, and severe infection, there was a delay of 2–3, 1–2, and 1 day, respectively, for symptom expression. In Gangotri valley, there was 2 day (light), 1 day (moderate), and 1 day (severe) delay in symptom expression under orchard conditions. Some of the differences between our findings and that of Mills and LaPlante (1951) could be attributed to the delay in ascospore release and germination, which is dependent upon leaf wetness period, temperature, and inoculum abundance. The delay in ascospore release and symptom development in Uttarakhand may also be due to differences in the apple cultivars, topography and microclimatic conditions. The data collected during the trials were analyzed to establish its relevance in scab prediction and rescheduling fungicide applications. The minimum time required for successful infection of a quality tissue on the tree was taken as the minimum infection time. The relationship of temperature and leaf wetness for release of ascospore and infection of the host are consistent with the finding of several other scientists (Olivier et al., 1983; Schwabe 1979; Stensvand et al., 1997; Sys and Soenen, 1970; Thakur and Khosla, 1999). Need for revision of Mills criteria for infection was also suggested by MacHardy and Gadoury (1989).

TABLE 1.3   Mills Table and Our University Data to Appear at Incubation Period Based on Temperature and Leaf Moisture

Average on a Daily Basis Temperature (°C)	Minimum Wetting Hours of Leaves for Infection (Approx. Hours)[a]		Days Mandatory (After Infection for Symptom Emergence)
	As Per Mills Table	As Per University Data	
25	11	9	
16 and 24	9	6	9
15	10	8	12
14	10	8	13
13	11	9	13
12	11–5	9	14
11	12	10	15
10	14	13	16
9	15	13	17

[a]The infection period is considered to start at the commencement of the rain.

1.9   SCAB PREDICTIVE AND WARNING SERVICE IN UTTARAKHAND HILLS

Apple scab forewarning service carried out under ICAR, NATP, NAIP, and UCOST projects is being followed in Uttarakhand. Such forewarning helps the orchardists in need-based fungicide use. It begins in the early spring and predicts the time of initial disease development. The maturity and discharge of ascospores from the mature pseudothecia coincide with the pink bud to petal fall stage of the host. The quantity of primary inoculums is measured as (1) ascospore discharge (productivity) based on the number of mature spores/cm^2 on overwintered leaf area and (2) ascospore dose, which is the number of spore/volume of air, (3) the infection period of the pathogen is predicted using modified Mills table by monitored by measuring ambient temperature leaf wetness period. Information on the maturation and release of ascospores, degree-days for the maturation of ascospores, and PAD values are also collected from different regions, forecasting time, and extend of primary infection. In most of the countries, greater reliance is given to the first aspect in the prediction of primary infection, whereas the second aspect is utilized to develop a disease prediction system (after initiation of infection) and for determining the threshold level of inoculums. All these pieces of information regularly collected from our experimental sites located at Auli, Tuni, Gwaldam, Joshimath, Harsil, Mukteshwar, Purola, Bona, Koti-Kanasar, Almora, and Ramgarh of Uttarakhand and are used for forecasting scab epidemic in the fruit growing belts. Ascospore dose measures the actual concentration of the inoculum in the orchard air at different stages of host phenology and this is dependent on (1) ascospore productivity and (2) factors that influence spore release, that is, air temperature, light, time of days, climatic date, and leaf wetting by rain/dew. Numbers of traps are available for monitoring of ascospores dose in the air. Studies by the large number of workers provide ample evidence of ascospore maturity coinciding with the dormancy break of apple trees. In Delicious cultivar, there observed an increase in the percentage of colored spores until the bloom to early petal fall stage and then they decreased gradually. Much reliance is given for spray program commencing at Green tip to early petal fall stage of apple trees and continuing the fungicidal spray at short intervals until the primary scab season is over. In several countries, protective spray in the form of single application technique or reduced doses in SAT is commonly practiced. By analyzing the 20-year data generated from our studies, it is clear that tree phenology tree phonological stages can be used for developing a predictive equation for improving chemical management program (Singh et al., 2012; 2015; 2016). In Uttarakhand, μMETOS and ASP were able to predict infection periods accurately as the tagged leaves showed new scab lesions accordingly. The minimum leaf wetting hours required for the initiation of primary infection by ascospores of the pathogen was indicated by the Revised Mills table. Except for ascospores released after the sunset due to rains, leaf wetting should be computed from the start of rain. A linear statistical model based on PAD and the accumulated degree-days from the ascospore maturation was developed by Singh and Kumar (2009). Even after the development of automatic weather data-based scab predictive systems, the majority of the orchardists still follow the traditional fungicide spray schedule, that is,

10-day spray schedule starting from the early petal fall stage in spring. The above information collected from experimental sites on the infection period is passed on to the orchardists by blowing a characteristic signaling, telephonic communication, SMS, local newspaper, government organization, and through personal contacts or messages flashed 4–5 times through "All India Radio, Najibabad" on the urgent need to undertake immediate spray or to reschedule already recommended spray program (Singh et al., 2012; 2015; 2016). Such forewarning has benefited the grower in minimizing damages due to scab and also reduced fungicide usage.

1.10 PHENOLIC COMPOUNDS

Phenolic compounds are considered as one of the most important parameters responsible for imparting resistance to diseases and are evident from their higher accumulation in the scab-resistant varieties (Singh et al., 2008). The estimation of phenolic compounds in plant samples was done through high-performance liquid chromatography (HPLC) (Shimadzu Corporation, Kyoto, Japan) according to Singh et al. (2002). The results indicated a wide variation in phenol and phenolic acid content between the susceptible and resistant apple cultivars. All the Delicious cultivar had four phenolic acids, in which benzoic acid was maximum (17.47–47.11 mg) followed by tannic (1.30–2.22 mg), caffeic (0.06–0.21 mg), and ferulic acid (0.05–0.11 mg) per gram fresh weight of the plant leaves. The benzoic acid content in "Golden Delicious" leaves was almost twice as high as in the Red Delicious and Royal cultivars. "Red Delicious" had the highest level of tannic and ferulic acid in the leaves; whereas caffeic and benzoic acids were highest in "Golden Delicious." The average total phenolic content in the leaves of three cultivars was the highest level of benzoic acid (34.155 mg) followed by tannic acid (1.460 mg), caffeic acid (0.143 mg), and ferulic acid (0.0766 mg). However, in fruits caffeic, ferulic. and benzoic acids were completely absent, Red Delicious cultivar had tannic (1.198 mg), caffeic (0.076 mg), ferulic (0.048 mg), and benzoic (17.473 mg) acids. The average tannic acid content was maximum in Red Delicious (1.198 mg per gm fresh weight of the fruit) that was followed by Golden (0.696 mg) and Royal Delicious (0.518 mg) cultivars (Singh et al., 2015). HPLC analysis of infected and healthy fruits and leaves revealed a wide variation in their number and type of phenolic acids. There was statistically significantly less phenolic acid in scab-infected leaves and fruits than in healthy ones. It varied greatly between the disease-resistant and susceptible cultivars as well. In Delicious cultivars, the content of phenolic acid was high both in healthy leaves and fruits. The relation between the induction of phenolic compounds and disease resistance has been reported by several scientists in other host–pathogen systems as well (Benhamon et al., 2000; Daayf et al., 2000; Prasad et al., 2010; Singh et al., 2000; Singh et al., 2008; van Peer et al., 1991). Among the three most important Delicious cultivars, Golden Delicious had minimum scab incidence and was also rich in phenolic acid. Other cultivars with low phenol content, viz., Royal and Red Delicious, had a high incidence of scab. Our findings were in accordance with the findings of Gunen et al. (2005) and Singh et al. (2008). Red Delicious, which is a highly susceptible cultivar (90%–95% disease intensity) had very low content of different phenolic acids like ferulic, tannic, benzoic, salicylic, and caffeic acid. An inherent variation was observed in the phenolic acid contents of cultivars that are highly and moderately susceptible to scab. The amounts of phytochemicals also vary from year to year, season to season, and region to region (Jill Max, 2005). The findings of our study further stress upon the importance of the role of phenolic compounds in imparting resistance to the apple scab pathogen *V. inaequalis.*

1.11   DISEASE MANAGEMENT

Scab is one of the most destructive diseases affecting apple in the whole apple growing areas of India, which requires an intensive fungicidal application for its management. The approach toward the control of apple scab differs from that of the steps followed for other diseases. There are three approaches: (1) interrupting the life cycle of the scab pathogen, (2) according protection against primary (biological and chemical control) and secondary infection (chemical control), and (3) resistant varieties (MacHardy et al., 2001).

1.11.1   BIOLOGICAL CONTROL

Isolation and evaluation of indigenous antagonists were carried to determine the effect of antagonists on pseudothecial density in relation to the progress of asci and ascospores. The conidial germination was significantly condensed by 50% by the six antagonistic isolates (*Trichoderma* sp., *Epicoccum nigrum, Chaetomium* sp., *Myrothecium* sp., *Fusarium* sp., *Botrytis cinerea*). The three bacterial isolates completely inhibited germination and scab formation on the leaf surface (UHCR16, UHCR 22, UHCR 47). These isolates provided the greatest scab suppression in experiments that was similar to that provided by Flusilazole (Singh, 2006, Singh et al., 2005, 2007, 2008).

Urea is one chemical that could be utilized to accelerate leaf litter decomposition significantly and earlier leaf fall better the rotting of apple leaf litter during the overwintering stages after urea treatment (Carisse and Rolland, 2004). Five percent urea and hundred percent cow urine spray not only suppressed the ascospore production entirely in the fallen apple leaves but also helped in the early decomposition of the leaves by increasing the microbial movement of fallen leaves, whereas no such apparent effect on leaves treated with Bordeaux mixture, copper, and carbendazim plus mancozeb was observed through these some equally effective fungicides in suppressing ascospore discharge (Prasad et al., 2018).

1.11.2   CHEMICAL CONTROL

Chemical control is a successful component of an overall management program and provides flexibility in disease management. Different manage-ment strategies have been developed to thwart the disease resourcefully through the scheduled application of fungicides in a protective squirt program (Gupta, 1985), besides, application of systemic and sterol-inhibiting fungicides (Schwab, 1980; Schwabe and Jones, 1983; Thakur and Gupta, 1990, 1992). The ergosterol biosynthesis inhibiting (EBI) fungicide had a prolonged inhibitory effect on the development and sporulation of the fungus, thereby preventing the development of distinctive apple scab symptoms. The EBI fungicides have been applied according to their activity at an appropriate time so that grower should adopt need-based spray program instead of unnecessary sprays.

1.11.2.1   PRELEAF FALL SPRAY

Primary source of inoculums overwintered on infected leaves in orchards that could give ascospores all through the growing season, foliar fungicides alone may not give acceptable control of the disease. The application of urea at 5% level of concentrations prior to abscission of leaves provided maximum (94.58%) inhibition of pseudothecial progress and suppression of ascospore productivity (95.50%). This urea spray not only controls the ascospore production but also promotes the decomposition of the leaves by enhancing the microbial activity. Carbendazim also exhibited 94.28%

reduction but dithianon, chlorothalonil, and dodine showed least pseudo-thecial inhibitory action. Efficacy of thiophanate-methyl, biteranol, dodine, chlorothalonil, and captan was observed here as well and confirmed earlier reports (Gupta and Lele, 1989; Gupta, 1989; O'Leary and Suttan, 1986; Gadoury et al., 1989, Prasad et al., 2018; Singh and Kumar, 1997, 1999a, b, and c). The spray schedules provided logical good management of primary and secondary infections of scab. Singh and Kumar (1999a, b, and c) recom-mended that fungicidal cover would provide a precautionary measure to manage the apple scab disease in those areas where it is being disseminated through airborne conidia.

1.11.2.2   PREINFECTION (PROTECTIVE) ACTIVITY

Systemic and nonsystemic fungicides at diverse concentrations (0.01%–0.30%) were sprayed at different phonological stages (full bloom, petal fall, fruit pea size, fruit walnut size, fruit development I, fruit development II, fruit development III, 20–25 days earlier than harvest) of the trees until run off with a foot sprayer during the upward season with a solitary application of 5% urea in autumn at preleaf fall stage. Seven to eight sprays per growing season were given for management of apple scab. EBI fungicides, Punch 40% emulsifiable concentrates (EC) (0.1%), Score 25% EC (0.015%), and Topas 10 EC (0.05%) were found effective with respect to foliage and fruit and were significantly superior over every one treatment throughout the period of study. The standard protectant, Indofil M-45 (0.30%) in a repeated spray schedule gave lower disease reduction with esteem to leaves (55.92%) and fruits (71.18%) infection. The differences among fungicides were not entirely consistent between years. In general, however, treatment with the EBI and sterol-inhibiting *fungicides* (SIF) fungicides resulted in a superior protective activity as compared to the conformist protectants (Thakur and Gupta, 1992; Singh and Kumar, 1997, 2001).

1.11.2.3   POSTINFECTION (CURATIVE) ACTIVITY

The spraying of fungicides after the incident of infection on terminal leaves at petal fall stage can be determined by a microprocessor-based orchard environment monitor (RSS-412 ASP and µMETOS). The efficiency of most of the conformist fungicides in controlling the disease at this stage is partial to only a few hours and therefore should be functional within 18–24 h of forewarning. Punch (0.2%), Score (0.015%), and Topas (0.05%) were very much effective in preventing the formation of scab lesions on leaves, when they are applied after 72 h of infection. These fungicides also gave total control of scab up to 120 h Bavistin (0.05%) was too efficient in controlling the scab after 72, 96, and 120 h, whereas contaf was effective only up to 72 h of infection periods. The sprays were not repeated in fewer than 9 days (incubation period) from the previous spray, even though the additional infection was recorded. Throughout the period of study 18–21 infection periods occurred, but only six to seven sprays were given (Singh and Kumar, 1997, 2001; Singh et al., 2001).

1.11.2.4   POSTSYMPTOM (ERADICATIVE) ACTIVITY

Discontinuity in executing the fungicide sprays can affect the establishment of scab infection both on the foliage and fruit of apple. Thus there is an impending necessity to apply fungicides that have a soaring level of eradicating or antisporulant activity to inhibit the spore production from the lesions produced by such infections. Perusal of data revealed that dificonazole (0.015%) was more effective than the other fungicides in reducing the recovery of the fungus from lesion 5 week after treatment and condensed the conidial production up to 86.32%. Flusilazole (0.01%) and penconazole (0.05%) were also found effective fungicides, inhibited the conidial production (Singh and Kumar, 1999a, b, and c).

1.11.2.4.1   Spray Schedule

Effective spray program for the management of apple scab disease was formulated/recommended on the base of the fungicides evaluation for their different means of action against the disease. Such spray strategies are in popular use in every apple increasing areas in India (Singh et al., 1997; Singh and Kumar, 1999a, b, and c; Sharma, 1995). The following protective spray schedule may be followed to protect trees: at silver tip to green tip stage spray mancozeb (0.3%), dodine (0.1%), or dithianon (0.1%); at pink bud stage spray bitertanol (0.05%) or mancozeb (0.3%) + sulfur (0.2%); at petal fall stage spray carbendazim or thiophanate methyl (0.05%); at pea sized fruits spray dodine (0.075%), fenarimol (0.04%), myclobutanil, mancozeb (0.3%), or bitertanol (0.05%). At walnut-sized fruits spray carbendazim or thiophanate methyl (0.025%) + mancozeb

(0.25%) or captan (0.3%). At fruit development spray captan (0.1%), biter-tanol (0.1%), mancozeb (0.3%), or fenarimol (0.05%). Spray mancozeb (0.3%) or captan (0.3%) 20–25 days prior to harvesting (Singh and Kumar 1999a, b, and c).

1.11.2.5   COMPARATIVE EFFICACY OF ANTISCAB SPRAY PROGRAMME

The apple orchards of Uttarkashi district are badly affected by the scab disease due to which the growers are incurring heavy losses. The incidence of scab was more among Delicious varieties of apples. The 75% orchardists applied 1–15 sprays of chemicals locally available at Harsil Government Department. Others applied three to five sprays after the appearance of the disease. Therefore an effort has been made to evaluate the loss incurred to apple growers owing to the occurrence of scab disease in their orchards. A village wise list of apple orchardists was obtained from the Harsil Centre of Horticulture Department. A sample of 20% of the apple orchardists was taken randomly from each listed village. In total, 60 orchardists were preferred for study. The primary data was collected from the selected orchardists during 1996–97, with the assist of a well-structured questionnaire, based on individual meeting and observations of the orchards. Secondary information was collected from the research workers and Horticulture Department of Uttarakhand Government. The data was tabulated and then analyzed in the desired manner (Kumar et al., 1998, 1999).

1.11.3   RESISTANT CULTIVARS

Breeding of disease-resistant fruit cultivars usually involves combining the best pomological characters of susceptible cultivars with disease resistance as an additional character. Genetical resistances are more often identified in primitive species or obsolete cultivars with mediocre pomological characters. This often requires the breeder to improve pomological characters simulta-neously while transmitting the desired disease resistance. Early attempts on apple improvement did not involve disease resistance. Though, by the end of the 19th century, researchers were becoming aware of the possibility of developing scab resistant varieties. Aderhold (1897), in Germany, presented data on the susceptibility of 160 apple varieties. While non showed absolute immunity, 11 including Antonovka proved to be noticeably resistant. The program for the expansion of apple cultivars resistant to scab was initiated in 1948 by Hough in Illinois University and Shay of Purdue University. Since then a formal cooperative program designed as PRI has been in effect between Purdue, Rutgers, and Illinois universities.

Twelve scab resistant varieties Prima, Priscilla, Sir Prize, Jonafree, Red Free, Coop-12, Coop-13, Mac Free, Liberty, and Freedom were introduced at two locations in Himachal Pradesh from western countries. Some promising cultivars Prima, Priscilla, Mac Free, Liberty, Coop-12, Red Free, and Sir Prize are under critical observations since their introduction in 1984 in Himachal Pradesh. The varieties have been interred/planted in the heart of scab prone areas along with scab susceptible cultivars. All the seven cultivars showed complete field resistant to apple scab both on foliage and fruits. Apple Prima and Priscilla are the earliest to break dormancy in the initial week of March in low hills and in the third week at high hills every year. The cultivars introduced from the United States are most appropriate for growing in Himachal Pradesh and Uttarakhand under the low chilling requirement. Apple Red Free, Coop-12 (Emra), and Priscilla are performing excellently in the mid hills but not in high hills. These have been recommended for commercial cultivation in the low and mid hills of Himachal Pradesh and Uttarakhand. The fruits mature early, develop a scarlet red color within 84–96 days, and are sweet (Thakur and Vaidya, 2006).

1.11.4   STORAGE SCAB IN COMMERCIAL APPLE CULTIVARS

Jammu and Kashmir, Himachal Pradesh, and Uttarakhand are the leading producer of fresh apples (*Malus domestica*) in India. Apple harvest starts in mid-August and may not finish until early November, depending on varieties. At harvest, the fruit is placed into wooden or plastic-coated boxes or plastic bins, and bins are then transported into storage and packing facilities. Spoilage of apples is an immense problem in Uttarakhand, provoked by the absence of appropriate storage amenities and scarcity of information on factors causing losses during storage. Scab infection that happens just before harvest may be symptomless at picking yet build up into storage scab lesions after harvest. The spots are small, generally less than 1 cm in size, brown or jet black color is depending on the variety of apple, and are often shiny because of the intact cuticle. They are circular and darker in color than orchard lesions and have definite borders. Late season scab noticeable at harvest is almost impossible to differentiate from storage scab. Such lesions often remain viable in storage and increase by forming fringe of olive-green mycelia around the edge of the scab spots. The fruit rot occurred during the normal storage life of apple cultivars susceptible to storage diseases based on the observance of samples harvested from the orchard without fungicides applications in 2005–2011. Shares of rotted apples increased very rapidly toward the ending of the normal storage life of the cultivars, reaching on average 7.57% of lost fruits at that period. However, economically significant losses caused by the occurrence of storage diseases of apples arose only during the last quarter of the storage life. Fruit samples that were harvested from the orchard through the integrated plant protection had 3.54% of rotted fruits at the end of their storage life on average. Economically significant losses caused by rotting of apples were usually recorded only through the last month of their normal storage life.

The efficacy of fungicides as preharvest spray-on storage scab was also tested. Penconazole (0.05%) and flusilazole (0.01%) were effective as preharvest fungicidal sprays in controlling scab during storage. While fungicides akin to carbendazim (0.05%), bitertanol (0.05%), difenoconazole (0.015%), and captan (0.30%) were efficient in controlling both scab and postharvest rotting. The cultural filtrate of *Gliocladium* and *Trichoderma* spp. were useful in controlling storage scab up to 30 and 60 days, respectively, while used as dip treatment (Anonymous, 1998). Garlic extract was least effective in checking storage rot to some extent, though Econeem at 0.2% and 0.4% proved quite effective. Maximum protection up to 2 months of storage was given by flusilazole (0.1%), carbendazim (0.025%), and difenoconazole (0.01%) is given in Table 1.4.

TABLE 1.4 Fungicidal Dip Treatments on the Progress of Storage Scab and Fruit Rot in Red Delicious Apple

Treatment	Conc. (%)	Dipping Time (min)	Infection at 0 day		Percent Fruit Rot After Days		Marginal Mean
			No. of Spots	Area	30	60	
Econeem	0.2	60	19.2	8.6	55.55	66.66	61.11
	0.2	30	17.5	4.2	11.11	33.33	22.22
	0.4	60	16.5	7.3	11.11	33.33	22.22
	0.4	30	19.1	6	11.11	55.55	33.33
Garlic extract (50:1)	0.2	60	23.90	8.90	33.33	44.44	38.88
	0.2	30	26.60	9.90	44.44	77.77	61.11

TABLE 1.4 *(Continued)*

Treatment	Conc. (%)	Dipping Time (min)	Infection at 0 day		Percent Fruit Rot After Days		Marginal Mean
			No. of Spots	Area	30	60	
Garlic extract (25:1)	0.4	60	32.20	11.80	55.55	77.77	66.66
	0.4	30	32.90	13.80	55.55	100.00	77.78
Flusilazole	0.01	10	22.80	8.80	11.11	22.22	16.66
Carbendazim	0.025	10	20.10	6.60	11.11	22.22	16.66
Difenoconazole	0.015	10	27.90	10.70	22.22	22.22	22.22
Control			32.90	13.90	44.44	100.00	72.22

KEYWORDS

- apple
- disease
- scab
- management

REFERENCES

Aderhold, R. (1897). Revision der species *Venturia chlorospora, inaequalis* and *ditricha autorum. Hedwigia* 36: 81.

Anonymous. (1998). Epidemiology and Integrated Management of Apple Scab. Status report 1987–1998, Dr. Y.S. Parmar University of horticulture and Forestry, Solan, 166pp.

Benhamou, N., Gagne, S., Quere, D. L., and Dehbi, I. (2000). Bacterial–mediated induced resistance in cucumber: beneficial effect of the endophytic bacterium *Serratia plymuthica* on the protection against infection by *Phythium ultimum. Phytopathology* 90: 45–56.

Carisse, O. and Rolland, D. (2004). Effect of timing of application of the biological control agent *Microsphaeropsis ochracea* on the production and ejection pattern of ascospores by *Venturia inaequalis*. Phytopathology 94: 1305–1314

Daaye, F., Ongena, M., Boulanger R., Hadrami, I. El., and Belanger, R. R. (2000). Induction of phenolic compounds in two cultivars of cucumber by treatment of healthy and powdery mildew-infected plants with extract of *Reynautria sachalinensis. J. Chem. Ecol.* 26: 1579–1593.

Gadoury, D. M. and MacHardy, W. E. (1982). A model to estimate maturity of ascospores of *Venturia inaequalis. Phytopathology* 72: 901–904.

Gadoury, D. M. and MacHardy W. E. (1986). Forecasting ascospore dose of *Venturia inaeualis* in commercial apple orchards. *Phytopathology*. 76: 112–118.

Gadoury, D. M., Mack Hardy, W. E., and Rosenberger, D. A. 1989. Interaction of pesticide application schedules for disease and insect control in apple orchards of the Northeastern United States. *Plant Dis.* 73:98–105S.

Gunen, Y., Misirli, A. and Gulcan, R. (2005). Leaf phenolic content of pear cultivars resistant or susceptible to fire blight. *Sci. Hortic.* 105: 213–221 .

Gupta, G. K. (1985). Recent trends in forecasting and control of apple scab (*Venturia inaequalis* (Cke.) Wint.). *Pesticides* 19: 19–31.

Gupta, G. K. (1989). Pre-leaf spray of chemical and cow urine in suppression of ascospore. *Pesticides* 23:23–24.

Gupta, G. K. and Lele V. C. (1989). Roll of urea in suppression of ascigenous stage, and comparative in vitro efficacy of fungicides against apple scab. *Indian J. Agric. Sci.* 50: 167–173.

Heald, F. D. (1963). Manual of Plant Diseases, 3rd ed. McGraw Hill Book Company Inc., New York, 953 pp.

Heye, C. C. and Andrews, J. H. (1983). Antagonism of *Athelia bombacina* and *Chaetomium globosum* to the apple scab pathogen, *Venturia inaequalis. Phytopathology* 73: 650–654.

Jill Max (2005). An apple a day keeps the cancer away. Scien Central News. Feb. 12, 1–3

Kumar, Jagdish, Singh, K. P., and Doshi, J. K. 1998. Apple scab-estimation of loss and economics of its control. *Indian J. Plant Path.* 16: 55–56.

Kumar, Jagdish, Singh, K. P., and Doshi, J. K. (1999). Factor affecting adoption of recommended spray schedules for the control of apple scab in Uttarkashi district of U.P. hills. (Abstr.) National symposium on challenges and prospects of plant pathology in the coming millennium, *NBRI.* p.42.

MacHardy, W. E. and Gadoury D. M. (1989). A revision of Mills's criteria for predicting apple scab infection period. *Phytopathology.* 79: 304–310.

MacHardy, W. E., Gadoury, D. M., and Gessler, C. (2001). Parasitic and biological fitness of *Venturia inaequalis*: Relationship to disease management strategies. *Plant Dis.* 85: 1036–1051.

Mac Hardy, W. E. (1996). Apple Scab: Biology, Epidemiology and Management. 545 pp. Academic Press, APS, St. Paul, Minnesota.

Mills, W. D. and LaPlante, A. A. (1951). Diseases and insect in the orchard. Extension Bulletin. Cornell University. 11: 1–5.

O'Leary, A. L. and Suttan, T. B. (1986). The influence of temperature and moisture on the quantitative production of pseudothecia of *Venturia inaequalis. Phytopathology* 76: 199–204

Olivier, J. M., Lambert, C., Lefeuvrt, M. (1983). Application Dur thermohumectoographe Kitinara. Etude des resques de tavelure due pommieralechella du maineet Loire (France). Bulletin OEPP. 13: 47–56.

Prasad D., Singh, A., Singh, K. P., Bist, S. and Singh, U. P. (2010). The role of phenolic compounds in disease resistance in Geranium. *Arch. Phytopathol. Plant Prot.* 43: 615–623.

Prasad, R.K., Singh, K.P. and Gupta, R.K. (2018). Pre leaf fall spray of chemical, cow urine and fungal antagonists on spring ascospore production of the apple scab pathogen, *Venturia inaequalis. Int. J. Curr. Microbiol. Appl. Sci.* 7(8): 575–586.

Prasad, R.K., Singh, K.P., Gupta, R.K. and Kumar, J. (2018). Prevalence and severity of apple scab in Uttarakhand hills. *Int. J. Curr. Microbiol. Appl. Sci.* 7(10): 1166–1184.

Prasad, R.K., Singh, K.P., Gupta, R.K. and Kumar, J. (2018). Symptoms, etiology and disease cycle of apple scab in Harsil, Gangotri fruit belt, Uttarakhand, India. *Int. J. Curr. Microbiol. Appl. Sci.* 7(11): 3554–3565.

Schwabe, W. F. S. (1979). Changes in scab susceptibility of apple leaves as influenced by age. *Phytophylactica.* 11: 53–56.

Schwabe, W. F. S. (1980). Curative activity of fungicide against apple leaf infection by *Venturia inaequalis. Phytophylactica,* 12: 199–207.

Schwabe, W. S., Jones, A. L. and Blerk, E. V. (1989). Relation of degree-day accumulation to maturation of ascospore of *Venturia inaequalis* in South Africa. *Phytophylactica,* 21: 13–16.

Schwabe, W. F. S., and Jones, A. L. (1983). Apple scab control with bitertanol as influenced by adjuvant addition. *Plant Dis.* 67: 1371–1373.

Sharma, J. N. (1995). Efficacy of fungicidal spray schedules for the control the apple scab. *Indian J. Mycol. Plant Pathol.,* 25: 250–253.

Singh, K. P. (2005). Integrated management of apple scab through development of scab warning system in Uttaranchal hills. In: Final report of NATP-CGP project. G B Pant University of Agriculture & Technology, Hill Campus, Ranichauri, p. 46.

Singh, K. P. (2006). Investigation on development of bio-control measures for the management of saprophytic stage of apple scab pathogen *Venturia inaequalis* in Uttaranchal Himalayas. ICAR, Report, New Delhi, p. 53.

Singh, K. P., and Kumar, J. (1997). Maturation and diischarge of ascospores of *Venturia inaequalis* in Central Himalayas of India (Abstr.) International Conference on Integrated Plant Disease Management for Sustainable Agriculture, organised by Indian Phytopathological Society, held at IARI, New Delhi from November 10–15.

Singh, K. P. and Kumar, J. (1999a). Severe incidence of apple scab in Bhatwari fruit belt of Uttar Pradesh Himalayas.(Abstr).National symposium on challenges and prospects of plant pathology in the coming millennium, NBRI, p. 24.

Singh, K. P. and Kumar, J. (1999b). Studies on ascospore maturity of *Venturia inaequalis,* the apple scab pathogen, in Central Himalayas of India. *J. Mycol. Plant Pathol.* 29: 408–415.

Singh, K. P. and Kumar, J. (1999c). Efficacy of different fungicidal spray schedules in combating apple scab severity in Uttar Pradesh Himalayas. *Indian Phytopathol.* 52: 142–147.

Singh, K. P., and Kumar, J. (2001). Environmental factor favouring development of pseudothecia and ascospores maturation of *Venturia inaequalis* in hills of Uttaranchal. International Symposium on "Frontiers of fungal diversity and diseases in South-East Asia. Organized by Department of Botany, DDU Gorakhpur University, Gorakhpur from Feb. 9–11, 2001.

Singh, K. P., and Kumar, J. (2004). Incorporation of weather forecasting in integrated biological- chemical management of *Venturia inaequalis.* In: 15th International Plant Protection Congress on "Plant Protection towards the 21st century" organized by The International Association for the Plant Protection Sciences, held from May 11–16, 2004 at Beijing, China

Singh, K. P. and Kumar, J. (2005). IPM of apple diseases. Technical Bulletin, GBPUAT, CFHA 6: 1–48.

Singh, K. P. and Kumar, J. (2007). Prediction of ascospore maturation of *Venturia inaequalis* in Central Himalayas. 3rd Asian Conference on Plant Pathology on "The role of Plant Pathology in rapidly globalizing economies of Asia" organized by The Indonesian Phytopathological Society and Gadjah Mada University from August 20–24, 2007 at Yogyakarta, Indonesia

Singh, K. P. and Kumar, J. (2008). Disease warning system for scab of apple: a field study. GBPUAT, CFHA 22: 1–18

Singh, K. P. and Kumar, J. (2009a). Forewarning scab caused by *Venturia inaequalis* in commercial apple orchards. 5th International Conference on Plant Pathology in the Globalized Era organized by Indian Phytopathological Society, IARI, New Delhi from November 10–13, 2009 at IARI, New Delhi.

Singh, K. P., and Kumar, J. (2009b). Potential ascospore dose of apple scab fungus, *Venturia inaequalis*, from Indian Himalayas. *Indian J. Agric. Sci.* 79: 184–189.

Singh, U. P., Prithiviraj, B., Singh, K. P., Sarma, B. K. (2000). Control of powdery mildew (*Erysiphe pisi*) of pea (*Pisum sativum*) by combined applicaton of plant growth-promoting rhizobacteria and Neemazal™. *J. Plant Dis. Prot* 107: 59–66.

Singh, K. P., Kumar, B. and Kumar, J. (2012). Development of computer-based weather monitoring and disease warning system for integrated management of apple scab in Uttarakhand hills. In: Eco-friendly innovative approaches in plant disease management, 682p., Singh, V. K., Singh, Y. and Singh, A. (eds.). International Book Distributors, Dehradun, pp. 113–146.

Singh, K. P., Kumar, J., Chauhan, M. S., Prasad, R. C., and Singh, Amitabh. (2005). Developing apple scab warning services in Uttaranchal Himalayas. (abst.). *J. Mycol. Plant Pathol.* 35: 493.

Singh, K. P., Kumar, J. and Kumar, B. (2010). Apple disease research in the Gangotri valley region of India. In: Microbial diversity and plant disease management, 625p., Singh, K.P. and Shahi, D.K. (eds). VDM Verlag Dr. Muller GmbH & Co. KG, Germany/USA/UK, pp. 276–301.

Singh, K. P., Kumar, J., and Pal, Ramesh (1997). Efficiency of fungicides for apple scab control in Garhwal Himlayas. (Abstr.) Society of Mycology and Plant Pathology, annual conference, Jan. 11–13, held at Department of Botany, Dr. Babasaheb Ambedkar Marathwada University, Aurangabad, Maharashtra.

Singh, K. P., Kumar, J., and Pal, Ramesh. (1997). Efficacy of different fungicidal sprays schedules in combating scab severity in apple in U.P. Himalayas. (Abstr.) Indian Phytopathological Society, National Symposium, February 15–17, held at Rani Durgavati Vishwavidyalaya, Jabalpur, M.P.

Singh, K. P., Kumar, J., Prasad, R. K., Chauhan, M. S., and Singh, A. (2006). A new integrated approach in forecasting primary apple scab, *Venturia inaequalis* (Cke.) Wint. Key note address in National Seminar "Current trends in crop disease management for improving productivity" organized by RRL, Jorhat and IPS, New Delhi from January 19–20, 2006, CTCDM-2006/04.

Singh, K. P., Kumar, J. and Singh, H. B. (2001). Curative and protective action of ergosterol-biosynthesis inhibiting fungicides in relation to infection periods against apple scab in Uttaranchal Himalayas. *Indian J. Plant Pathol.* 19: 34–38.

Singh, K. P., Prasad, Dinesh and Kumar, J. (2008). Devising an integrated apple disease management program through the use of antagonists, need based fungicides and advisory services in Uttarakhand hills. (abst.). *Indian Phytopathol.* 61: 398.

Singh, K .P., Prasad, D., Srinivas, P. and Kumar, J. (2007). Apple pests in organic orchards of Gangotri valley. (abst.). *Indian Phytopathol.* 60: 399.

Singh, K. P, Kumar, J., Singh, A., Prasad, R. K., Singh, R. P. and Prasad, D. (2015). Predicting potential ascospore dose of *Venturia inaequalis* (Cks) Wint in farmers apple orchards in Central Himalayas of India. *Plant Pathol. J.* 14: 189–195.

Singh, K.P., Kumar, J., Singh, A., Prasad, R.K., Singh, R.P. and Prasad, D. (2016). Maturation, ascospores discharge pattern and relevance of Mills criteria for predicting apple scab infection period in India. *Plant Pathol. J.* 15: 108–123

Singh, Amitabh, Singh, K. P., Rajwar, G. S., Dimri, D. C., and Kumar, J. (2006). Integrated control of powdery mildews by the mycoparasite *Ampelomyces quisqualis* and plant products. National symposium on microbial diversity and plant health problems, organized by DDU Gorakhpur University, Gorakhpur and IPS-M E Zone, New Delhi from December 18 and 19, 2006, TS. IV-P 12, p. 91.

Singh, Amitabh. Singh, K. P., Rajwar, G. S. and Singh, U. P. (2010). Phenolic acid content–a criterion for selection of resistant apple cultivars against *Podosphaera leucotricha* (Ell. and Ev.) Salmon. *Arch. Phytopathol. Plant Prot.* 43: 1138–1143.

Stensvand, A., Amundsen, T. and Semb, L. (1992). Apple scab (Venturia inaequalis) in Norway. *Acta Phytopathol. Entomol. Hungarica* 27: 593–598.

Stensvand, A., Gadoury, D. M., Amundsen, T., Semb, L., and Seem, R. C. (1997). Ascospore release and infection of apple leaves by conidia and ascospores of *Venturia inaequalis* at low temperature. *Phytopathology* 87: 1046–1053.

Sys, S., and Soenen A. (1970). Investigations on the infection criteria of scab on apple with respect to the table of Mills and La Plante. Agriculture, Heverlee, Belgium. 18: 3–8.

Thakur, V. S. and Gupta, G. K. (1990). Evaluation of pre symptom expression conidia production and viability of *Venturia inaequalis*. *Indian Phytopathol.* 43: 520–526.

Thakur, V. S. and Gupta, G. K. (1992). Post-infection fungicidal inhibition of apple scab (*Venturia inaequalis*) sporulation. *Indian J. Agric. Sci.* 62:629–36.

Thakur V. S. and Khosla K. (1999). Relevance of Mills infection periods to apple scab (*Venturia inaequalis*) prediction and rescheduling fungicide application in Himachal Pradesh. *Indian J. Agric. Sci.*, 69: 152–156.

Thakur, V. S. and Vaidya, S. (2006). Status and prospects of integrated pest management strategies in selected crops: Apple. In: A. Singh, O.P. Sharma and D.K. Garg (eds.) Integrated Pest Management: Principles and Applications vol. 2. pp. 448–477.

Van Peer, R., Nieman, G.J., and Schippers, B. (1991). Induced resistance and phytoalexin accumulation in biological control of Fusarium wilt of carnation by Pseudomonas WCS417r. *Phytopathology*, 81: 728–734

Yan, Z., Reddy, M. S., Ryu, C. M., McInroy, J. A., Wilson, M., and Kloepper, J. W. (2002). Induced systemic resistance against tomato late blight elicited by plant growth-promoting rhizobacteria. *Phytopathology*, 92: 1329–1333.

CHAPTER 2

IMPORTANT DISEASES OF APPLE (*Malus domestica* L.) AND THEIR MANAGEMENT

ANAMIKA JAMWAL[1,*], J. N. SRIVASTAVA[2], and UPMA DUTTA[3]

[1]Krishi Vigyan Kendra—Kathua, Sher-e-Kashmir University of Agricultural Sciences & Technology of Jammu, Jammu and Kashmir, India

[2]Department of Plant Pathology, Bihar Agricultural University, Sabour, Bhagalpur, Bihar, India

[3]Division of Microbiology, Sher-e-Kashmir University of Agricultural Sciences & Technology of Jammu, Jammu and Kashmir, India

[*]Corresponding author. E-mail: annajamwal@gmail.com

ABSTRACT

Apple (*Malus pumila* Mill) Syns. (*Malus domestica* L.) temperate fruit is broadly created on the planet after banana, orange, and grapevine. The biggest apple delivering nation on the planet is China. It is a deciduous tree and has a place with Rosaceae family. Fruits are sweet and pomaceous. Apple tree originates in central Asia, is a wild progenitor, and *Malus sieversii* is still found today. Apple trees and parts of their plants are powerless for various infections and insect parasites and bacteria that cause severe diseases in apple plant. These are scab disease, powdery mildew, phytophthora collar rot, sooty blotch and flyspeck, black rot of apple, and fire blight of apple. The chapter will deal with these diseases, their causes, and management in different sections.

2.1 SCAB DISEASE

Introduction/Economic Importance

Scab disease was first time reported in1819 from Sweden and furthers other apple-growing regions of World but in India, scab disease was first time reported in 1935 from Kashmir region of Jammu and Kashmir on the native cultivar Ambri (Nath, 1935). The disease is reported in almost all growing states of India. Scab disease occurred as epidemics in Kashmir region of Jammu and Kashmir in the year 1973 and infected about 70,000 acres apple orchard and further resulted in a loss of rupees about 55 lakhs (Joshi et al., 1975). Scab plant sickness in apple has major economic significance. It happens each year causing noteworthy monetary misfortunes in yields. Losses are mainly due to fruit infection, resulting in low yield, poor fruit quality, and poor storage. Infection also leads to defoliation with poor tree vigor and winter survival.

Causal Organism: (*Venturia inequalis*) *(*Cke) Went.
Imperfect stage: *Spilocaea pomi* Fr.

The disease is caused due to *V. inaequalis* fungal pathogen. The pathogen survives two distinct phases of life cycle. Perfect stage/sexual stage of pathogen survives on diseased fallen leaves in winter season and imperfect stage (asexual stage) survives on leaves and fruits during the growing season as a parasite. The respective pathogen host plants are blooming crab apples (*Malus* spp.), hawthorn (*Crataegus* spp.), and loquat (*Eriobotrya japonica*). Pear (*Pyrus* spp.) trees and their parts of the plant are also infected with a related fungus and *Venturia pirina*, causing almost identical symptoms (Gupta, 1985).

Symptoms:

Scab disease indications appeared on the upper ground plant parts of the apple tree, including leaves, petioles, blossoms, sepals, pedicels, juvenile shoots, bud scales, and fruits. Indications on blooms mostly occur in small, dull green sores at the blossom base. Previously and during blossoming, it happens on the sepals and stem pedicel resulting in fruit drop. The lesions are initially seen in the spring season, usually at the underside of expanding leaves. Later it spreads on the upper leaf surfaces also. Lesions may appear quite large (more than 1 cm in diameter) on young leaves and small in size on expanded and older leaves because older leaves are often resistant to infection. The lesions are circular and small in the early stages. With time it increases in size and due to the production of asexual spores (conidia) it becomes olive and velvety in color. Impacted tissues gradually become blurred and puckered. Regularly, the leaf injuries get broken and torn. Injuries on the leaves and fruits are shown in appearance as large rankled and "scabby," with a certain edge. The symptom on fruits in water-soaked areas is velvety, green to olive-brown lesions (Gupta, 1990, 1985).

New fruit infections may exhibit signs of fruit distortion. Severely infected leaves and fruits drop frequently from the tree. Defoliation of leaves, thanks to scab disease infection, continues for 2 or 3 years during a row, the tree becomes more vulnerable to freeze damage, insect injury, and other disease infections. The symptoms of scab disease in fruits and leaves are depicted in Figures 2.1 and 2.2.

FIGURE 2.1   Scab lesions on fruits.

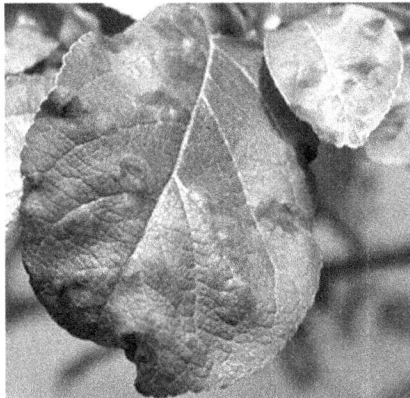

FIGURE 2.2   Secondary scab lesions on leaves.

Disease Cycle and Epidemiology:

Apple scab organism (*V. inaequalis*) is live to the region between the host cuticle and the epidermis in living leaves. When diseased leaves have fallen and passed on, they are totally colonized by the contagious mycelium. In leaf debris, most of the pseudothecial initials form within a month after leaves fall and enter in a dormancy period. Development in contaminated buds in mellow atmospheres will deliver conidia that fill in as essential inoculums. Mating occurs in the leaves in the Spring. When pseudothecia grow wet the asci extend through the pseudothecium's highest level. The ascospores are persuasively released into the air and dispersed through the downpour of breeze and wind. If there is enough humidity due to rain or dew, the ascospores infect and cause lesions on apple flowers and young leaves (Gupta and Lele, 1976).

The primary lesions developed on apple blossom and young unfolding leaves produce conidia after 9–30 days, which are scattered to different leaves and fruits and also causing secondary contaminations. A single lesion can be produced up to 100,000 conidia. The rate of ascospore or conidia lesion infection depends on temperature, humidity and, in part, on the tissue host, genotype, and plant age characteristics. Lesions grow more gradually during a cold season or resistant apple cultivars, and they may be smaller in size, resulting in fewer secondary cycles.

Moistness influences the creation of conidia and it requires 60%–70% moistness for spore creation. Contamination of ascospores and conidia is highly dependent upon the wetness of leaves and fruits as well as on average temperature. The Mills table relating the term and temperature of the leaf wetness is used to decide on conidia contamination. For instance, if leaves are wet at an average temperature of 18 °C for 9 h, the infection will be low, but if at the same temperature the leaves are wet for 18 h or more, the infection will be severe. Lesions will produce conidia after 9 days if the average temperature is 18 °C but it takes 17 days at a lower temperature (8 °C) (Gupta and Lele, 1976).

Integrated Disease Management:

Cultural Control:

- Control of apple scab sickness can be overseen by cultural activities, planting of malady safe assortments, utilization of synthetic compounds, and utilization of natural control operators.

- Follow cultural operations, the sanitation (picking and disposing of infected fallen leaves on the ground) can be an effective method of disease control. But it is not practical for big orchards (Gupta, 1985).
- Using urea solution splash applied to apple trees, not long before leaf drop or on infected fallen leaves on the ground, reduces the amount of ascospore inoculum during the coming season.
- Control of apple scab sickness is often done by pruning the trees regularly. This process will help in proper air movement and more sunlight penetration to the soil surface in orchards.
- Proper air movement and sufficient sunlight getting to the ground will help in fast drying of fallen infected leaves, resulting in poor growth of scab ascospore.
- Extra irrigation must be avoided, especially when climatic conditions are favorable to the production of scab (Gupta, 1985).

Choose Resistant Cultivars and Rootstocks:

- Enormous quantities of scab infection safe assortments of apples are accessible. Planting of scab safe assortments is the best alternative for the disease.
- The scab-resistant varieties of apples are exempt from Capitalism, Independence, Gold Rush, Jonafree, Liberty, Pure, and Red. The varieties get acceptance from both farmers and consumers. Apple varieties David, Harvest Gold, Mary Potter, and Prairie Fire are crab apples that are scab resistant to blooming. These varieties have greater bloom potential, thanks to no risk of defoliation by apple scab infection and delightful landscape plants.
- Varieties susceptible to scab disease, namely, Delicious, Fuji, Gala, Golden Delicious, McIntosh, and Winesap are the chosen varieties of apples for consumers to develop in dry regions where scab disease is not a restriction of apple cultivation (Gupta and Lele, 1976).

Chemical Control:

- Scab sickness of apple can be control with the use of preventive (protectant) and curative chemical fungicides.
- Preventive chemicals are applied on the apple leaves and fruits before the disease. On the off chance that ascospores or conidia are available on plant surfaces, fungicide application keeps the parasitic spores from sprouting or entering the host tissue.

- Spraying of chemicals at 7 days regular intervals in the early season is a necessity to prevent the infection on newly emerging tissues (Gupta and Lele, 1976).
- A spray scheduled in protective mode is adopted on apple trees with 7 sprays of nonsystemic and systemic during the growing season of the crops.
- Therapeutic chemicals have constrained foundational action. Whenever applied, not long after contamination, they are equipped for entering the leaf and controlling further improvement of the growth.
- Growers apply curative fungicides only when there is a need. During the dry spring season, application of fungicides may control initial infection. At apple harvest, the residual effect of fungicides is either absent or present at a safe level (Gupta and Lele, 1976).

Biological Control:

Biological control agent or natural microbial antagonist is a good substitute of chemical fungicides in the control of scab disease of apple; it is a viable option in the future (Ellis et al., 1997).

2.2   POWDERY MILDEW

Introduction/Economic Importance:

Powdery mildew disease was first time reported in 1871 from Iowa, United States on apple seedlings and further around the world in the apple-growing regions (Bessey, 1877). The disease is reported all most all growing states of India. This disease is primarily loss to seedling stocks and nursery. It is a big problem altogether Victorian apples. The disease is more vulnerable to winter squash, Golden Delicious, Buckingham, White dotted Red (Maharaji), Granny Smith and Ambary cultivars. Powdery mildew disease can cause poor growth of a tree, deadly buds, reduced area of leaf, reduced shoot growth, and lower quality of fruit. Heavy infection of mildew in plants reduces yield up to 80% (Glawe, 2008).

Causal Organism: *Podosphaera leucotricha* (El. Ev.) Salmon
Imperfect stage: *Oidium farinosum* (Cooke)

Powdery mildew is caused by *P. leucotricha* [(El. Ev.) Salmon], fungal organism. The occurrence of *Erysiphe heraclei* and *Podosphaera oxycanthae* can also cause the disease. The host plant is crab apple (*Malus* spp.) and pear (*Pyrus* spp.).

Symptoms:

Symptoms showed on newly shoots, newly and mature leaves, blooms, and also on fruits. Symptoms indications are generally observed on the leaves and fruits. First of all disease symptoms appeared during the spring season is 3–4 days delay in opening new buds. Symptoms of the disease appeared as on leaves, blooms, and fruits covered with a white to light dim powdery mass, these powders mass are spores of the pathogen (Glawe, 2008). The disease rapidly spreads from leaf to other parts of the plant. White to light dim powdery mass on twigs of apple tree disappear they entire covering with dim earthy colored or earthy colored fruiting bodies of the pathogen. The infected foliage turns out to be hard and fragile.

The disease signs are observed as white spots on the top and lower surfaces on the new shooting leaves. The disease might cause some leaves to twist, break, or become disfigured. If disease appeared on succulent leaves and growing shoots pathogen cause reduce the dimensions of the whole shoot and also leaves and shoots may turn brown in midsummer season. Flowers do not develop normally, they turn greenish-white, and become fail to produce fruits (Hickey and Yoder, 1990). The symptom of powdery mildew on dormant shoot, blossom, leaves, and fruits are shown in Figures 2.3–2.5.

FIGURE 2.3   Symptoms on dormant shoot.

FIGURE 2.4   Symptoms on blossom and leaves.

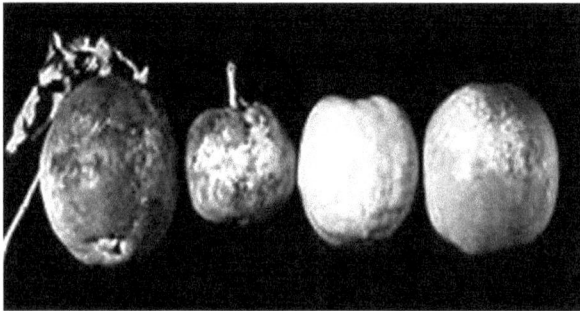

FIGURE 2.5   Symptoms on fruit.

Infection in Apple Shoots:

In the spring season the terminal buds are started to develop. The fungus colonizes green tissue that is emerging young. These infected flag shoots showed color as silver-gray, and may also have shown defoliation, stunted growth, and symptoms of dieback. The primary contaminations of the shoots produce inoculums in the advance stage of the developing season, which cause secondary contamination on leaves, blooms, and even fruits. Intensely infected trees are debilitated, and mycelium and various earthy-colored fruiting bodies (ascocarps) are required to strike secondary inoculums (Hickey and Yoder, 1990).

Infection in Leaves and Fruits:

Secondary infection occurs on expanded young leaves with the fall of wind-borne spores of powdery mildew. Fungal colonies consisting of mycelium/hypha and spores on leaf surface show white patches and chlorotic. Leaves infected along the margin become curled, crinkled, or folded longitudinally. Impacted tissues produce the powdery, silver-greyish appearance characteristic of powdery mildews during disease progression. Infections on blossom receptacle or young fruit during ripening fruits will cause net resetting and discoloration. Fruits exhibit distorted and small in size, causing low yield and lower quality (Hickey and Yoder, 1990).

Infection in Apple Blossoms:

Infected flowering buds develop silver-grayish color and opened within 5–8 days later after healthy buds. Petals are blurred and color-impose pale or light green. Blossoms develop shrivelling and struggling to bear fruit. The buds stay lethargic till following spring. The heavily contaminated buds with pathogen at the hour of opening may wipe out the yield of following season (Turechek et al., 2004).

Disease Cycle and Epidemiology:

Apple flower and shoot buds are infected by the mycelium of *P. leucotricha* in the previous year. The mycelium stays dormant in wintertime. Onset of spring season, breaks in dormancy and fungus resume growth, colonizing on the development of young leaf shoots and tissue. Asexual conidia are produced from these primary infections on conidiophores and are dispersed by wind.

Pathogen conidia germinated at high humidity (>70%) and between 10 °C and 25 °C temperature. Conidia grow to frame hyphal outgrowths that traverse the surface of the leaf, grow, and then straighten into oppressor form. These structures release chemical substances that allow parasitic contamination pegs to penetrate the epidermal cells of the plant and then extend to frame haustoria. Haustoria are specific organs shaped within living plant cells, assimilating supplements, and remaining the parasite. The contamination protocol (hyphal outgrowth > appressorium > contamination peg > haustorium) is rehashed as the mold state progresses or as optional diseases lead to new settlement production until helpless tissue is not accessible at this stage. Development in the late season may cause an unexpected increase in mold action. Despite contributing to the production of rapid inoculums, auxiliary disease cycles are also responsible for tainting horizontal and terminal buds that will carry the organism during the winter. (Hickey and Yoder, 1990; Turechek, et al., 2004).

Integrated Disease Management:

Cultural Control:

- The primary contamination is also regulated by eliminating the first sources of inoculum (i.e., infected flower and shoot buds in the preceding year).
- Infected white terminal shoots must be pruned out during winter or early spring. Pruning in big orchards is not economically feasible. If cheap manpower is out there, it is going to help to great extent (Yoder, 1992).

Choose Resistant Cultivars and Rootstocks:

- Resistant cultivars are accessible that exhibit characteristic protection from buildup and need control just under high sickness tension. Resistant cultivar of apple are Jonafree, Prima, and Enterprise, however, they are not broadly developed.
- Choice of the cultivar is impacted more by business bid, fruits size, fruits attractiveness, and pollination characteristics than by ailment obstruction.
- Cultivars like Golden Delicious and Granny Smith are widely grown but are moderately to highly susceptible to mildew and can require the management of chemical diseases (Hickey and Yoder, 1990). Maharaja Chunth and Golden Chinese cultivars are resistant to powdery mildew (Verma and Gupta, 1988).

Chemical Control:

- Secondary infection and fruit infection can be controlled by foliar sprays of fungicides. In commercial orchards, fungicides are almost always used to control mildew, as well as other apple diseases.
- Fungicides are usually applied at 7–10 days intervals from the tight-cluster stage until terminal shoot growth ends during midsummer. Chemical sprays are most effective during the growth

of the leaf, and thus the postbloom period. Sometimes, this could be 18 sprays in highly susceptible cultivars for disease management (Yoder et al., 1997).

- Most effective chemicals that are used for the control of mildew are sulfur, fenbuconazole, myclobutanil, and strobilurins. Apple growers do not apply just one fungicide, but they apply different action groups of fungicides, alternately. For instance, sulfur is applied on less susceptible cultivars at low-risk time. Benzimidazoles were effective against mildew but, thanks to the widespread resistance developed in *V. inaequalis* (apple scab), their application was reduced.
- Powdery mildew gives chronic effect on tree vigor and yield and the advance effects are longevity and profitability of the trees. Successive management of mildew is also dependent on apple grower knowledge about management strategy.

2.3 PHYTOPHTHORA COLLAR ROT

Introduction/Economic Importance:

One disease that commonly afflicts apples is *Phytophthora* collar rot. Apple cultivar "Grimes Golden" are most susceptible to this disease. *Phytophthora* collar rot is highly intense on apple trees planted on rootstocks of Malling Merton 106 (MM.106). Practice of unpolluted cultivation of apple and overhead irrigation with contaminated surface water also help in spreading the disease. The host plants of causal organism are many other tree fruits, including nut trees.

Causal Organism:

The causal organism of this disease is *Phytophthora cactorum,* but sometimes *Phytophthora megasperma, Phytophthora syringae, Phytophthora cambivora*, and *Pythium ultimum* are also involved in the development of collar rot (Rana and Gupta, 1984).

Symptoms:

Phytophthora disease occur around the crown, spread down to the rootlets, and a soil-line lesion develops. If scion cultivar is defenseless, the disease can spread to the association of the bud as well. When the bark dries out, the canker frames a distinct shape, and the callus tissues build at the edges. Equally, the *Phytophthora* organism can attack vulnerable cultivar fruits. Contaminated apple trees develop a solid, light tan spoil on or near the ground, as well as rot (Rana and Gupta, 1984).

Phytophthora collar rot assaults the apple trunks approximately up to 76 cm from groundline of soil. The disease exhibits contamination between the soil line and the root-crown layer. Contaminations often begin at the lateral root intersection of the trees. New canker problem to define at ground level. As they enlarge, they become black, depressed and, with an unusual margin, roughly oval shape. Contaminated bark gives earthy-colored, often delicate, soft, or thin symptoms. The earthy- to ruddy earthy-colored wood staining and a sticky exudate under the ulcer's dead bark indicate *Phytoph- thora's* closeness. Dark streaks also occur near to the shift and extend out past the edge of the canker. On the off chance that an infection expands for quite a while, just the minor regions show the run of the mill shading and surface of recently slaughtered tissue. The canker grows quickly in both even and vertical bearings. A definitive impact of neckline spoil is to support the influenced appendage roots, or trunk, coming about death of influenced organs or the whole tree. The main signs of the infection are declining vigor, poor terminal bud development, and the premature creation of small, scanty yellow leaves in summer or rosy bronze leaves in early harvest. Early fruit maturation and the development of extra colored fruits are common symp- toms of the disease. The foliar effects can occur only on branches, above the canker, and severely infected trees die eventually. *Phytophthora* collar rot is usually associated with cancers caused by sporadic or secondary invaders (Rana and Gupta, 1984). Contamination of collar rot disease in apple tree and apple fields is depicted in Figures 2.6 and 2.7, respectively.

Disease Cycle and Epidemiology:

The pathogen mycelium survives as thick-walled oospores or as chla- mydospores in trunk cankers during winter and in the soil for many years. In the soil of old apple plantations, oospores and chlamydospores occur in the greatest numbers. These thick-walled oospores resist unfavorable environment for a long period of time, namely drought and freezing temperatures. They too are largely resistant to chemical therapy. Oospores may grow in bark hole during late winter, or the mycelium can invade the bark at the soil groundline. The numerous zoospores are formed in the presummer or late-spring from sporangia in the soil. The lower part of the trees is reached with zoospores, swims in rain flow of water or in the water of irrigation. The fruits that established near the ground are additionally contaminated by rain splash and infect new branches and cause fruit decay. With the help of few zoospores *Phytophthora* fungus can produce large numbers of spores (Anonymous, 1989).

FIGURE 2.6   Phytophthora collar rot at the base of apple tree.

FIGURE 2.7   Apple orchard infected with collar rot.

Moist soil and natural temperatures are preferred for fungal growth and infection. *Phytophthora cactorum* pathogen grows at 13 °C and stays active up to 21°C. Apple's "Midwest" susceptible cultivars tend to be affected most often during the cycle of blossom, again in late August or early September, and in October at the time of commencement or dormancy. Fungus is most active in soil during these times and host is most susceptible.

Integrated Disease Management:

Phytophthora collar rot disease is very hard to control and difficult to track. Draw out long, cool season and wet environment before blooming, and planting can be seriously affected by the disease. Defensive measures are necessary to control the illness. If infections are late noticed, control of the disease is too late and the apple tree is saved (Anonymous, 1989).

Cultural Control:

- *Phytophthora* collar rot is generally prevalence in damp and poorly drained soils or high clay content soil.
- New planting of apple should be planted on slopes land with good drainage. The moderate slope should be created in fields before apple planting on ridges. It will help in smooth water flow.
- Tiling well-drained orchard wetlands improve internal soil drainage and control the disease.
- Flooding and soil saucer around the apple tree trunks should be avoided.
- Planting of apple trees should be shallow as within the nursery.
- High-density planting will cause most contamination of disease.
- Fertilization should not be applied in orchards during late summer or early fall. Application of fertilizers during these periods induces late growth and low cold tolerance. The young apple trees must be given support.

Choose Resistant Cultivars and Rootstocks:

- There should be no planting of susceptible rootstocks where the soil is heavy and poorly drained.
- Apple, Midwest, Dutch, Grimes Golden, and Lodi cultivars are highly vulnerable to rotary collar disease.
- Golden Beautiful, Jonathan, McIntosh and Rome Beauty Cultivars seedling rootstocks are fairly resistant to collar rot disease.
- The cultivars Melba, Red Wonderful, Rich, and Winesap are highly resistant to collar rot disease.
- Disease-free nursery stock should be planted as contaminated nursery stocks are a vital disease source.
- The susceptibility of the Dwarf rootstocks to *Phytophthora* collar rot varies.

- The rootstocks East Malling (M), viz., M-4 and M-9 are recognized as being reasonably resistant. Ottawa-3 has also been recognized as relatively resistant to Canadian rootstock viz.
- The M-25, M-26, Malling-Merton (MM) 103, MM-104, MM-106, MM-107, MM-109, MM-110, MM-111, MM-113, and MM-115 rootstocks are highly susceptible.
- The M-2, M-7, and MM-112 rootstocks are intermediate type resistors. Yet the rootstocks M-9 and M-26 are moderately susceptible.
- Provides best management practices for the production of resistant apple cultivars on resistant rootstocks (Anonymous, 1989).
- Dwarf rootstocks, notably MM-106, are vulnerable to winter injury.

Chemical Control:

- Disease can be controlled by drenching with Blitox/fytolon/Blue Copper @0.1% or Dithane M45 @0.03% in 30 cm radius around the apple tree trunk (Agrwal, 1970).
- Foliar sprays by Fosetyl–aluminum completely control the disease and also increase the growth of trees and fruit yield (Utkhede and Smith, 1995).

If the infection of collar rot occurs, some success in arresting and eradicating the disease may be obtained if the grower follows the following points:

- Inarch resistant varietal 1-year-old whips into the trunk well above the cankered region. If 25%–50% or more of the trunk circumference of trees is infected this treatment has value.
- Expel soil from the base of a late contaminated tree to reveal the entire cancer zone.
- Remove every sick tissue and leave the storage compartment zone open to allow drying of the contaminated territory. It will forestall further malady improvement. In late fall, top off the territory around the storage compartment with great new soil. (Agrwal, 1970).
- *Phytophthora*-specific fungicide should be used before growth begins at planting time or in the spring.
- Application of chemicals in the fall after harvest has been shown to be successful in combination with good cultural practices and tolerant rootstock planting (Byrde and Jordan, 1977).

2.4   SOOTY BLOTCH AND FLYSPECK DISEASES

Introduction/Economic Importance:

Sooty blotch and flyspeck are different diseases and both diseases generally appeared together on same fruit superficial, darkly pigmented blemishes and smudges, or discoloration. Both diseases degrade the market price of fruits. They often appeared on the same fruit superficial at the same time. Sooty blotch and flyspeck diseases reduce market value/market price of fruits. These diseases spread widely in the apple cultivar "Midwest." Both diseases are managed in the same way, and are usually taken together.

   Causal Organism of Sooty Blotch: *Gloeodes pomigena, Peltaster fructicola, Geastrumia polystigmatis*, or *Leptodontium elatius*

   Causal Organism of Flyspeck: *Schizothyrium pomi* or *Zygophiala jamaicensis*

   The fungus *G. pomigena* (Schw) causes sooty blotch, and the fungus *S. pomi* (Mont and Fr.) (formerly *Microthyriella rubi*; anamorph *Z. jamaicensis*) causes flyspeck (Saha, 2002).

   In the last of 20th century three fungal species had been recognized as an easygoing creature of dirty smudge infection on North Carolina apples on the base of their morphology. These are *P. fructicola, G. polystigmatis*, and *L. elatius* (Johnson, et al., (1997).

   The host plants of sooty blotch and flyspeck are ash blackberry, citrus, crab-apple, dogwood, grape, hawthorn, leatherwood, magnolia, maple, mountain ash, oak, orchids, peach, pear, persimmon, prickly-ash, raspberry, orange, tulip poplar, willow, and wintergreen. The host plants of flyspeck are American elderberry, banana, blackberry, carnation, apple, and grapes.

Symptoms:

A. Sooty Blotch:

Symptoms of sooty blotch appeared as "smudge" appearance or as sooty on the apple fruit surface (Figure 2.8). Symptoms of sooty blotch turned up on the spherical of fruits as dirty or overcast blotches. The blotches with indefinite margins are olive green. They are a fourth of an inch or massive in distance across and can be combined to cover a large portion of the fruit. This fungus produces pycnidia that have a large number of spore during late May/early June to autumn and move by air or rain splash. The pycnidia are found with some characteristic, namely, thick-walled, joined dull hyphae, and breach cell-like fragments. The presence of "smudge" results from the proximity of many moments, dim pycnidia, which is interconnected by a mass of free, joined dull hyphae. The growth of the dingy smear is usually limited to the outside surface of the skin of the fingernail. The hyphae infrequently enter between the epidermal cell dividers and the skin of the fingernail. The fungus grows and develops with a wide range of temperature, namely, 18 °C–27 °C.

B. Flyspeck:

Symptoms of flyspeck appeared as shiny, black, and round shape dots on the apple fruit surface that resembling by the fly excreta appear on the apple fruit skin (Figure 2.8). The individual "fly spots" are more widely distributed and larger than the sooty blotch pathogen pycnidia. The pathogen flyspeck develops sexual fruiting bodies (pseudothecia) and is interconnected by fine hyphae. Overwhelming scouring or blanching will remove the imperfections. Symptoms of sooty blotch and flyspeck diseases are depicted in Figures 2.8–2.10.

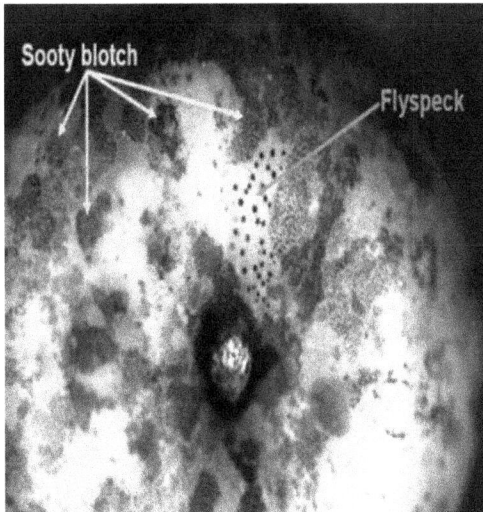

FIGURE 2.8   Sooty blotch and flyspeck on apple fruit.

FIGURE 2.9   Sooty blotch on apple fruit.

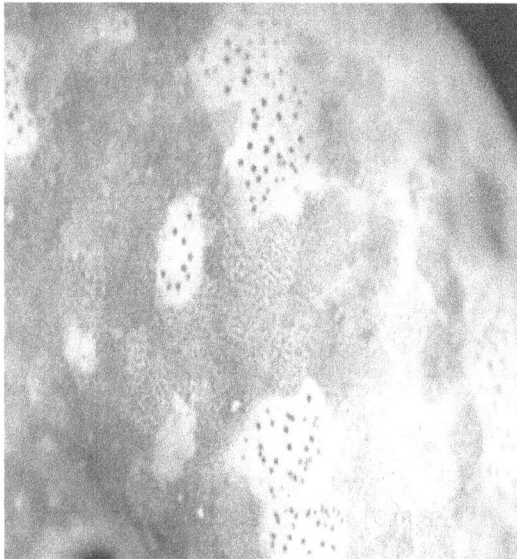

FIGURE 2.10   Flyspeck on apple fruit.

Disease Cycle and Epidemiology:

G. pomigena and S. pomi both fungi overwinter as pseudothecia on infected apple twigs and woody reservoir hosts. The cycles of the disease involve favorable temperature and humidity conditions, which both diseases require for infection.

A. Sooty Blotch:

The pathogen pycnidia on infected plants formed huge amounts of spores (conidia) that flow out of diseases and gather in a coagulated mass. The conidia are transmitted from late May or early June to harvest time via air and a wind-blown splash of rain through plantations. The thick-walled, dim hyphae framed on apple fruits and various twigs frequently split into pieces like cells. They produce subsequent infection on spherical fruit after spread. The pathogen growth ranged from 18 °C to 27 °C at broad temperatures. Cool environment followed by July and August, the hot environment in May and June, sooty blotch as dirty smudge hasn't occurred on spherical fruits. Illness episodes are usually extreme with cool spring blustery conditions coupled with summer rains and preharvest low temperatures. The length of incubation from the infection to the onset of symptoms may be as short as 5 days under ideal conditions. In any case, this period in the plantation, as a rule, lasts 20 to over 60 days. Contamination of the fruits can occur whenever the petal drops, but it is usually prevalent in the middle to prefall phase.

B. Flyspeck:

This fungus formed both ascospores and conidia during the late spring season. Ascospores and conidia are spread from plant to plant by air currents. The incubation period of fungus is about 15 days in cool weather (18 °C).

Integrate Disease Management:

Sooty blotch and flyspeck diseases are prevalent in damp, shaded areas of apple plantation. Management practices that help in air movement and making a dry condition in plantation areas are best in disease control in orchards. Fungicides must be applied, starting shortly after petal fall (calyx) and continuing when cooler weather occurs in the autumn prior to harvest is good disease management practice. Mostly growers spray preventive fungicide at 10–14 days interval for regulating sooty blotch/flyspeck disease. Wide range of integrated pest management options, however, are available to growers to scaleback pesticide utilization.

Cultural Control:

A. Pruning

- Sooty blotch/flyspeck can be controlled by pruning the trees. Annual pruning is the best practice for controlling the disease.

Fruit surface gets dry by pruning and hence less chances of disease infection. Sooty blotch/flyspeck pathogen requires wet or moist condition to grow.

- Additionally, well-pruned trees allow better penetration of fungicides. Even, frequent sprays of fungicides are not ready to control the damage done by sooty blotch/flyspeck, if fruits are in inner canopy of inadequately pruned trees.

B. *Bramble Management:*

- Different species of *rubus*, including blackberries and raspberries, are a big sooty blotch/flyspeck reservoir. Therefore cultivation of those berries near apple planting areas should either be avoided or planted with distant, using windbreak or hedgerow between apple orchards and berries.
- Wild brambles on the sides of orchard blocks should be planted in the spring and summer months. This helps in managing the volume of sooty blotch spores and flyspeck.

C. *Thinning:*

- In the early season, the soothy blotch smudge/flyspeck dots formed on the fruit surface. The apples are too little to even think about touching one another.
- High-density planting gives damp condition and favor the two sooty blotch smudge/flyspeck dots.
- Proper fruit thinning will reduce cluster formation, resulting in effecting disease control by fungicides application.

D. *Mowing:*

- Mowing is a technique to advance better wind current and quick dry condition in plantation areas.
- Tall grassy apple trees with low-hanging branches can stay wet with dew until late morning.
- Regular mowing allows low-draping apples to dry quickly and discourages multiplication of pathogenic sooty blotch/flyspeck.

Choose Resistant Cultivars and Rootstocks:

- There are no apple cultivars immune to both sooty blotch/flyspeck disease.

- Some cultivars have physiological characteristics that reduce the possibility of damage to sooty blotch/flyspeck.
- Short maturing cultivars show less damage as sooty blotch/flyspeck pathogen gets less time to multiply on the surface of the apple. Therefore planting of short duration maturing cultivars, viz., red apple cultivars can reduce the damage caused by sooty blotch/flyspeck.

Chemical Control:

- Chemicals can be used effectively and safely to manage sooty blotch/flyspeck.
- The sooty blotch/flyspeck infection warning system has been established in recent years. The wetness from rain or dew during this phase must be assessed for 175 h between first cover spray and second-cover fungicide spray. In second-cover spray fungicides are preventively applied (at intervals of 10–14 days) until harvest. In primary and second cover sprays, a mixture of benzimidazole fungicide (i.e., thiophanate-methyl) and a contact fungicide (captan) should be used with a disease warning system.
- In addition, a personal weather sensor can be used to measure trees wetness. For information assortment, the sensor is mounted in the lower shelter of a delegate tree at the plantation. The information is downloaded to a PC as regularly as fundamental (for the most part once per week first and foremost and all the more frequently at moving toward the limit level). With the wetness estimation framework, information assortment starts after the principal spread shower. Wetness hours are included every day till 175 complete hours. The second spread splash is applied after 175 h. A standard interim timetable of 14 days is followed for the rest of the period (Agrwal, 1967).
- Strobilurin chemicals, kresoxim-methyl and trifloxystrobin, were tested for sooty blotch/flyspeck control. Alternate sprays of kresoxim-methyl + trifloxystrobin and thiophanate-methyl (Topsin-M) + captan were as effective in controlling sooty blotch/flyspeck disease as full-season sprays of thiophanate methyl plus captan. Strobilurin fungicides are less detrimental to human health and climate. That also provides the same control level as traditional sprays (Agrwal, 1967).
- Potassium bicarbonate ($KHCO_3$) may be a new, moderately effective chemical for sooty blotch/flyspeck disease management in apples. But it the simpler treatment for apple disease management, particularly mildew in organic planting areas.

- Flyspecks are sexual fungal fruiting bodies (pseudothecia) and interconnected with very fine hyphae. The imperfections are often removed through vigorous rubbing or bleaching.

2.5 BLACK FRUIT ROT DISEASE

Introduction/Economic Importance:

The causal organism of disease is *Botryosphaeria obtusa* (Schwein.) and *Diplodia seriata*. Shoemaker pathogens often cause black fruit rot, frog-eye leaf spot, and branch canker and dieback of shoots in the apple tree. The disease is severe on summer and fall varieties. The disease is often completely controlled within the orchard by ablation the limb cankers and follow the spray schedules.

Causal Organism: *B. obtusa* (Schweinitz)

The pathogen was firstly detail described by Schweinitz (1832) as *Sphaeria obtusa Schwein.* Cooke (1892) identified the pathogen as *Physalospora obtusa* (Schwein.). Shoemaker (1964) was considered the pathogen a species of *Botryosphaeria* Ces. & De Not. fungus and introduced the new combination *Botryosphaeria obtuse.*

Symptoms:

Black fruit rot disease symptoms appeared on fruit, leaves, and bark caused by the pathogen *B. obtusa.* Commence checking your apple trees for the sign of infection a few weeks after the petals.

(a) Symptoms as Fruit Rot:

Infected fruit with disease stand rotten tissue with concentrated brown and black rings may cover large areas of the fruit. Pycnidia fungal spore is seen as small black spots on older infections. Some fruits mummify and stay tied to the tree. Early infection of fruits leads 1-week advancement in fruit ripening than normal harvest. Infected fruits are rotten at the core.

(b) Symptoms as Frogeye Leaf Spot:

Infected apple leaves showed frogeye leaf spot symptoms and later than black rot disease develop. On the leaves are found the circular lesions with violet or red outer borders and the light tan interiors.

(c) Symptoms as Branch Cankers:

Symptoms appeared on infected limbs as a sunken, reddish-brown region sometimes formed with rough-looking or broken bark. Those cankers may not be evident readily. If the plants have fruit rot or frogeye leaf spot on them, care should be taken to spot the canker on shoots. Symptoms of fruit rot, frogeye leaf spot, and branch cankers in apple are depicted in Figures 2.11–2.13.

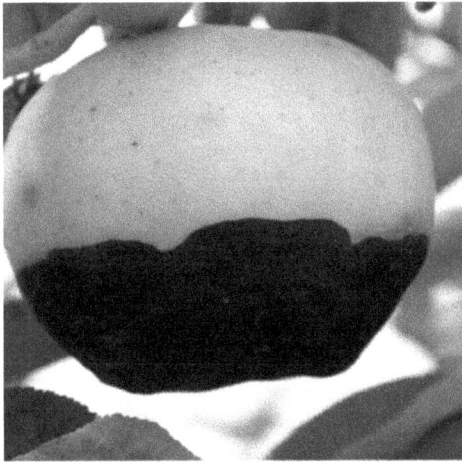

FIGURE 2.11   Fruit rot.

FIGURE 2.12   Frogeye leaf spot.

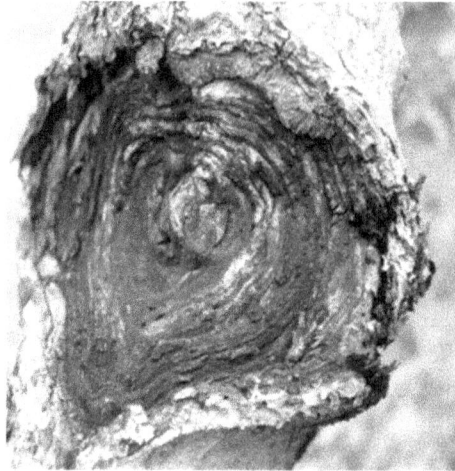

FIGURE 2.13   Branch cankers.

Disease Cycle and Epidemiology:

Throughout the winter season the fungus is present in cankerous, mummified fruits, and deadwood bark. The black fungal fruiting bodies (pycnidia and perithecia) release conids and ascospores within the season. These spores propagate the disease into healthy leaves, fruit, and wood. The utmost discharge of spores occurs around the time of blossom, but conidia assembly may continue throughout the summer during wet periods. The conidia will remain viable for at least a year.

Infection with the leaf usually occurs during the petal-fall period. Conidia are attached to the leaf and should germinate within 5 or 6 h with a wet condition. Using natural openings below the surface or by insects, hail, or other wounds, they enter the leaves after conidia germination. The spread of the pathogen, as well as disease, is additionally favored by RH (relative humidity) and rain. Spore germination and infection are most favored at about 32 °C temperature.

Integrated Disease Management:

- The blend of cultural practices and fungicidal control is the best control practice of disease of apple.
- The inoculum may be reduced by expelling preserved apples and by pruning deadwood in the tree or on the ground surface of the plantation.

- Punned shoots, twigs, are a reservoir source of disease for next season.
- Pruning of trees every year and utilization of adjusted composts on soil test premise help in lessening the sickness contamination.
- Cancer commonly develops if trees are week and stress. Distinctive social practices/activities must be followed to keep up solid apple trees.

2.6   FIRE BLIGHT DISEASE

Introduction/Economic Importance:

Fire blight pathogen was introduced first time in northern Europe with bacterial ooze from fruit containers that imported from Northern America during 1950–1960. *Erwinia amylovora* bacteria was firstly described as the cause of a disease (Burrill, 1883). Finally, the causes of fireside blight in apple were reported in 1995–1996 from Hungary, Romania, Northern Italy, and Northern Spain. *E. amylovora* has been reported in more than 40 countries as a cause of a disease.

Causal Organism: *E. amylovora.*

The host plants of the causal organism are pear, quince, loquat, cotoneaster, pyracantha and hawthorn. Symptoms are similar in host plants and can be easily identified.

Symptoms:

Symptoms showed as brownish coloure, necrotic appearance of leaves, twigs and flowers. Because symptoms also appeared as burned appearance or fire and hence the name "fire blight." The typical symptoms are infected leaves showed brown to black color and infected twigs/branch, the assembly of exudate, and therefore the characteristic "shepherd's crook" of terminal shoots. Accordingly, the disease produces blossom blight, fruit blight, limb or trunk blight, or collar or rootstock blight (Van der Zwet and Keil, 1979; Van der Zwet and Beer, 1995).

(a) Symptoms as Blossom Blight and Spur Blight:

Symptoms of blistering blight and blight occur during the spring season. As well as new shoots, bacteria can enter the trees through the blooms. Infected blossoms give symptoms of brown water-soaked, wilt and switch. Inside the cluster, bacteria spread rapidly into other flowers and move down within the spur. Infected spurs on apples and pear turn brown (Van der Zwet and Keil, 1979; Van der Zwet and Beer, 1995).

(b) Symptoms as Shoot Blight:

Shoot curse begins at shoot tips and moves quickly down to more established twig segments. Initially, contaminated twigs were shown to be water-soaked, and turned dull earthy or black colored. As infected shoots shrivel, the branches curl at the point of growth and look like the evildoer of a shepherd, or a "J." Contaminated leaves remain attached throughout the midyear to dead branches. Beads of velvety white microscopic organisms will overflow during warm and humid climate (Van der Zwet and Keil, 1979; Van der Zwet and Beer, 1995).

(c) Symptoms as Twig/Branch/Stem Cankers:

If infection-causing bacteria move from infected twigs to the most branches, often the bark splits along the contaminated areas surface, creating a noticeable canker. Younger trees bark shows a water-saturated environment and dark brown to purple cankers. Cankers can girdle most tree branches and trunk causes extra dieback (Van der Zwet and Keil, 1979; Van der Zwet and Beer, 1995).

(d) Symptoms as Fruit Blight:

Apple fruits may be infected with fire blight. Pathogen caused fruit rot and rotted areas turned brown to black and coated with whitish-colored bacterial ooze droplets. Apple fruits ultimately dry out and shrive into mummies.

(e) Symptoms as on Rootstock:

Symptoms of fire blight similar to those of stem cankers that grow near the union of the graft. Fire blight infections in rootstocks can rapidly kill the tree by girdling the rootstock (Van der Zwet and Keil, 1979).

Disease Cycle and Epidemiology:

During the winter season, fire blight bacteria live as cancers in living tissue on the trunk or main limbs, and even on mummified fruit. Primary infection by splashing rain, pollinating insects (i.e., bees, wasps of pollen, flies, ants) or by pruning (Paulin, 2000) during production. During spring cankers become active and bacteria are reproduced. Many bacteria can be produced by the one active canker, enough to infect a whole orchard. These bacteria rapidly multiply within the nectar of the blossom, and spread to the spurs (blossom-bearing twigs), new shoots, and branches, resulting in secondary infections. Shooting infections can also be caused by wounds caused by sucking insects (aphids, leafhoppers, or tamed plant bugs), freezing or frost damage, wind blowing, wind-driven rain, or hail. When a shoot is infected, the bacteria in the heart blight quickly multiply, and ooze droplets are often seen within three days. Shooting remains highly vulnerable to infection until vegetative growth stops and thus the end bud is made (Thomson, 2000).

The pathogen growth is optimal at 27 °C. It can also grow in temperature ranging from 5 to 31 °C. At high temperatures (5 °C or below, 31 °C and above), the growth of diseases and the extension of lesions slow down due to unfavorable conditions (Thomson, 2000).

Integrated Disease Management:

Management of fire blight disease of apple is difficult to control. Though the management is not easy, the application of integrated management practices may minimize the damage and losses fire. The following recommended practices may help the grower in controlling fire blight disease to great extent.

(a) Cultural Control:

- Pruning of blight infected twigs and cankers during winter (Dormant season) removes fire blight bacteria from the orchard and checks the infection in spring.
- Summer pruning of infected shoots, done carefully, reduces the movement of bacteria within the orchard.
- Contaminated shoots can only be pruned during the dry season. Cuts cannot be made in wood that is diseased in good tissue below 12–15 in. Pruning tools must be sanitized after each cut. Tools are often sanitized by dipping them into the 10% bleach solution (1 volume of bleach to 9 volumes of water) containing a couple of drops of soap. When the answer seems dirty, the new bleach solution should be prepared.
- Sections of the contaminated plant should be collected from the orchard and burned or thrown into the garbage.
- Proper sanitation practices are very effective in stopping infestation of the disease.

(b)Chemical Control:

- Sprayed with a copper-based pesticide to rescale the bacteria build up on buds and bark. Copper can be added to green tips at the silver tip (when buds start swelling) but not later.
- Streptomycin is an antibiotic that destroys bacteria in the pear blight before they reach the fruit tree.

(c) Control Sucking Insects:

Sucking insects create wounds through which bacteria from the fire blight may enter. Such pests should be managed in the growing season. Insecticides should not be applied to protect bees, during bloom.

KEYWORDS

- apple
- diseases
- symptoms
- management

REFERENCES

Agarwal, R. K. (1961). Problem of root rot in Himachal Pradesh and prospect of its control with antibiotics. *Himachal Hortic.,* 2: 171–178.

Agrwal, R. K. (1967). Relative efficacy of fungicides for the control of *Gloeodes pomigena* (Schw) causing Sooty blotch disease of apple. *Proc. Acad. Sci.* 37: 171–178.

Anonymous, (1989). Pakage and Practices for fruit crops. Dr Y. S. Parmar University of Horticulture and Forestry, Nauni, Solan, Himachal Pradesh.

Bessey, C. W. (1877). On Injurious Fungi: The Bligh (*Erysiphe*). Iowa State College of Agriculture. Bienn. Rep., pp. 185–204.

Burrill, T. J. (1883). New species of Micrococcus (bacteria). *Am. Nat.,* 17: 319.

Byrde, R. J. W. and Jordan, V. W. L. (1977). Systemic Fungicides (R. W. Marsh, Ed.) Longman, London and New York. 280 pp.

Ellis, M. A., Chatfield, J., and Draper, E. (1997). Scab of Apple and Crab Apple. The Ohio State University Extension Factsheet. HYG-3003-94.

Glawe, D. A. (2008). The powdery mildews: a review of the world's most familiar (Yet poorly known) plant pathogens. *Annu. Rev. Phytopathol.,* 46: 27–51.

Gupta, G. K. (1985). Apple Scab (*Venturia inequalis) (*Cke) Went. E. Merck (India) Ltd., Bombay, 36 pp.

Gupta, G. K. (1990). Apple scab – a review. *Indian J. Agric. Sci.* 18; 157–165.

Gupta, G. K. and Lele, V. C. (1976). Morphology, physiology and epidemiology of the apple scab fungus (*Venturia inequalis*) (Cke) Went., in Kashmir valley. *Indian J. Agric. Sci.* 18; 157–165.

Hickey, K. D. and K. S. Yoder. (1990). Apple powdery mildew. Pages 9–10 in A. L. Jones and H. S. Aldwinckle, Eds. *Compendium of Apple and Pear Diseases*. American Phytopathological Society, St. Paul, MN, USA.

Johnson, E M., Sutton, T. B., Hodges, C. S. (1997) Etiology of apple sooty blotch disease in North Carolina. *Phytopathology*, 87: 88–95.

Joshi, N. C., Malik, A. G., Kaul, M. L. and Anand S. K. (1975). Some observation on epidemics of scab disease of apple in Jammu and Kashmir during 1973. *Indian Phytopathol.*, 28: 288–289.

Nath, P. (1935). Studies on the diseases of Apple in northern India. II A short notes on apple scab due to *Fusicladium dendriticum* Puck. *J. Indian Botany*, 14: 121–124.

Paulin, J. P. (2000). *Erwinia amylovora*: general characteristics, biochemistry and serology. In J. Vanneste (Ed.) Fire Blight: The Disease and Its Causative Agent, *Erwinia amylovora*, pp. 87–116. Wallingford, UK, CABI. 370 pp.

Pfeiffer, D. G., Bergh, J. C., Fell, R. D., Yuan, R., Walsh, C. S., Yoder, K. S., Biggs, A. R., Kotcon, J. B., Derr, J. F., Chandran, R. S., Weaver, M. J., Baniecki, J. F., Brown, A. and Parkhurst, J. (2010). *Spray Bulletin for Commercial Tree Fruit Growers.* Virginia Cooperative Extension Publication.

Rana, K. S. and Gupta, V. K. (1984). Occurrence of pythiaceous fungi in collar rot affected apple soil of Himachal Pradesh. *Indian Phytopathol.*, 37: 39–42.

Saha, L. R. 2002. Hand Book of Plant Diseases. Kalyani Publisher, Ludhiana, pp. 260–270.

Shoemaker, R. A. (1964). Conidial states of some *Botryosphaeria* species on Vitis and Quercus. *Can. J. Botany*, 42: 1297–1301.

Thomson, S. V. (2000). Epidemiology of fire blight. In J. Vanneste, Ed. Fire Blight: The Disease and Its Causative Agent, *Erwinia amylovora*, pp. 9–36. Wallingford, UK, CABI. 370 pp.

Turechek, W. W., Carroll, J. E. and Rosenberger, D. A. (2004). Powdery mildew disease of apple. Tree Fruit Factsheet, Cornell University.

Utkhede, Raj and Smith, Emmerson (1995). Control of *Phytophthora* crown and root rot of apple trees with fosetyl-aluminium in new plantings. *Pest Manage. Sci.*, 45 (2): 117–122.

Van der Zwet, T. & Beer, S. (1995). Fire blight: its nature, prevention and control. A practical guide to integrated disease management. USDA *Agricultural Information Bulletin,* No. 631.

Van der Zwet, T. & Keil, H. L. (1979). Fire blight: a bacterial disease of rosaceous plants. United States Department of Agriculture (USDA) Handbook 510. Washington, DC, USDA.

Verma, K. D. and Gupta, G. K. (1988). Field reaction of apple germplasm to powdery mildew. *Indian J. Agric. Sci.*, 58: 223–234.

Yoder, K. S. (1992). Powdery mildew of apple. In: J. Kumar, H. S. Chaube, U. S. Singh, and A. N. Mukhopadhyay (eds.) Plant Diseases of International Importance, pp. 66–89, Vol. 3. Diseases of Fruit Crops, Prentice Hall, NJ.

Yoder, K. S., and Biggs, A. R. (1997). Apple cultivar susceptibility to the powdery mildew fungus. *Biol. Cult. Tests Control Plant Dis.,* 12: 42–43.

CHAPTER 3

AONLA (EMBLICA OFFICINALIS GAERTN.): KEY DISEASES AND THEIR MANAGEMENT

S. K. MAHESHWARI*, HARE KRISHNA, and LOKESH KUMAR

ICAR-Central Institute for Arid Horticulture, Bikaner, Rajasthan, India.

Corresponding author. E-mail: maheshwariskciah@gmail.com

ABSTRACT

Aonla is known by different names in various regions across the globe for example, Emblic myrobalan, Indian gooseberry, *Aonla, amla* (India), *ganlanshu* (Chinese), and *kimalaka* (Indonesian). This crop is quite hardy, prolific bearer, and is remunerative for the farmers having marginal land. There is an increase in effective and ecologically aware uses of natural resources, rehabilitation of wastelands, and improved economic return to farmers coupled with enhanced employment. *Aonla* is quite admired for its nutritional attributes. The edible aonla fruit tissue has 3 times more protein and 160 times more ascorbic acid than an apple. It is rich in polyphenols, tannins, minerals, and contains plentiful amount of ascorbic acid, which is next to only Barbados cherry (*Malpighia glabra* L.). *Aonla* or Indian gooseberry (*Emblica officinalis* Gaertn.) is crippled by many diseases mediated by fungi, bacteria, viruses, nematodes, and parasitic plants. These diseases as well as pathogen either alone or in combination causes substantial damage to crop resulting in heavy economic losses every year. This communication deals with important diseases of *Aonla* or Indian gooseberry and their management strategies.

3.1 INTRODUCTION

Arid region in India is spread over 38.70 million ha and covers about 12% of the total area of the country. Growing of annual crops during the most congenial part of the year is the only source of livelihood for the people of the arid region. These crops often encounter abiotic stresses like drought, hot, and desiccating winds, nutrient deficiency during their growth period leading to poor productivity.

Aonla (*Emblica officinalis* Gaertn.), also known as Indian gooseberry, is an ancient, indigenous fruit of India. It is cultivated in India since the *Vedic* era and is linked with country's ancient tradition, culture, and heritage. Its importance to ancient Indian civilization can be well understood with the fact that it is often regarded as *Divya* and *Amrut* or *Amrit Phala* in Sanskrit, which means fruit of paradise or fruit of nectar. It is widely hailed as a wonder fruit for health-conscious population. The popularity and nutritional value of *aonla* can be explained from changing the Western proverb "an apple a day keeps the doctor away" to "an *aonla* a day keeps the doctor at bay." *Aonla* is known by different names in various regions across the globe, for example, emblic myrobalan, Indian gooseberry, *Aonla*, *amla* (India), *ganlanshu* (Chinese), and *kimalaka* (Indonesian). This crop is quite hardy, prolific bearer, and remunerative for the farmers having marginal land. There is an increase in effective and ecologically aware uses of natural resources, rehabilitation of wastelands, and improved economic return to farmers coupled with enhanced employment.

Aonla is quite admired for its nutritional attributes. The edible *aonla* fruit tissue has 3 times more protein and 160 times more ascorbic acid than an apple. It is rich in polyphenols, tannins, minerals, and contains plentiful amount of ascorbic acid, which is next to only Barbados cherry (*Malpighia glabra* L.). Its fruits are loaded with appreciable amount of minerals, phosphorous, calcium, iron, carbohydrates, and vitamins (Table 3.1).

TABLE 3.1 Nutritional **Value of Fruit of** *Aonla* **(per 100 g edible portion)**

Edible Portion (%)	Moisture (%)	Protein (g)	Fat (g)	Minerals (g)	Fiber (g)
89	81.8	0.5	0.1	0.7	3.4
Carbohydrates (g)	Calcium (%)	Phosphorous (%)	Iron (mg)	Vitamin C (mg)	Nicotinic acid (mg)
14.1	0.05	0.02	1.2	600	0.2

3.2 USES OF *AONLA*

All parts of the plant including the fruits of *aonla* are medicinally rich and are used in the preparation of various Ayurvedic medicines. *Aonla* is an essential ingredient of more than 175 formulations in Ayurveda. The active ingredient, which is being contributed toward significant pharmacological action of *aonla*, is attributed to *Phyllemblin* by Indian scientist. Fruits are commercially used for preparation of *chayanprash and triphala*, which is used for curing biliousness, chronic dysentery, and other health-related disorders. *Aonla* is helpful in the treatment of diabetes, dyspepsia, and so on. Fruits are acrid, cooling, refrigerant, diuretic, and laxative, and dried fruits effectively treat diarrhea, hemor-rhages, dysentery, peptic ulcer, jaundice, anemia, cough, and dyspepsia. The fruits have been found to have antioxidant, hepatoprotective, cyto-protective, antimutagenic, antitumor, and antimicrobial attributes. Ash of *aonla* fruits has minerals like chromium, zinc, and copper. *Aonla* fruits are regarded as adaptogenic, which improves immunity. *Aonla* plants are used as an efficient antiseptic for dressing wounds. *Aonla* leaves are employed for dyeing matting, bamboo wickerwork, silk, and wool into brown colors. Matting can be colored dark hues with a infusion made from the bark. The fruits are also utilized for preparing hair dye and black ink.

Fruits have substantial requirement in the industries for the prepara-tion of a variety of healthcare products like hair oils, dyes, shampoos, face creams, and toothpowders. *Aonla* fruits are edible but rarely eaten fresh as they are astringent and sour; hence, they are commonly used for prepara-tion of pickles, candy, jelly, jam, and so on. The *aonla* preserve (*murabbas*) is one such delicacy of the Indian fruit preservation industry for which *aonla* is well known for. These days, several *aonla*-based health beverages are being offered by food industries to woo the consumers.

Aonla or Indian gooseberry (*Emblica officinalis* Gaertn.) is crippled by many diseases mediated by fungi, bacteria, viruses, nematodes, and parasitic plants. These diseases as well as pathogen either alone or in combination cause substantial damage to crop resulting in heavy economic losses every year. This communication deals with important diseases of *aonla* or Indian gooseberry and their management strategies.

3.3   RUST DISEASE

Introduction/Economic Importance

This is a major and economically significant disease of *aonla*, especially in Rajasthan. Severity of rust disease was first reported from Rajasthan by Tyagi (1967). At Faizabad in Uttar Pradesh, the rust symptoms on fruits were noted during the month of October. This disease has been reported from Uttar Pradesh by Misra (1988). In Assam, Borah et al. (1998) have also reported this disease for the first time. Subsequently, it was also reported from Nagpur (Maharashtra), Tanjore (Tamil Nadu), and Bihar. The disease caused considerable losses in major *aonla* growing belts of the state of Uttar Pradesh (Rawal, 1993).

Symptoms

Reddish spots are developed on plant leaves from the starting of August. Conspicuous brown rust pustules arise on the *aonla* leaves and fruits, which later turn dark brown to black. Under severe condition, many pustules coalesce together and occupy a large fruit area. Severely affected fruits prematurely drop. The reddish brown uredo pustules can be seen on upper as well as lower surfaces of the leaflets. On infected leaves, pinkish brown pustules come into being.

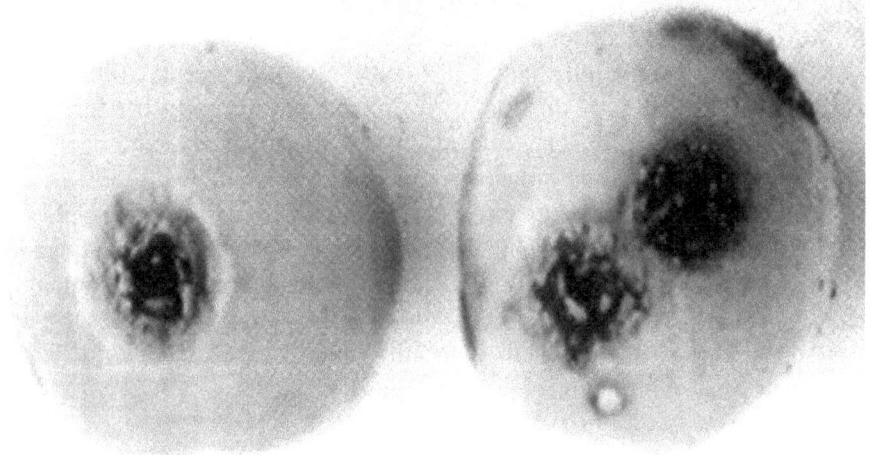

FIGURE 3.1   Rust disease on *aonla* fruits.

Casual organism: *Ravenelia emblicae*
Taxonomic Position
Divisionc—Eumycota
   Subdivision—Basidiomycotina
      Class—Teliomycetes
         Order—Pucciniales
            Family—Reveneliaceae
               Genus—*Ravenelia*
                  Species—*emblicae*

Disease Cycle:

Uredia of this pathogen are typically subepidermal erumpent bearing smooth walled uredospores produced singly on pedicel. The uredosori ruptures the epidermis and uredospores are exposed. Growth stages of uredospores and the reactions host tissue and fungal colony can be noticed. Telia are subepidermal in origin, erumpent bearing pedicellate teleospores. The uredospores are easily blown away by air which infect the new leaflets and fruits during growing season.

Epidemiology:

Under conditions in Rajasthan, disease development was more when relative humidity increases, while temperature decreases (Anonymous, 1989). This disease requires live host for infection by pathogen (obligate parasite) and its establishment under favorable conditions.

At Faizabad, the rust symptoms on fruits were observed in second week of October (40th meteorological week) and reached on peak in the month of December (52nd meteorological week). Temperature showed highly significant negative correlation whereas relative humidity (morning) and sunshine hours/day showed significant positive correlation with % disease intensity (Anonymous, 2015).

Management:

Cultural Control:

- Sanitation and clean growing of *aonla*.
- Destruction of infected plant parts and weeds from *aonla* orchard.
- Pruning should be done properly in orchard for reducing humidity.

Chemical Control:

- The use of fungicides is imperative to save crop losses and also increase productivity for better yield.

Three sprays with Dithane Z-78 (0.2%) at monthly interval from July to September is effective against this disease (Rawal and Saxena, 1989).

- This disease can also be controlled by wettable sulfur (0.25%) during July to September.

Use of Resistant Varieties:

- Resistant varieties of this disease are required for reducing this disease.
- Cultivar Francis and NA-6 have been found % disease index with 5.50% and 15.50%, respectively, against this disease under Faizabad conditions (Anonymous, 2015).

Integrated Disease Management Strategies:

- Integrating various managing practices for the containment of this disease seems to be the best option. Combining variety resistant with application of cultural and chemical control can be useful for reducing this disease.

3.4 ANTHRACNOSE DISEASE

Introduction/Economic Importance:

This disease is the second serious disease in *aonla*-growing areas. It was observed mostly in Rajasthan and Uttar Pradesh.

Symptoms:

This disease appears on leaflets and fruits in August to September (Misra and Shivpuri, 1983). At the beginning, symptoms are minute, circular, brown to gray spots with yellowish margin in leaflets. Leaves are then dried up at advance stages. In fruits, pin-lead like spots appear with dark brown to pink having yellow halo, while on fruits, depressed lesions develop which later turn dark. At advance stages, many spots coalesce with each other to develop into bigger lesions. The central areas of spots stay grayish elevated with minute fruiting bodies—the acervuli arranged in rings. Under severe condition, plenty spots result in the fruits becoming shrivel and rot.

Causal organism: Colletotrichum gloeosporioides

Taxonomic Position:

Division—Eumycota
   Subdivision—Deuteromycotina
      Class—Coelomycetes
         Order—Melanconiales
            Family—Melanconiaceae
               Genus—*Colletotrichum*
               Species—*gloeosporioides*

Disease Cycle:

The fungus perpetuates in infected crop debris and in the soil. Spores are released from last year's diseased tissue. Conidia germinate by germ tube which forms appressorium after coming in contact with the host. The infection peg arises from the appressorium and penetrates the host. The fungal spores are carried out by rain splash or wind and cause secondary infection to soft and succulent tissues. The fungus obtains nutrients from plant cells and these cells are killed by leaf lesion. The lesion expands as the fungus spreads.

Epidemiology:

Temperature and moisture are one of the most vital governing factors which influence the disease development. The infection is favored under hot and humid weather. This disease responds positively to elevated temperature along with cloudy weather.

Management:

Cultural Control:

- Discard affected plant parts such as leaves and fruits at initial stage.
- Proper pruning to be followed in orchard so as to improve air circulation within the crown.
- Plant trees at wider spacing and keep the adjoining area clear of vegetation.

Chemical Control:

- Spraying of carbendazim (0.1%) or difolatan (0.2%) can suppress the disease (Bhardwaj and Sharma, 1999).

Integrated Disease Management Strategies:

- This disease can be contained by adopting suitable strategies. Combined application of cultural and chemical means of control is recommended.

3.5 *ALTERNARIA* FRUIT ROT

Introduction/Economic Importance:

This rot was noticed in Allahabad (Uttar Pradesh) during the month of January to February. Pandey et al. (1984) found *Alternaria alternata* in dropped fruits of *aonla*.

Symptoms:

Initial disease symptoms appear as small brownish to black spherical necrotic spots on fruits and expansion in circular fashion with the advancement of the disease. In advance stage, the spots turn dark brown to black and adjoining spots coalesce together. The middle part of affected tissues happens to be soft and pulpy.

Causal organism: *Alternaria alternata*

Taxonomic Position:

Subdivision—Deuteromycotina
   Class—Hyphomycetes
      Subclass—Sporomycetidae
         Order—Moniliales
            Family—Dematiaceae
               Genus—*Alternaria*
                  Species—*alternata*

Disease Cycle:

The fungus perpetuates as mycelium or spores in infected debris of plant parts. Conidia can remain viable up to few months at ambient temperature. The infection begins from the fruits, where innumerable conidia are formed.

These are spread by wind, water, rain splash, and cause secondary disease. The conidia germinate owing to moisture availability at temperature nearly 25 °C –30 °C, which in turn gives rise to germ tubes that enter the host tissue.

Epidemiology:

The presence of high moisture, followed by warm and dry weather conditions is most favorable for disease occurrence. Very high temperatures along with continued dry spells check this disease. Dew and frequent rains accompanied with moderate- to high-temperature result in disease progression.

Management:

Cultural Control:

- Use of disease-free planting material.
- Removal of infected plant debris from the *aonla* orchard.
- Harvesting to be done quite carefully to avoid any injury to fruits.

Chemical Control:

- Mancozeb (0.25%) spray at 10 days interval to reduce this fruit rot.
- One spraying of carbendazim (0.1%) should be done 15 days prior to fruit harvest.
- Borax (0.5%) treatment or sodium chloride (1%) application checks the rot.

Biological Control:

- Use of antagonistic (*Trichoderma* spp. and *Pseudomonas* spp.) could be beneficial in managing the disease.

Besides control measures indicated above, full sanitary measures should be adopted during storage as well as in transit.

KEYWORDS

- **Aonla**
- **diseases**
- **symptomatology**
- **management**

REFERENCES

Anonymous (1989). Epidemiological studies. Annual Report, All India Co-ordinated Research Project on Arid Zone Fruits.

Anonymous (2015). Epidemiological studies of aonla. Annual Report, All India Co-ordinated Research Project on Arid Zone Fruits. pp. 90.

Bhardwaj, S. S. and Sharma, L. M. (1999). Diseases of minor fruits. *In*: Diseases of Horticultural Crops-Fruits (Eds. Verma, L. R. and Sharma, R. C), Indus Publishing Company, New Delhi, pp. 540–562.

Borah, R. K., Dutta, D. and Hazarika, P. (1998). Some new records of fungi from north east India. *Van-Vigyan* 36 (1): 41–43.

Misra, A. K. (1988). Studies on diseases of fruit crops. Annual Report, CIHNP. Lucknow.

Misra, A. and Shivpuri, A. (1983). Anthracnose, a new disease of aonla. *Indian Phytopathology* 36: 406–407.

Pandey, R. S., Bhargava, S. N., Shukla, D. N. and Dwivedi, D. K. (1984). Two new fruit diseases of aonla caused by *Alternaria* species. *International Journal of Tropical Plant Disease* 2: 79–80.

Rawal, R. D. (1993). Fungal diseases of tropical fruits. *In*: Advances in Horticulture 3: Fruit Crops (Eds. Chadha, K. L. and Pareek, D. P.), Malhotra Publishing House, New Delhi, pp. 1255–1273.

Rawal, R. D. and Saxena, A. K. (1989). Evaluation of different fungicides against black leaf spot of ber. *Indian Journal of Horticulture* 46: 413–414.

Tyagi, R. N. S. (1967). Morphological and taxonomical studies on the genus *Revenelia* Berk. Occurring in Rajasthan. Ph.D. Thesis, University of Rajasthan, Udaipur.

CHAPTER 4

AVOCADO (*PERSEA AMERICANA MILL.*) DISEASES AND THEIR MANAGEMENT

PRITI S. SONAVANE* and V. VENKATARAVANAPPA

Central Horticultural Experiment Station, ICAR-IIHR, Chettalli, Kodagu, Karnataka, India

Corresponding author. E-mail: Priti9883@gmail.com

ABSTRACT

Avocado, (Avocado) belongs to the family Lauraceae commonly called as butter fruit. The crop is basically native to Central America suitable for tropical and subtropical climatic conditions. The fruit is well known as super food not only because of its high nutritional value but also for its role in the cosmetic and health care industries. Though the fruit is native to Central America it was introduced to India from Sri Lanka during trade transit. In India, butter fruit is well known in southern states like Tamil Nadu, Kerala, Maharashtra, Karnataka, and in the eastern Himalayan state of Sikkim. It is majorly grown as intercrop or in multiple cropping systems. Now efforts are being made to exploit this crop commercially in the states of Kerala, Tamil Nadu, and Karnataka. However, the avocado crop notable suffers from a number of diseases both under field and after the harvest is major limiting factor for the successful remunerative avocado cultivation. Not much work has been attempted on the diseases of Avocado in India and information available is scanty. In order to provide effective and economic management of the diseases, the important diseases of Avocado and there management are discussed.

4.1   INTRODUCTION

Avocado from the family of *Lauraceae* is commonly known as butter fruit in India. The crop is basically native to Central America suitable for tropical and subtropical climatic conditions. The fruit is well known as super food not only because of its high-nutritional value but also for its role in the aesthetic and pharmaceutical industries. The fruit contains 4.6 g of protein, 19.62 g of carbohydrate, 1166 mg of potassium, 23.0 mg of vitamin C, 175 mg beta-sitosterol, and 0 mg cholesterol (USDA National Nutrient Database). The fruit was well known as fresh food 9000 years ago by American communities and later semidomesticated by Mayan and Aztec civilizations (Perez-Jiménez, 2008). Though the fruit is native to Central America, it was introduced to India from Sri Lanka during trade transit. In India, butter fruit is well known in southern states like Tamil Nadu, Kerala, Maharashtra, and Karnataka, and in the eastern Himalayan state of Sikkim. It is majorly grown as intercrop or in multiple cropping systems.

Avocados cultivation is not recommended in soils with high clay content to avoid Phytophthora root rot incidence. Proper drainage and soil pH in the range of 5 to 7 is extremely important for proper crop cultivation. The avocado cultivars can be grown in the region with subtropical conditions wherein the average temperature ranges from 20 °C to 25 °C, which is favorable for flowering and fruit set.

Avocado is an evergreen tree, flowers appear together in a compound panicle or raceme and the fruit is usually green in color with single seed, sometimes during ripening they turn into purple color, and are highly variable in shape and size. Three different types of avocado are widely grown, namely, West Indian, Guatemalan, and Mexican type, respectively. Guatemalan is of medium size, ovoid or pear-shaped green fruit that turn dark green when ripe (Bergh, 1969). Mexican varieties are small size with delicate skin that turns glossy green or black when ripe. The flesh of avocados is light green near the skin, becoming yellowish as moving toward the seed. The flesh is whitish when unripe but slowly turns yellow when softens with buttery texture. "Cukes" are seedless, pickle-shaped fruit. All three horticultural races of Avocado (West Indian, Guatemalan, and Mexican) have well adapted to the tropical and subtropical conditions of India. West Indian race, hybrids of Guatemalan, and Mexican race are grown in localized pockets of Maharashtra, Tamil Nadu, and Karnataka. Both the Mexican and Guatemalan races are grown successfully in Sikkim. In avocado-growing areas of Sikkim, the temperatures ranged from 12 °C to 30 °C with an average annual rainfall of 2000 mm. The Mexican race is cultivated on mid-altitude hills (pH 5–6) (Ghosh, 2000).

Now efforts are being made to exploit this crop commercially in the states of Kerala, Tamil Nadu, Karnataka, Maharashtra, and Northeast states. However, the avocado crop notable suffers from a number of diseases both in field condition, storage, and during transport which is a major limiting factor for the successful remunerative avocado cultivation. Though the crop is still being explored in India, there is no much information available about diseases occurring in India. In this chapter, an attempt is made to describe the important diseases of Avocado and their management in order to provide effective and economic management of the diseases.

4.2 FUNGAL DISEASES

4.2.1 *PHYTOPHTHORA* ROOT ROT

Introduction/Economic Importance:

Phytophthora root rot severely hampers the plant growth in all the stages of the crop. It occurs from seedling stage in nursery to trees in fields. The disease is reported from throughout the world causing severe losses in fruit yield and production (Allen, 1985).

Symptoms:

The pathogen infection starts from the root zone, the feeder rot turn black, decayed with disintegration of tissues. From the roots, the infection spreads to other parts of the trees. Lower truck of tree shows stem canker (Darley and Zentmyer, 1957; Pegg et al., 2009). Once the disease spreads to leaves the leaf margin turns brown, leading to severe leaf fall. Since the roots are infected, the nutrients and water movement is affected making the tree weak (Figure 4.1). Heavy leaf fall exposes the branches and major limbs to sunburn. After severe decline, death of the tree takes places after a few months to several years. Though the avocado trees are hardy, the trees cannot withstand waterlogged condition, therefore proper drainage is essential requirement.
Casual Organism: *Phytophthora cinnamomi* Rands

Etiology:

Sporangia in *P. cinnamomi* formed only in aqueous solution. Their size and shape vary depending on environmental conditions and nutritional status

(Ribeiro, 1978). Therefore, it is not a reliable characteristic for identification. The sexual oospore is spherical, pleuritic with smooth wall. Antheridia is mostly unicellular and amphigynous, although some are paragynous (Waterhouse and Waterson, 1966a, Zentmyer, 1977; Gerrettson-Cornell, 1983 and Daniel et al., 2003). Chlamydospores and oospores are abundantly formed on vegetable juice, V8 broth, but not on PDB (potato dextrose broth). V8 agar is the most appropriate media for oospore production. In soil, germination of chlamydospores is stimulated by root exudates containing sugars and amino acids (Zentmyer, 1980).

FIGURE 4.1 Phytophthora root rot.

Disease Cycle:

Primary source of inoculum in *Phytophthora cinnamomic* is infected soil and plant tissues. Presence of motile zoospores helps easily dissemination through soil and water. In absence of host, it survives by forming dormant chlamydospores. Under favorable conditions, chlamydospores form mycelia and sporangia. Sporangia, once mature, releases abundant zoospores. These zoospores swim in water to reach the root zone and enter through newly formed root tips which are delicate.

Epidemiology:

Maximum mycelial growth of *P. cinnamomi* can be seen around the temperatures 24 °C and 27 °C, respectively. The temperature falls below 10 °C or above 33 °C to 34 °C will restrict growth. Further, it is also showed that there is no correlation between the geographic or host origin of the isolates and the growth response to different temperatures (Shepherd and Pratt, 1974; Zentmyer et al., 1976; Zentmyer, 1981).

Integrated Disease Management:

- Strict quarantine has to be followed and disease-free planting material should be collected from certified nursery.
- Integrated disease management with bioagent fungicides and growth regulators as well as planting resistant rootstocks is very important.
- Mechanical injury while carrying out cultural practices to be avoided as it gives free entry to the pathogen. Avoid direct contact of irrigation water to the truck region.
- Soil drenching or Bordeaux pasting at the truck region helps in maintaining pH and also suppresses the gummosis occurrence.
- In grafted plants, the bud union should be 30 to 45 cm above the base.
- Physical injuries to the trees near the collar region should be avoided.
- The trunk should be painted with 10% Bordeaux paste up to 50 to 75 cm from the soil in premonsoon and postmonsoon seasons.
- After scraping the diseased parts with a sharp knife, paste 10% Bordeaux or copper fungicides followed by 3 to 5 spraying with Fenamidone 10% + Mancozeb 50% WDG (1 g/L) or Fenamidone 4.44% + Fosetyl-Al 66.66% WDG (1g/L) or Famoxadone 16.6% + Cymoxanil 22.1% SC (1.5 g/L) at 15 days interval, reduces disease spread effectively.
- Soil drenching with Fenamidone 10% + Mancozeb 50% WDG (2 g/L) or Fenamidone 4.44% + Fosetyl-Al 66.66% WDG (2 g/L) or Famoxadone 16.6% + Cymoxanil 22.1% SC (2 g/L) with 0.5% *Trichoderma* commercial formulation is also effective in managing the disease (Martin-Sanchez et al., 2007).

4.2.2 ANTHRACNOSE

Introduction/Economic Importance:

Anthracnose disease will appear usually when fruits start ripening. It is endemic in most producing areas and is the most severe disease in heavy rainfall areas (Darvas and Kotze 1981; Fitzel 1987; Fucikovsky and Luna 1987). The mature avocado fruit is rotted, consequently, very important losses occur during commercialization. In a market survey in South Africa, Sanders and Korsten (2000) recorded incidences on overripe avocados up to 80%. In other studies, losses of 71% on unripe fruits (Fitzel, 1987) or 36% on mature fruits have been recorded (Korsten et al., 1991).

Symptoms:

First, small round spots in large number may appear on the leaf surface. These spots under high humidity increase rapidly in size forming necrotic areas with abundant spore mass. Young leaves are more susceptible to this infection.

Anthracnose disease produces peculiar sunken concentric spots on the avocado fruits. It can be said that the pathogen mostly hibernates in the fruits and once the fruits start to ripen, the sunken spots with slimy pink bodies can be observed on fruits. These are nothing but numerous amount of spore mass. These spots often coalesce, and lead to blight, withering, and necrosis of infected plant tissues. Pathogen spreads from leaves to fruits during peanut stage or after flowering stage (Perez-Jemenez, 1997).

Initial symptoms are small black circular spot appearing on the skin of the fruit and spreads as the infection progresses. Infection in peanut stage of fruiting with heavy moisture and ambient temperature results in fruit drops and total crop loss. Fruit rot, anthracnose, leaf blight, wither tip, and die back of shoot, all are caused by *Colletotrichum gloeosporioides*. Disease development after harvest is the result of quiescent infection of fruit on trees before harvest (Figure 4.2). The fungus may remain dormant in green fruit for many months.

Casual Organism: *Colletotrichum gloeosporioides*

Etiology:

Pure culture of pathogen produced whitish narrow mycelium but later took on a slight dark color. Acervuli are formed in abundant on affected host surface with orange spore masses. Conidia in masses are pinkish but are hyaline individually, single celled, (8–20 × 2.5–5 μm) either oval to oblong with obtuse ends or ellipsoidal with rounded apex and a narrow, truncate base and sometimes contain 1 to 2 oil globules. Conidiophores are light brown in irregular acervuli, appearing orange in mass.

FIGURE 4.2 Anthracnose on fruits.

Disease Cycle:

Diseased plant debris around the trees acts as primary source of perennation and fresh infection. The pathogen survives in this plant debris as saprophytes for long duration (Fitzel, 1987). High humidity and high winds favor conidial dissemination from diseased to healthy plants. Conidia served as the primary source of inoculums. On fruits, conidium germinate forms a germ tube and a terminal appressorium from which an infection peg emerges and penetrates the outer wax layer and cuticle of the fruit skin (Binyamini and Schiffmann-Nadel, 1972). Nevertheless, the presence on unripe fruit of antifungal compounds (i.e., epicatechin) limits the growth of the infection peg, which remains in latency until fruit ripening (Prusky et al. 1983). During ripening, levels of these compounds decrease by enzymatic action (i.e., laccase) (Guestsky et al., 2007). Then, latent infections become reactivated and the hyphae invade the flesh until most of the fruit is rotten. In advanced stages, acervuli are produced beneath the fruit surface. These condia are dispersed in air when the cuticle ruptures.

Epidemiology:

C. gloeosporioides required for promote infection with warm temperature of 18 °C to 28 °C, pH range 5.8–6.5, and high humidity (above 80%) for conidial germination and appressorium formation. In dry condition, the pathogen remains dormant and under favorable conditions it retains its active stage.

Integrated Disease Management:

The critical phases for disease control are flowering, fruit set, and after harvest. Plant health is important for keeping the infection of twigs away. Tree and field sanitation before monsoon showers very essential to keep the disease in check. Fallen plant debris should be collected and burnt to remove inoculum (Korsten, et al., 1998).

Dieback showing branches should be pruned followed by pasting with Copper oxychloride 50% WP (0.25%) or Copper hydroxide (0.15%) are effective when floral buds begin to swell and throughout the fruit development period (intervals of 14–28 days) (Peterson and Inch, 1980; Darvas and Kotze, 1987). Pruning of diseased twigs and its protection with copper fungicide (Copper oxychloride 50% WP or 10% Bordopaste) is quite effective in checking this disease. Spraying of Hexaconazole 5% EC (0.5 mL/L) is very effective against this disease.

In South Africa, Korsten et al. (1991) evaluated the strain of *B. subtilis* A6, isolated from avocado phylloplane, against the most important postharvest diseases (anthracnose, fruit rot, and stem-end rot), and confirmed that antagonistic water dip treatments (suspensions of *B. subtilis* at 1 × 10 or 2.1 × 109 cells mL) of avocado fruits were as least as effective as Prochloraz 45%EW in controlling these diseases. In Australia, Stirling et al. (1995) selected *Bacillus* spp. and yeasts from avocado phyllosphere that reduced lesion development and size on detached avocado when applied (at 10–109 spores mL) on artificially *C. gloeosporioides* inoculated avocado fruits. In Mexico, Vidales-Fernandez et al. (2007) have demonstrated the potential of *Trichoderma harzianum* (Amicus-L, 10 monthly treatments at 0.1%–0.4%) to control the disease.

4.2.3   CERCOSPORA SPOT DISEASE

Introduction/Economic Importance:

Cercospora spot is also known as *Pseudocercospora* spot; blotch or black spot was reported in Florida by Stevens (1922). He described that the disease caused by *Cercospora* leads to inferior quality of fruit. Subsequently, Zentmyer (1984) referred to it as most severe disease in avocados grown in Florida.

In South Africa, it is most important preharvest disease of avocado and can cause losses up to 69% in untreated orchards (Darvas and Kotze, 1987). The disease is also important in Florida, several areas of Mexico, and Australia, since warm and humid climates are conducive for the disease (Darvas and Kotze, 1979; Ploetz et al., 1994; Teliz, 2000a, 2000b).

Symptoms:

The fungus causes small (3–6 mm in dia.) angular spots on margins of leaf, initially the spots are brown in color and later turn into purple. Many of the spots are surrounded by yellow halos. These spots enlarge and coalesce, infecting the whole leaves, which become deformed. Similar small spots appear on fruit, which may crack and permit the entry of secondary pathogens. Sometime the fungus can invade the flesh. Under high humidity, the spots will show presence of greyish mycelium and spores (Darvas and Kotze, 1979; Ploetz et al., 1994).

Casual Organism: *Pseudocercospora purpurea*

Etiology:

The mycelium consists of prostrate hyphae and more or less erect conidiophores. The mycelium develops as greyish-brown to dark brown mat in and on the plant parts and in culture. Conidia are club-shaped to cylindrical, with a truncate base, pale olive, 9 to 11 septate, straight or curved.

Disease Cycle:

The fungus can survive in soil, plant debris, or infected fruits up to one year. Under high humid and warm condition, conidia are dispersed by wind rain splash. The conidia lands on susceptible host through stomata or natural opening or wounds in the plant. Once infection is successful, the symptoms appear to be visible lesions on the leaves.

Epidemiology:

Infection mainly occurs through conidia, which develop on infected organs at high humid and warm conditions. Subsequently, water, wind, or insects spread them to the infection points. Penetration can be direct or through wounds. After entry into susceptible host, the pathogen remains dormant for about three months. Small fruits and mature fruits are immune, whereas intermediate size fruits are susceptible (Ploetz et al., 1994).

Integrated Disease Management:

- Among cultivars, susceptible genotypes are 'Fuerte' and 'Ryan' compared to 'Hass' and 'Edranol' (Darvas and Kotze, 1987).
- In South Africa, where Cercospora leaf spot is considered as the most problematic preharvest disease of avocado (Darvas and Kotze, 1979), chemical control has been widely developed and it is currently recommended that *Cercospora* spot be controlled, as a component of an integrated disease control program, by monthly applications from October to January of Mancozeb 75% WP (0.2%), Copper oxychloride 50% WP (0.2%–0.3%) alone or in combination with systemic fungicides such as Azoxystrobin 23% SC (0.03%) (Willis and Mavusco, 2007).
- Severity of *Cercospora* spot is reduced when avocado trees are treated with preharvest sprays of *B. subtillis* (isolate B246, at a concentration of 107 cells ml) alone or in combination with different fungicides (Korsten et al., 1992).

4.2.4   SCAB DISEASE

Introduction/Economic Importance:

The disease appears in the regions with significant rainfall (Jenkins 1934; Ploetz et al. 1994; Teliz, 2000a, Teliz, 2000b).

Symptoms:

On the infected fruit, the spots are roughened or crust-like lesion or to a freckled appearance of the diseased organ (Pohronezny and Simone, 1994). Further, the spots crack and become leathery. Often, there is intersecting raised ridges (Figure 4.3). The lesions are superficial but may facilitate the entry of other fruit pathogens. On leaves, small (3 mm) dark brown spots develop along leaf veins (Jenkins, 1934). Severe infections may crinkle and distort the leaves. Scabby large lesions may occur on leaf veins, pedicels, and twigs.

FIGURE 4.3   Scab.

Casual Organism: *Sphaceloma perseae*

Etiology:

Conidia are as hyaline, usually continuous or with single septation elliptical to oblong, mycelium white to dark grey. Stroma embedded into host tissue, and acervuli are small, with white, cream to olive masses of clustered conidiosphores and bear conidia on top and sides (Jenkins, 1934).

Disease Cycle:

S. perseae enters in early stage of the crop but waits for infection for suitable growth stage of the host and high humid condition for colonization. During low temperature and high humidity, acervuli are produced containing clusters of conidia on conidiophores. Injuries caused by insect create entry wounds for *S. perseae* and greatly aggravate scab development.

Epidemiology:

The pathogen grows in low temperature and high humidity; in contrast, conidia may be formed throughout the whole year, although they are more abundant at the end of the infection period. Conidia are dispersed and spread by wind, rain, or insects. In Mexico, populations of thrips (*Thysanoptera*) correlate with the number of lesions on fruits, since wounds caused by thrips favor the establishment and development of the fungus (Teliz, 2000a, Teliz, 2000b). Young fruits and leaf tissues are susceptible whereas mature ones are resistant (Ploetz et al., 1994).

Integrated Disease Management:

- Field sanitation is essential as infected plant debris are primary source of inoculum.
- Trees should be aerated by pruning and infected stems removed. As scab and thrips populations are positively correlated, thrips control may help to control the disease (Teliz, 2000a, 2000b, Palmateer, 2006).
- Chemicals should be applied to protect the fruit, particularly when humidity is above 60%. Sprays the crop with Propaconazole 10% EC (0.5 mL) or Hexaconazole 5% EC (0.5ml/lit) or Copper oxychloride 50% WP (2.5 g/L) are helpful in controlling the disease (Lopez-Herrera et al., 2003).

4.2.5   FRUIT ROT DISEASE

Introduction/Economic Importance:

Fruit rot is caused by many fungi, some fungi can be major cause of the disease or it may also be due secondary infection (Zentmyer and Mircetich, 1965), although the species of genus *Botryosphaeria* have been reported as primary pathogen. Though the infection may be in early stage, the symptoms on fruits are expressed at ripening stage thus, the symptoms of fruit rot are indistinguishable from those caused by anthracnose. Therefore, the disease is reported or referred as the *Dothiorella/Colletotrichum* fruit rot complex (Darvas and Kotze, 1987).

Symptoms:

The symptoms only develop on softening fruit after ripening, thus fruits at early stage are asymptomatic carrier. On unripened surface, small reddish-brown irregular spots may be observed. As the fruit ripens, the spots gradually enlarge, sink, and turn black and the lesions enter into the flesh and develop unpleasant odor. Unlike *Collectotrichum*, fruit rot lesions are usually superficial with no sporulation. In later stage, fruits are covered with grey mycelium (Darvas and Kotze, 1987).

Casual Organism: *Botryosphaeria dothidea*

Disease Cycle:

The primary source of inoculums is conidia and ascospores abundantly produced on diseased fruits and plant debris. They are normally disseminated by wind and rain (Darvas, 1982). After initial establishment, the fungus remains dormant and rotting is initiated at fruit-ripening stage.

The fungus survives as pycnidia (small dark "pimple-like" structures) on diseased wood. Pycnidium produces and releases conidia (spores) through the growing season. Following hydration, inoculum spreads by wind and rain splash, disseminating the fungi from plant to plant, and from one part of the vine to another.

Epidemiology:

Conidia and ascospores germinate at 28 °C to 32 °C. Warm and wet weather favored the germination of conidia.

Integrated Disease Management:

- Emphasis should be placed on preventive measures such as eradication of weeds, sanitation of fields, use of certified seedlings, and removal of dead wood or leaf tissues from trees (Zentmyer and Mircetich, 1965).
- Trees stressed by rot diseases, drought, or nutrient deficiency are more susceptible than healthy trees. It is important to avoid saline conditions because the fungus will survive on dead portions of leaves caused by salinity stress.
- Preharvest sprays of the trees with Copper fungicides (Zentmyer et al., 1965) give some control (Darvas, 1982; Darvas and Kotze, 1987).
- Postharvest chemical treatments with wax alone reduce the occurrence of the pathogen or associated with *Colletotrichum* spp. (Darvas and Kotze, 1981). Dipping the fruit before storage in biological agent such as *Bacillus subtilis* or wax treatments with significantly reduce the disease (Korsten et al., 1991).

4.2.6 SEEDLING BLIGHT

Introduction/Economic Importance:

The disease can occur in all the stages of crop. *Phytophthora* seedling blight is the most serious and important disease of avocado worldwide and avocado growing areas of India.

Symptoms:

Water soaked, brown to black lesions are observed on leaves. Under high humid weather, a mycelial growth can be seen at infected region. The terminal bud may be killed.

Casual Organism: *Phytophthora* spp.

Disease Cycle:

Primary source of inoculum in *Phytophthora cinnamomic* is infected soil and plant tissues. Presence of motile zoospores helps easily dissemination through soil and water. In absence of host, it survives by forming dormant chlamydospores. Under favorable conditions, chlamydospores form mycelia and sporangia. Sporangia once mature, releases abundant zoospores. These zoospores swim in water and reach the host and enter the plant through newly formed root tip.

Epidemiology:

Heavy rainfall and high humidity aggravates the disease when seedlings are young and succulent.

Integrated Disease Management:

- In order to avoid soil borne primary inoculums, field sanitation is important.
- Proper selection of seeds is also necessary to avoid seed-borne infection.
- Fungicides in combination with bioagents will be helpful to manage disease.

4.2.7   STEM-END ROT

Introduction/Economic Importance:

Stem-end rot is caused by many primary and secondary pathogens or it may be integration of different diseases with similar symptoms. This disease is an economically important postharvest disease in most avocado producing areas (Darvas and Kotze, 1981; Hartill, 1991; Ploetz et al., 1994).

Symptoms:

Stem-end rot can be seen at various stages of fruit development. If the disease is observed in early stage, there will be premature dropping of fruits, severely reducing the yield. The infected stem will be discolored with dark-brown to black lesion at the stem end of avocado fruit. The rot progresses when fruit ripens toward the fleshly parts of the fruit. Some fungal species causing stem-end rot may discolor vascular tissues in the advance front of the rot.

Casual Organism: *D. aromatica, Colletotrichum, Botryosphaeria* and its associated anamorphs, *Fusarium, Nectria* or *Phomopsis*):

Epidemiology:

The species of fungi associated with the disease remain dormant as endo-phytes or phellophytes in avocado stems, thus they can infect fruit from endophytically colonized tissues. Pycnidia or ascomata may be produced on dead wood, leaves, or fruits. Conidia or ascospores are spread by rain or wind to fruit or pedicel, initiating the infection. Infection can also be initiated during harvesting through the cut surface (Hartill and Everett, 2002). In general, the infection develops when the fruit ripens (Darvas and Kotze, 1981; Hartill, 1991; Ploetz et al., 1994).

Integrated Disease Management:

- Preharvest sprays with Copper oxychloride 50WP (0.25%) and Copper hydroxide 77% WP (0.15%) alone or in combination with other fungicides Carbendazim 12% + Mancozeb 63% WP (0.2%) give some control of stem-end rot.

4.2.8   ROSELLINIA (DEMATOPHORA) ROOT ROT

Introduction/Economic Importance:

Dematophora root rot or white root rot of avocados was first reported in California by Raabe and Zentmyer (1955). The pathogen is heavily damaging the crop under favorable environmental conditions thus, limiting factor for the production of avocado in southern Spain (Perez-Jiménez et al., 2005, Perez-Jiménez, 2006).

Symptoms:

Symptoms of plants infected by *R. necatrix* or *Armillaria* spp. are quite similar. The first observable symptom of *Rosellinia* root rot is the existence on root surfaces of a white-cottony mycelium and mycelia strands colored white or black. In the first case, that is, apoplexy, the tree suddenly declines in vigor; the leaves wilt and dry, and the tree dies during the next few weeks (Ruano-Rosa, 2006).

Casual Organism: *Rosellinia necatrix*

Etiology:

Conidia produced on the synnemata are one-celled, solitary, elliptical to ovoid, colorless to pale brown, smooth. Teleomorph is spherical, black, 1 to 2 mm in diameter perithecia that have papillate ostioles (Hanlin, 1990). They are embedded in a mat of septate, brown hyphae.

Disease Cycle:

The infective cycle of *R. necatrix* is basically underground, mostly through the contact between diseased and healthy roots of trees and further penetration into the roots by mycelial strands. Chlamydospores, conidia, and ascospores seem to play a secondary role in the infective cycle of the fungus.

Epidemiology:

Rosellinia root rot requires high soil moisture (near to field capacity) and temperatures of 20 °C to 25 °C for growth and development.

Integrated Disease Management:

- It is essential to remove diseased plants stumps and roots completely where the fungus has been present. The pathogen is already present in the soil for longer duration. Therefore, sanitation of field by removing all old root and stem debris and extensive exposure of soil for solarization is essential (Lopez-Herrera et al., 1998, Pliego et al., 2007).
- After planting, it is important to avoid water logging and excessive irrigation.

4.2.9 *VERTICILLIUM WILT*

Introduction/Economic Importance:

The disease was first reported by Zentmyer (1949). It was also reported from South Africa, Chile, Ecuador, Spain, and Australia. Initially, in California the disease was called apoplexy or asphyxiation. It is not a serious disease although the appearance of disease is sporadic in several growing areas.

Symptoms:

In initial stage, the leaves suddenly wilt from one branch, and slowly entire tree is covered. The leaves rapidly start drying but remain hanging on affected branches for several weeks. Brown to grey-brown discolored streaks can be observed on the tissue beneath the bark.

Casual Organism: *Verticillium dahliae*

Etiology:

It produces clusters of sticky conidia at nodes of elongated phialides on vertically branched conidiophores. The mycelium is hyaline, multinucleated, and septate. The phialospores are colorless, elongated-elliptical to cylindrical, and one-celled. The fungus also produces dark brown to black microsclerotia.

Disease Cycle:

During dormant stage, *verticillum* survives by production of microsclerotia which survives in soil for long period of time under adverse conditions. The microsclerotia later germinates when exposed to root exudates of favorable host. As the root tip gets infected, the pathogens colonizes the root cortex and later subsequently infect the vascular tissue. The mycelia invade upwards the xylem producing abundant conidia and the vessels may be plugged causing canopy wilt (Zentmyer, 1984).

Epidemiology:

These fungi can affect various host plants and for several years overwinters inside soil surface. Temperature between 21 °C and 25 °C generally favors disease development. *Verticillium* enters plants through physical injury, root tip, and nematode feeding.

Integrated Disease Management:

- It is recommended not to plant avocados in fields where susceptible crops of *V. dahlia*, that is, tomato, potato, straw-berry, or olives have been cultivated (Zentmyer, 1984).
- Avocado rootstocks have been evaluated for their sensitivity toward *Verticillium* wilt and it was recorded that Mexican rootstocks are more resistant to the disease than Guatemalan (Halma et al., 1954; Ben-Yaacov and Frenkel, 1973).
- Soil fumigation, soil solarization, maintaining adequate field sanitation, deep ploughing, avoiding over watering, and using resistant cultivars are the possible way out to control the disease effectively.
- Soil application of biocontrol agent like *Trichoderma viride* @ 2.5 kg/ha mixed with well-rotten farmyard manure or neem seed kernel cake of about 100 kg. After mixing the bioagent with manure, incubate it for 21 days in shed before application.

4.2.10   SOOTY MOLD

Introduction/Economic Importance:

Sooty mold appears in orchards infested with sucking pests. They do not infect avocado, but grow on the sugary substance excreted by sucking insects (Pohronenzy et al., 1994).

Symptoms:

Black velvety coating can be seen on the leaves, twigs, and fruits (Teliz, 2000a, 2000b). Leaf surfaces can be covered to such an extent that photosynthesis is hampered and the leaves become chlorotic.

Casual Organism: *Capnodium* spp.

Etiology:

Conidia are dark, thick-walled, and multiseptated. When they contact honey dew, they germinate and produce dark, velvety mat that contains ascocarps and conidia.

Disease Cycle:

A densely populated orchard where there is low-light intensity provides favorable conditions for sooty mold disease. Sugary substance secreted by the insects is stated to be a condition favorable for development of sooty mold. Incidence of insects on the shoot is directly associated with disease severity (Pohronenzy et al., 1994).

Epidemiology:

High humidity and high incidence of sucking pests proved to be favorable for disease development.

Integrated Disease Management:

- Washing the fruit in the packing house with water or a Ca-hypochlorite solution usually removes these fungi.
- Indian oil formulation (tree spray oil) of 1% at 15 days interval could control sooty mold effectively (Prakash, 1988; Misra and Prakash, 1993).
- Spraying of Elosal (900 g/450 L) at 10 to 15 days interval proved quite effective control over sooty mold disease.
- The insects and aphids that produce honeydew should be controlled. Reduction of dust along roadways may reduce the problems.

4.3   BACTERIAL DISEASES

4.3.1   *BACTERIAL SOFT ROT*

Introduction/Economic Importance:

Bacterial soft rot rarely occurs under the field condition but usually is a major problem in storage and transit. The fruits are transformed into partially or totally into a soft decayed pulpy mass with metallic shine (Figure 4.4). When the fruit is cut opened, the flesh is brownish red to brown-black often liquefied and has a putrid odor.

Casual Organism: *Erwinia herbicola* and *E. Carotovara*

The soft rot causing species of Erwinia are rods shaped occurring singly, in pairs, or in chain and produce no spores or capsule. The organism lives saprophytically on decaying plant debris including diseased leaves and fruits left behind in the field. Once the organism enters the fruits by means of pectic enzymes, the soft rot occurs. Fruits with thick skin are less infected than thin skinned fruits. The organism becomes systemic and spreads to other fruits.

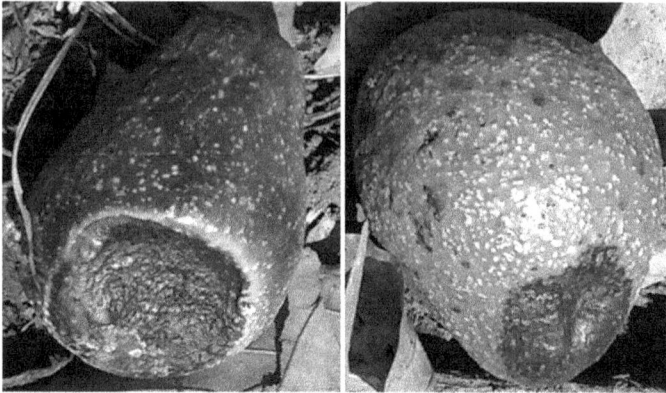

FIGURE 4.4   Bacterial soft rot.

Integrated Disease Management:

Bacteria colonies are usually present on the surface but once exposed to injury or opening or when plants are under stress the damage is caused. The disease will appear more on fruits, immediately spraying streptomycin sulphate 500 to 1000 ppm or Phytomycin antibiotic @1500 ppm in combination with Copper oxychloride 50% WP @ 2.5 g/L of water for effectively management of the disease.

4.4   ALGAL DISEASE

4.4.1   ALGAL LEAF SPOT

Introduction/Economic Importance:

Algal leaf spot disease of avocado is caused by an alga and that alga reduces the activity of photosynthesis of leaves.

Symptoms:

The algae are visible on upper surface of leaves as orange to red circular spots. It turns grey to black in later stage. Sometimes, yellow halo can be seen surrounding the spots (Pegg, 1991). Algal spores are disseminated from older leaves to younger leaves by wind and rain splash (Figure 4.5).

FIGURE 4.5 Algal leaf spot.

Casual Organism: *Cephaleuros virescens* (Syn. *C. parasiticus* Karst,).

Pathogen:

Sporangiophores are long unbranched filaments that are reddish in color, filaments measuring 52.05–235.96 × 10.41–15.61 µm, emerging through the thalloides disk, 5 to 12 stalked. The sporangia are subspherical to oval, purple brown in color measuring 19.08–27.76 × 15.61–20.82 µm.

Disease Cycle:

The fresh infection is initiated by zoospores formed by the sporangia. Initial entry into the host is achieved through inflicted wounds or washed into stomata during rains. Humid condition is congenial for forming fruiting bodies.

Epidemiology:

Disease infection increases during the rainy season; the rain water splash serves as mode of spread for alga from one infection site/plant to another. Temperature of 30 °C maximum and 25 °C minimum with high RH and wind velocity and frequent moderate rains are favorable for the growth, development, and rapid dispersal of *Cephaleuros virescens*.

Integrated Disease Management:

- Reducing canopy humidity by judicious pruning can help in disease management (Pegg, 1991; Vock, 2001).
- Spraying of Copper Oxychloride (0.3%) effective in managing the disease.
- Spraying Bordeaux mixture (2:2:250) followed by copper oxychloride to be effective against disease.

KEYWORDS

- **Avocado**
- **diseases**
- **symptoms**
- **management**

REFERENCES

Allen, R. N. (1985). Avocado diseases. Agfact H6. AB.5. Department of Agriculture, New South Wales.

Ben-Yaacov, A., Frenkel, M. (1973). Sensitivity of avocado rootstocks to Verticillium wilt. *California Avocado Society Yearbook* 57:111–113.

Bergh, B. O. (1969). Avocado (*Persea Americana* Miller.). In: Fewerda, F. P., Witt, F. (Eds) Outlines of perrennial crop breeding in the tropics, Miscellaneous papers no 4, Landbouwhoge School (Agricultural University) Wagen in gen, The Netherlands, pp. 23–51.

Binyamini, N., Schiffmann-Nadel, M. (1972). Latent infection in avocado fruit due to *Collectotrichum gloeosporioides. Phytopathology*, 62:592–594.

Daniel, R., Wilson, B. A., Cahill, D. M. (2003). *Phytophthora cinnamomi* in native vegetation communities of Southern Victoria: morphological variation and paragyny among isolates. *Aust. Pl. Patho.* 32:403–409.

Darley, E. F., Zentmyer, G. A. (1957). Oak root fungus on Avocados. *California Avocado Society Yearbook* 41:80–81.

Darvas, J. M. (1982). Etiology and control of some fruit diseases of avocado *(Persea Americana* Mill) at Westfalia Estate. DSc (Agric) thesis, University of Pretoria.

Darvas, J. M., Kotze, J. M. (1981). Post harvest diseases of avocados. *South African Avocado Growers'Association Yearbook* 4:63–66.

Darvas, J. M., Kotze, J. M. (1987). Avocado fruit diseases and their control in South Africa. *South African Avocado Growers'Association Yearbook* 10:113–116.

Darvas, J. M., Kotze, J. M. (1979). Cercospora spot of avocados. *South African Avocado Growers'Aassociation Research Report* 3:38–39.

Fitzel, R. D. (1987). Epidemiology of anthracnose disease of avocados. *South African Avocado Growers'Association Yearbook* 1:113–116.

Fucikovsky L., Luna, I. (1987). Avocado fruit diseases and their control in Mexico. *South African Avocado Growers'Association Yearbook* 10, 119–121.

Gerrettson-Cornell, L. (1983). A compendium of the morphology of *Phytothora cinnamomi* Rands from Australia. *Acta. Bot. Hung.* 29:91–105.

Guestsky, R., Kobiler, I., Avilla-Quezada, G., Prusky, D. (2007). Metabolism of epicatechin by laccase of *Collectotrichum gloeosporioides. Book of Abstracts of the VI World Avocado Congress*, Vina del Mar, Chile, p. 98.

Ghosh, S. P. (2000). Avocado production in India, In eds: Minas K. Papademetriou, Avocado production in Asia and Pacific, http://www.fao.org/3/X6902E/x6902e06.htm.

Halma, F. F., Zentmyer, G. A., Wihelm, S. (1954). Susceptibility of avocado root stocks to Verticillium wilt. *California Avocado Society Yearbook* 38:153–155.

Hanlin, R. T. (1990). Illustrated genera of ascomycetes: American Phytopathological Society Press, *St. Paul. Minn.*, p. 63.

Hartill, W. F. T. (1991). Post-harvest diseases of avocado fruits in New Zealand. *New Zealand Journal of Crop and Horticultural Science* 19, 297–304.

Hartill, W. F. T., Everett, K. R. (2002). Inoculum sources and infection pathways of pathogens causing stem end rots of 'Hass; avocado (*Persea americana*). *N Z J Crop Horti. Sci.* 30:249–260.

Jenkins, A. E. (1934). A species of *Sphaceloma* on avocado. *Phytopathology* 24:84–85.

Korsten, L., De Villers, E. E., Jager, E. S., Cook, N., Kotze, J. M. (1991). Biological control of avocado postharvest diseases. *South African Avocado Growers' Association Yearbook* 14:57–59.

Korsten, L., Lonsdale, J. H., De Villeirs, E. E., Kotzé, J. M. (1992). Effect of *Bacillus subtilis* and fungicide sprays for control of preharvest disease of avocado. *South African Avocado Growers' Association Yearbook* 15:9–11.

Korsten, L., Towsen, E., Claasens, V. (1998). Evaluation of Avogreen as post-harvest treatment for controlling anthracnose and stem-end-rot on avocado fruit. *South African Avocado Growers' Association Yearbook* 21:83–87.

Lopez-Herrera, C. J., Perez-Jimenez, R. M., Zea-Bonilla, T., Basallote-Ureba, M. J., Melero-Vara, J. M. (1998). Soil solariastion in established avocado trees for control of *Dematophora necatrix. Pl. Dis.* 82:1088–1091.

Lopez-Herrera, C. J., Zea-Bonilla, T., Perez-jimenez, R. M. (2003). Evaluacion de diferentes fungicides para el control de la podredumbre blanca del aguacate. *Proceedings of the World Avocado Congress*, Granada-Malaga, Spain, pp. 543–547.

Martin-Sanchez, P. M., Gonzalez, M. A., Zea-Bonilla, T., Perez-Jimenez, R. M. (2007). Control biologico de *Phytophthora cinnamomi* y *Rosellinia necatrix* con cepas de Pseudomonas Chlororaphis aisladas de rizosfera de aguacate. *Book of Abstracts of the VI World Avocado Congress*, Vina del Mar, Chile, p. 96.

Misra, A. K., Prakash, O. (1993). Host range and efficacy of different chemicals for the control of sooty mould of mango. *National Academy of Sci.* 63(B)II:233–235.

Munnecke, D. E., Kolbezen, M. J., Wilbur, W. D. (1981). Interactions involved in controlling *Armillaria mellea. Plant Dis.* 65:384–389.

Palmateer, A. (2006). Florida plant disease management guide: avocado (Persea Americana). University of Florida IFAS Extension. Retrieved December 1, 2015.

Pegg, K. G. (1991). Causes of disease, In: Broadly, R.H. (ed.) *Avocado Pests and Disorders. Queensland.* Department of Primary Industries. Brisbane, pp. 1–7.

Pegg, K., Lindy C., Elizabeth D. (2009). Avocado: In Tony Cooke, Denis Persley and Susan House (Ed.), *Diseases of Fruit Crops in Australia* (pp. 49–64) Collingwood, Australia: CSIRO Publishing.

Perez-Jemenez, R. M. (1997). Podredumbres radiculares del aguacate (*Persea americana* Mill.) en el suur de Andalucia. PhD Thesis, Universidad de Malaga, Malaga, Spain, p. 370.

Perez-Jiménez, R. M., Zea-Bonilla, Lopez Harrira. (2005). Avocado root rot in Andalocia: A review of South African Avocado grower Association Yearbook, 28: pp. 10–13.

Perez-Jiménez, R. M. (2006). A review of the biology and pathogenicity of *Rosellinia necatrix*; the cause of white root rot disease of fruit trees and other plants. *J. Phytopath* 154:257–266.

Perez-Jiménez, R. M. (2008). Significant avocado diseases caused by fungi and oomycetes. *Euro J Plant Sci Biotechnol* 2(1):1–24.

Peterson, R. A., Inch, A. J. (1980). Control of anthracnose on avocados in Queensland. *Queensl. J. Agric. Anim. Sci* 37:79–83.

Pliego, C., Cazorla, F. M., González-Sánchez, M. A., Pérez-Jiménez, R. M., De Vicente, A., Ramos, C. (2007). Selection for biocontrol bacteria antagonistic against *Rosellinia necatrix* by enrichment of competitive avocado root tip colonizers. *Res. Microbio.* 158:463–470.

Ploetz, R. C., Zentmyer, G. A., Nishijima, W. T., Rohrbach, K. G., Ohr, H. D. (1994). *Compendium of Tropical Fruit Diseases*, APS Press, St Paul, Minnesota, USA, p. 118.

Pohronenzy, K. L., Simone, G. W. (1994). Compendium of tropical fruit diseases. St. Paul: APS Press. p. 81. ISBN 978-0-89054-162-3.

Pohronezny, K. L., Simone, G. W. (1994). Scab In: Ploetz, R. C., Zentmyer, G. A., Nishijima, W. T., Rohrbach, K. G., Ohr, H. D. (eds.) Compendium of tropical fruit diseases. APS Press, St Paul, Minnesota, USA, p. 81.

Prakash, O. (1988). Sooty mould disease of mango and its control. *Int. J. Trop. Plant Diseases.* 9:277–280.

Prusky, D., Keen, N. T., Eaks, I. (1983). Further evidence of the involvement of a preformed antifungal compound in the latency of *Collectotrichum gloeosporioides* on unripe avocado fruit. *Physiol. Plant Pathol.* 22:189–198.

Raabe, R. D., Zentmyer, G. A. (1955). Susceptibility of avocados to Dematophora root rot. *California Avocado Society Yearbook* 39:172–174.

Ribeiro, O. K. (1978). Sporangia and zoospores production. In: Cramer, J. (Ed) A Source Book of the Genus *Phytophthora*, Ganther Verlag A.R., Vaduz, Germany, pp. 101–133.

Ruano-Rosa, D. (2006). Control biologic, caracterizacion y deteccion molecular de Rosellinia necatrix Prill., agente causal de la podredumbre blanca del aguacate. *PhD Thesis*, Universidad de Cardoba, Cardoba, Spain, p. 361.

Sanders, G. M., Korsten, L. (2000). Market survey of post-harvest diseases and incidence of *Colletotrichum gloeosporioides* on avocado and mango fruit in South Africa. *Trop. Sci.* 40:192–198.

Shepherd, C. J., Pratt, B. H. (1974). Temperature-growth relations and genetic diversity of A2 mating-type isolates of *Phytopthora cinnamomi* in Australia. *Aust. J. Bot.* 22:213–249.

Stevens, J. E. (1922). Avocado diseases. Florida Agric Exp Stn Bulletin, No. 161.

Stirling, A. M., Coates, L. M., Pegg, K. G., Hayward, A. C. (1995). Isolation and selection of bacteria and yeast antagonistic to preharvest infection of avocado by *Collectotrichum gloeosporioides. Aust. J. Agric. Res.* 46:985–995.

Teliz, D. (2000a). Enfermedades del aguacate. In: Teliz-Ortiz, D., Gonzalez-Hernandez, H., Rodriguez-Velez, J., Dromundo-Salazar, R. (Eds) *El Aguacate y su Manejo Integrado*, Mundi Prensa Mexico SA de CV, Mexici DF, Mexico, pp. 137–182.

Teliz, D. (2000b). El Aguacate Y Su Manejo Integrado. Mexico: Mundi-Prensa. pp. 219–221. ISBN 9789687462431.

Vidales-Fernandez, J. A., Tapia, V. M., Coria Avalos, V. M., Anguiano, J., Ochoa, S., Guilnlen-Andrade, H., Lara Chavez, B.N., Gutierrez, M., Sanabria, M.A., Chavez, J. (2007). Biological control of antracnose (*Collectotrichum gloeosporioides* Penz.) in the avocado (*Persea americana* Mill cv. Hass) growing in Michaocan, Mexico. *Book of Abstracts of the VI World Avocado Congress*, Vina del Mar, Chile, p. 99.

Vock, N. (2001). Avocado information Kit. Agrilink series Qal. Dpi, Queensland.

Waterhouse, G. M., Waterson, J. M. (1966a). CMI descriptions of pathogenic fungi and bacteria no 113 *Phytopthora cinnamomi*, *CAB International*, Wallingford, UK, p. 2.

Willis, A., Mavusco, Z. (2007). Evaluation of alternative fungicides for control of *Cercospora* spot on 'Fuerte'. *Book of Abstracts of the VI World Avocado Congress*, Vina del Mar, Chile, p. 109.

Zentmyer, G. A. (1949). Verticillium wilt of avocado. *Phytopathology* 39:677–682.

Zentmyer, G. A. (1977). Morphology of *Phytopthora cinnamomi*. *Mycologia* 69:701–713.

Zentmyer, G. A. (1980). *Phytophthora cinnamomi* and the diseases it causes, *The American Phytopathological Society* (Ed), Monograph No 10, St Paul, Minnesota, USA, p. 96.

Zentmyer, G.A. (1981). The effect of temperature on growth and pathogenesis of *Phytophthora cinnamomi* and on growth of its avocado host. *Phytopathology* 71:925–928.

Zentmyer, G. A. (1984). Avocado diseases. *Trop Pest Manage* 30:388–400.

Zentmyer, G. A., Leary, J. V., Klure, L. J., Grantham, G. L. (1976). Variability in growth of *Phytophthora cinnamimi* in relation with temperature. *Phytopathology* 66:982–986.

Zentmyer, G. A., Mircetich, S. M. (1965). Testing for resistance of avocado to *Phytophthora cinnamomi* in nutrient solution. *Phytopathology* 55:487–489.

CHAPTER 5

IMPORTANT DISEASES OF BAEL (*AEGLEMORMELOS* L.) AND MANAGEMENT STRATEGIES

K. JAYALAKSHMI[1,*], J. RAJU[2], S. RAGHU[3], and PRITI S. SONAVANE[4]

[1]National Bureau of Agriculturally Important Microorganisms (NBAIM), Kushmaur, Maunath Bhanjan, Uttar Pradesh 275103, India

[2]Plant Quarantine Station, Ministry of Agriculture and Farmers Welfare, Government of India, Mangalore, Karnataka 575011, India

[3]Crop Protection Division, National Rice Research Institute, Cuttack, Odisha 753006, India

[4]Indian Institute of Horticultural Research, Hesaraghatta, Bangalore, Karnataka, India

[*]Corresponding author. E-mail: jayalakshmipat@gmail.com

ABSTRACT

Bael is a domestic fruit tree of India having religious importance. It is also known as Bengal quince, Indian quince, golden apple, holy fruit, and stone apple. Bael (*Aegle mormelos* L.) is affected by many diseases caused by fungi, bacteria, and viruses. Nursery plants suffer from number of fungal foliar diseases namely root rot caused by *Fusarium solani*, leaf spot caused by *Myrothecium roridum* and *Alternaria altrnata*, die back and leaf spot caused by *Fusarium pallidoroseum*. These diseases as well as pathogen either alone or in combination causes substantial damage to crop resulting in heavy economic losses every year. This communication deals with important diseases of Bael (*Aegle mormelos* L.) and their management strategies.

5.1   INTRODUCTION

Bael (*Aeglemormelos* L.) is known for its nutritional and medicinal values. Bael tree is a national fruit tree of India and gaining religious importance, also called as golden apple, holy fruit, stone apple, Indian quince, and Bengal quince. From prehistoric times it is known in India also in the world. It is a deciduous tree that bears greenish-white and sweet-scented flowers and fruits are oblong or pyriform in shape. It is used in human drug preparation against respiratory infections, constipation, dysentery, diarrhea, and peptic ulcer. Beal having important medicinal properties such as antidiabetic, antimicrobial, anti-inflammatory, antipyretic, analgesic, antispermatogenic, cardioprotective, and anticancer (Sharma et al., 2011). Bael is grown in Uttar Pradesh, Uttaranchal, Madhya Pradesh, Deccan Plateau, Bihar, Chhattisgarh, Jharkhand, and along the foothills of the Himalayas and East Coast (Teaotia et al., 1963). There is no organized orchard/plantation of bael in India (Jauhari and Singh, 1971).

Bael trees are affected by many diseases caused by fungi, bacteria, and viruses. The trees suffer from a number of fungal foliar diseases, namely, root rot, leaf spot, dieback, and leaf spot and also some disorders (Anonymous, 2010). Beal trees suffering root rot (*Fusarium solani)*, leaf spot (*Myrothecium roridum, Alternaria altrnata)*, dieback, and leaf spot (*F. pallidoroseum)* disease (Anonymous, 2010). These diseases as well as pathogens either alone or in combination cause substantial damage to crops resulting in more economic loss. This communication deals with important diseases and different management strategies.

1.2   DISEASES

5.2.1   ALTERNARIA LEAF SPOT DISEASE

The disease is very common in bael trees. Bael trees is quite resistant to diseases but some disease occurred in little instance on beal. Leaf spot caused by *A. alternata* (Fr.) Keissler (Madan and Gupta, 1985).

Symptoms:

Alternaria leaf spot appears initially as small light or dark brown indefinite size spots appears on leaves with concentric rings. Later infected leaves become blight and fall. The disease is incited by seedlings that could also get infected. In most cases, only lower leaves are infected, but a severe infection can defoliate the plant (Maurya et al., 2002). On ripe fruit, sunken areas near the calyx are covered by a dark green, velvety growth. Grayish-green mold may appear on the stem scar or calyx end and spread over the entire berry on stored fruit. Infected fruit becomes soft and shriveled (Ansari et al., 1989).

Causal Organism: *A. tenuissima, A. alternata.*

Disease Cycle and Epidemiology:

The fungus persists in old twigs and in plant debris on the ground during winter. During winter conidia spread through the wind. Leaf infections occur in the spring during cool and wet weather periods. Fruit infections occur as berries start to ripen. The optimum temperature for disease development is 20 °C (Pandey et al., 1984).

Management:

  (1) Sanitation of bael orchards is very effective in disease control.
  (2) In the process, the infected plant debris should be removed and destroyed.
  (3) Foliar sprays of 0.25% mancozeb twice 15 days interval or 0.1% propi-conazole at 10–15 days interval is more effective in disease control.
  (4) Foliar application of 3 g/L of chlorothalonil 75WP or 1 g/L of carbendazim 50WP.

5.2.2   STALK END ROT AND FRUIT ROT DISEASE

This disease infects the bael plants during rainy season, in the month of June–July along with *Aspergillus* rot disease.

Symptoms:

Darkening of infected young shoots, immature drooping of fruits takes place during the rainy season. Rind becomes soft under-laying pulp. The affected rind turns dark brown at a later stage of infection. Fruits neither shrivel nor lose their shape until press. Infected fruits fall when there is heavy rain or strong wind. Cottony growth of fungal mycelium (*F. moniliforme*) was observed just beneath the hard shell. Later *Aspergillus* was grown on whole entire fruit and becomes soft and pulpy (Naqvi, 2002).

Causal Organism: *F. solani and F. moniliforme.*

Disease Cycle and Epidemiology:

Fusarium is extremely common in soils with saprophytic nature, which means it can grow and develop dead or dying tissues. Fungus produces chlamydospores which persist several years in soil. The fungus can invade in stems at the nodes or at the soil line or wounds created by pruning. Active growing stage, succulent parts, and ripening fruit are susceptible to infection as compared to green fruit. In fruits, infection is especially seen around the calyx and rotting in storage. The fungus may colonize fallen or aborted fruit and senescent flowers (Pandotra and Ganguli, 1964).

About 26 °C–28 °C temperature and high relative humidity favor disease development. Low nitrogen dose, high potassium dose, and acidic nature of soil predispose the plant to this disease. Survival of the pathogen in plains of India is rare as soil temperature above 37 °C kills the fungus.

Management:

(1) The infection of disease can be reduced by removal and destruction of affected fruits.
(2) Soil drenching with captan or thiram at 2g/L is effective.
(3) Spraying with benomyl 1 ml/L at 15 days intervals should be done to check the disease.
(4) Avoid damage during harvest and packing, good storage, and transit.

5.2.3 ASPERGILLUS ROT/DRY ROT

This disease infected the fruits during summer months (May–June). The intensity is severe and leads to 100% loss of fruit kept in storage since the infection was latent.

Symptoms:

Aspergillus rot disease is occurs most commonly in rainy season during flowering and early fruit development stages. Similar to *Alternaria* fruit rot disease, the fungus grows within the fruit often without showing external symptoms. However, infected fruit generally shows light brown to yellowish to reddish-brown discoloration. Brown discoloration which gradually becomes blackish and slimy, later, gets slightly depressed and covered by green conidial heads of the fungus. Shell of fruits become soft and rotten. Inner pulp produces black sooty growth of the fungus and emits fermented odor (Purohit and Vyas, 2005).

Causal Organism: *Aspergillus niger, A. fumigates, A. luchuensis.*

Disease Cycle:

Fungus colonizes with hemipteran insect infestation causes cracking on fruits.

Management:

   (1) Orchard sanitation, removal of old fruit and dead branches should be reducing the pre and postharvest incidence of disease.
   (2) Apply preharvest sprays are generally ineffective, but healthy-appearing fruits to the ground.
   (3) Insecticides spray to control insect pests.
   (4) Gently shaking of trees before harvest, can help in dropping infected fruits, but healthy appearing fruits to the ground.
   (5) Avoid water stress and over watering that may result in fruit cracking.

5.2.4   LEAF BLIGHT

M. roridum causes leaf blight. During favorable environmental conditions, soil born saprophyte that is considered to be a weak pathogen invades wounded or stressed plant tissue.

Symptoms:

Symptoms appeared as small, dark brown colored circular lesions or target spots on leaves. Under high humidity, black-colored fruiting bodies (sporo-dochia) develops and it can be covered with a white tuft of fungus mycelium (Verma and Sharma, 2005).

Causal organism: *M. roridum.*

Disease Cycle and Epidemiology:

Fungus survives in plant debris as mycelium. It spreads through wind-borne conidia released by sporodochia. The infection of disease is favored by wet weather conditions. The relative humidity of 80% and temperatures ranging between 21 and 23 °C are the most suitable conditions for disease spread.

Management:

  (1) Sanitation of orchard is very important to control the disease.
  (2) Periodical removal and destruction of affected debris
  (3) As early as, disease infection is noticed, in 15 days interval provide two to three sprays of mancozeb or copper oxychloride @ 2g/L three sprays are advisable at 15 days interval.

5.2.5   LEAF SPOT

The disease is very common in bael trees.

Symptoms:

During infection at the tip or margin of the leaves, necrotic lesions appear and turn to ashy gray and produce black fruiting structure acervullus (Elhadi and Yahia, 2011).

Causal Organism: *Colletotrichum gleosporioides.*

Disease Cycle and Epidemiology:

Air-borne conidia help in secondary spread of disease. Moist leaves due to dew and rains provide congenial environment for disease growth. Temperature ranging from 25 to 30 °C with relative humidity more than 80% encourages spore production and causes infection.

Management:

  (1) Removal and destruction of affected plant debris.
  (2) 2–3 sprays of mancozeb 2.5 g/L or carbendazim 1 g/L or blitox/blue copper at 3 g/L found effective.

5.2.6   POWDERY MILDEW DISEASE

Similar to other horticultural crops, bael trees are also infested by powdery mildew disease (Thampman, 1993).

Symptoms:

White or gray powder-like growth takes place on the tender leaves due to infection of powdery mildew disease. Generally, the infection starts from leaves and spreads downwards covering the stem. Affected leaves show twisting and curling finally defoliate (Yarwood, 1978; Elhadi and Yahia, 2011).

Causal Organism: *Oidium tingitanium*
Mycelium branched, hyaline, superficial, septate, haustoria lobate. Conidia are hyaline unicellular, elliptical, borne single or rarely in chains of two. Conidiophores are simple, erect with two or more basal cells (Yarwood, 1978).

Disease Cycle and Epidemiology:

Primary source of inoculums is through conidia in infected plant debris. Secondary spread takes place through air-borne conidia. Spread of disease is favored by warm humid weather condition combining with cool nights (Yarwood, 1978).

Management:

(1) Powdery mildew can be checked by spray of 0.3% wettable sulfur or Karathane of 0.1%.
(2) Dusting of fine sulfur, 2–3 times will also control the disease.

5.2.7   *PHOMA* LEAF BLIGHT AND *PHOMA* FRUIT ROT

The bael plants are quite often infected with *Phoma* leaf blight and *Phoma* fruit rot diseases (Thampman, 1993).

Symptoms:

As soon as the blossoms opened, they start turning brown due to infection of pathogen. This discoloration also goes to on the pedicels and adjacent leaves immediately.

Initially the small circular yellow to light brown spots produced on leaf lamina. As the lesions enlarge color changes from light brown to dark brown becomes irregular. In severe infection, such spots coalesce lead blightening, withering, and defoliation. Affected young fruits completely turn to brown and remained hanging on tree until dry (Morgan-Jones, 1967; Elhadi and Yahia, 2011).

Causal Organism: *Phoma glomerata.*

Disease Cycle and Epidemiology:

Infected leaves and fruits fallen on soil act as source of inoculums. Mainly, they spread through wind-borne conidia. Dry warm weather with intermittent rain and temperature ranging from 25 °C to 30 °C favor the development of disease.

Management:

Spraying benomyl @ 0.2% followed by copper oxychloride + zineb at 0.3% at 15 days interval after first appearance of the disease.

5.2.8 SOOTY MOLD

Sooty mold or sooty blotch common in all beal tree growing orchards found wherever honeydew or sugary substance present which is secreted by insects like, hopper, coccids, mealy bugs, and scales (Thampman, 1993).

Symptoms:

Superficial dark thread like mycelium growth produced on leaf surface on insects secreted sugary substance. On that secretion black encrustation is formed by pathogens which affect the photosynthetic activity (Kala, 2006).

Casual Organism: *Capnodium* spp.

Disease Cycle:

Infected leaves serve as primary inoculums.

Management:

(1) Insects should be controlled by spraying systemic insecticides monocrotophos or methyl dematon.
(2) Spraying with a mixture of sulfur + chlopyriphos/methyl parathion + gum @ 0.2% + 0.2% + 0.3% for the control of disease
(3) Spraying starch solution (1 kg starch/maida in 5 L of water, boiled, and dilute to 20 L) should be on the tree after the spraying of

insecticides. Starch dries and forms flake, which are removed along with the fungus.

5.2.9   BACTERIAL SHOT HOLE AND FRUIT CANKER

The disease was first time reported in 1953 by Patel et al. from Maharashtra (India).

Symptoms:

Symptoms appeared as round water-soaked spots surrounded by clear halon leaves, which later increases in size and form brown lesions with saucer-like depression in the center, surrounded by oily raised margin finally leads shot holes symptoms. The pathogen infects fruit, twig, and thorn. On fruits, the lesions are round, raised water-soaked without any halo. With growth of fruits, crater-like depression is formed in the center of spots surrounded by irregular, oily raised margins which become corky, irregular, and chocolate brown in color. The pathogen infects thorns also. Large lesions are commonly formed on twigs. The infected tissue ruptures in the center giving a rough corky appearance. Bacterial exudates are commonly formed on all infected parts (Sarah et al., 2008).

Causal Organism: *Xanthomonas bilvae.*

Disease Cycle and Epidemiology:

The bacteria confined in the twigs and thorns during winter. During spring, the bacteria spread by rain splash to leaves, twigs, and fruit. Secondary spread takes place through bacterial ooze from affected lesions or cankers from leaf or fruit. The systemic spread of the bacteria from leaves and twigs leads canker formation.

Warm environment with temperatures above 25 °C and rains are needed for bacteria multiplication, its exposer, and dissemination.

Management:

(1) Disease infection can be checked by taking precautions.
(2) The bael fruits should not be hurt during plucking and transportation. Packing of fruits should be tight to avoid injury.
(3) Spray of 500 ppm streptomycin sulfate solution controls the infection.

5.3   PHYSIOLOGICAL DISORDERS

Fruit cracking and fruit drop before ripening are two important physiological disorders which occurred in bael.

5.3.1   FRUIT DROP

Degeneration of the embryos in the initial stages of their development is another factor of fruit drop. Just after fruit setting, fruit drop observed due to nutrient deficiency leads.

5.3.2   FRUIT CRACKING

Fruit cracking is caused due to boron deficiency in young fruits. Lack of soil moisture and humidity may lead to cracking in developed fruits.

Management:

(1) Fruit cracking can be managed by providing good irrigation facility, making wind breaks around the orchard.
(2) At full bloom and after fruit set stage spraying 0.1% borax controls the fruit cracking.
(3) Spraying of 2, 4-D, and GA3 growth regulators reduce fruit dropping and fruit cracking.

KEYWORDS

- Bael
- diseases
- symptoms
- management

REFERENCES

Anonymous. The Cause of Wilting/Drying in Bael Plants. *Annual Report of AICRP on Arid Zone Fruits (ICAR)*; 2010; 196.

Ansari, N. A.; Khan, M. W.; Muhert, A. Effect of Some Factors on Growth and Sporulation of *Alternaria* Brasicae Causing *Alternaria* Blight of Rapeseed and Mustard. *Acta Bot. Indica.* **1989**, *17*, 49–53.

Elhadi, M. Y. Postharvest Biology and Technology of Tropical and Subtropical Fruits. Woodhead Publishing Ltd: Cambridge, UK, Vol. 2, 2011; pp. 110–115.

Ellis, S. D.; Boehm, M. J.; Coplin, D. Bacterial Diseases of Plants. *6th Fact sheet Agriculture and Natural Resources.* The Ohio State University; PP401.06; pp 2–4.

Jauhari, O. S.; Singh, R. D. Bael-A Valuable Fruit. *Indian Hort.* **1971**, *16(1)*, 9–10.

Kala, C. P. Ethnobotany and Conservation of *Aeglemarmelos* (L.) Correa. *Indian J. Traditional Knowl.* **2006**, *5(4)*, 537–540.

Madaan, R. L.; Gupta, P. C. A Leaf Spot Disease of *Aeglemarmelos* L. Caused by *Alternaria alternata* (Fr.) Keissler. *Indian J. Plant Pathol.* **1985**, *3*, 239.

Maurya, S.; Kumar, R.; Kumari, A.; Choudhary, J. S. First Report of *Alternaria* Leaf Blight in Bael (*Aeglemarmelos* (L.) Corr.) from Eastern Plateau and Hill Region of India. *J. Agric. Search*, **2016**, *3(4)*, 248–250.

Morgan-Jones, G. Phomaglomerata. In *CMI Descriptions of Pathogenic Fungi and Bacteria No. 134;* Commonwealth Mycological Institute: Kew, United Kingdom, 1967.

Naqvi, S. A. M. H. Diseases of Fruits and Vegetables: Volume I: Diagnosis and Management. Naqvi, S. A. M. H., Ed.; Kluwer Academic Publishers, New York, Boston, Dordrecht, London, Moscow. 2004; p. 679. eBook ISBN: 1-4020-2606-4, Print ISBN: 1-4020-1822-3.

Pandey, R. S.; Bhargava, S. N.; Shukla, D. N.; Dwivedi, D. K. Two New Fruit Diseases Caused by *Alternaria alternata*. *Int. J. Trop. Pl. Diseases*. **1984**, *2*, 79–80.

Pandotra, V. R.; Ganguli, D. Fungi of Medicinal and Aromatic Plants in the North West Himalaya. II. *Mycopath. Mycol. Applicata*, **1964**, *22*, 106116.

Patel, M. K.; Allayyanavarmath, S. B.; Kulkarni, Y. S. Bacterial Shot Hole and Fruit Canker of *Aeglemarmelos* Correa. *Curr. Acince*. **1953**, *22*, 216–271.

Purohit, S. S.; Vyas, S. P. *Medicinal Plant Cultivation-A Scientific Approach*. Agrobios: India, 2005; pp. 282.

Sharma, G. N.; Dubey, S. K.; Sharma, P.; Sati, N. Medicinal Values of Bael (*Aeglemarmelos*) (L.) Corr.: A Review. *Int. J. Curr. Pharm. Rev. Res.* **2011**, *1(3)*, 12–22.

Teaotia, S. S.; Dayal, K.; Asthana, M. P. Propagation of Jackfruit by Budding. *Sci. Cult.* **1963**, *29*, 46–47.

Thampman, P. K. *Trees and Tree farming*; Thampman, P. K. Ed. Peekay Tree Crops Development Foundation: Kerala, India, 1993.

Verma, K. S.; Sharma, A. Challenging Disease Problem in Tropical and Subtropical Fruits and Their Management. In *Challenging Problems in Horticultural and Forest Pathology*; Sharma, R. C.; Sharma J. N., Eds.; Indus Publishing Company: New Delhi, 2005; pp. 72–89.

Yarwood, C. E. *The Powdery Mildews*. Spencer D. M. Ed.; Academic Press, Inc. London: United Kingdom; 1978.

CHAPTER 6

CURRENT STATUS OF BANANA (*MUSA PARADISIACA* L.) DISEASES AND THEIR MANAGEMENT

KAVI SUMI, SUSANTA BANIK* and PEZANGULIE CHAKRUNO

Department of Plant Pathology, School of Agriculture Science and Rural Development (SASRD), Nagaland University, Medziphema 797106, Nagaland, India

*Corresponding author. E-mail: susanta.iari@gmail.com

ABSTRACT

Banana is the most important fruit crop commercially grown in a number of countries worldwide for its utilization as dessert and as staple food in certain parts of world. Among the fruit crops, it is one of the most important crops in international trade for earning the foreign exchange in many countries. Banana is the second highest fruit crop in production in India next only to mango. Although, banana occupies less than 12% of the area under fruit crops, it contributes nearly 32% of the total fruit production in India. India ranks first in the world, with 11% of the total world banana production. Today, banana is the much sought-after tropical fruit in India with a highly organized marketing network. Biotic stress caused by diseases and insect pests wreaks havoc on the production of banana in the state. Major diseases and insect pests of banana are discussed below.

6.1 INTRODUCTION

Banana (*Musa paradisiaca L.*) is a significant commercially cultivated fruit grown in almost all the countries of the world. It is the second bulkiest food crop belonging to the *Musaceae* family in the order *Scitamineae*.

Banana is considered a key food crop for more than 40 crores inhabitants of the emerging countries in the tropics (Molina and Valmayor, 1999). It has a multifaceted usage and is hence designated to as "Kalpatharu" (Plant of Virtues) in India. Apart from consumption as a dessert fruit, the leaves, pseudostem, and flower blossoms find various usages for human needs and for animal purposes. The fruit is wholesome and includes a high quantity of vitamin C, potassium, and vitamin B6 (Samson, 1986; Robinson, 1996). Banana was the foremost cultivated fruit and its domestication is considered to have initiated after the Pleistocene era (approximately 12,000 B.P.), with archeological evidence that shows that it was domesticated at least 6500 B.P. (Denham et al., 2003; 2004). The area of cultivation is 4.81 Mha with 100.9 MT of production on an average during 2012–13 worldwide (FAO). Banana (*Musa paradisica* L.) is one of the most important tropical fruit accounted by Pawar et al. (2010). India is the biggest producer of banana globally, yielding 29.22 million tones from 0.821 million hectares area with 34.2 MT ha^{-1} productivity (NHB, 2015). Among all states in India, Gujarat, Tamil Nadu, and Maharashtra accounts for production of 13.4%, 27.7%, and 14.4%, respectively with 8.25 m. MT, 4.3 MT, and 3.98 MT of production, respectively from 0.13 m ha, 0.08 m ha, and 0.06 m. ha area and 65.8 t/ha, 52.5 t/ha, 61.5 t/ha productivity in the country (National Horticultural Mission, 2011). In India, Banana is more prone to diseases than insect pest which amounts to huge losses in the total yield of the crop.

Lists of Banana Diseases

Sr. No.	Name of the Disease	Casual Organism
	Major Diseases	
1.	Panama Wilt	*Fusarium oxysporum* f. sp. cubense
2.	Sigatoka leaf spot	*Mycosphaerella musicola*
3.	Bunchy top of banana	Bunchy top virus or Banana virus—1 or Musa virus—1
	Minor Diseases	
1.	Moko disease	*R. solanacearum*
2.	Bacterial Soft rot	*Erwinia carotovora*
3.	Anthracnose disease	*Colletrotrichum musae*

6.2 FUNGAL DISEASES

6.2.1 *PANAMA WILT/FUSARIUM WILT/VASCULAR WILT DISEASE*

Introduction and Economic Importance:

Panama wilt in banana is one of the most destructive disease which affects commercial production as well as subsistence production all over the areas producing banana in the world (Ploetz, 2005). Originally, Panama was one of the primary countries to encounter the global epidemics of fusarium wilt that severely affected the dessert banana, Gros Michel that had dominated the export trade since its early days (Brandes, 1919). Although Panama wilt almost certainly started off in Southeast Asia (Ploetz and Pegg, 1997), the discovery of the disease was first made at Eagle Farm, Australia in 1876 in banana crop var. sugar (Bancroft, 1876). In Asia, the first account of wilt was described by Basu in 1911 in Chinsurah, West Bengal (Basu, 1911).

In the world, the Panama wilt disease is positioned as one of the top six most important plant diseases (Ploetz and Pegg, 1997) causing an annual yield loss of 60%–90% in many countries (Bhuvanendra, 2010). In India alone, this disease was projected to cause losses up to 30%–40% and in South India, the losses ranged from 2% to 90% (Thangavelu, 1999). The vicious nature of the disease is now prevalent in about all the banana-growing states of our country causing an incidence of about 30% in the fruit crop and up to 85% in ratoon crop except for Red Banana (AAA) and Nendran (AAB; Thangavelu and Mustaffa, 2010). The damage inflicted by Fusarium wilt can be very harmful in susceptible cultivars with fatalities soaring up to 100% (Thangavelu et al., 2001).

Symptoms:

The exterior symptoms produced by Panama wilt of banana are: "green leaf complex" and "yellow leaf complex." (Stover, 1962; Pérez-Vicente, 2004). The most prominent internal symptom is vascular browning of the rhizome and pseudostem (MacHardy and Beckman, 1981). The fungus penetrates the plant through the roots and inhabits the xylem vessels thereby blocking the course of water and the nutrients (Ploetz and Churchill, 2011). The fungus after invading the xylem, produces exterior symptoms like wilting of the crop gradually, the leaves progressively become yellow which proliferates from the leaf margins, and begins from older to younger leaves and ultimately crumple at the leaf petiole, leading to the longitudinal opening of the external leaf sheaths in pseudostem of the crop (Yin et al., 2011). The interior symptoms can be distinguished even after a month or two of appearance of suckers (Moore et al., 1995).

Causal Organism: *Fusarium oxysporum* f. sp. *cubense* (Foc)

Fusarium oxysporum f. sp. *cubense* (Foc), a soil borne fungus causes a fatal disease in banana, the Panama wilt. The fungus was initially isolated from a diseased banana plant by Bancroft (1876) and the fungus association with the diseased banana plant suffering from wilt was noted by Higgins (1904). The species was named *Fusarium cubense* by Smith (1910) whereas the first specific narrative of the fungal organism in in vitro culture was given by Ashby (1913). The first proof of the disease was done by Brandes (1919) using Koch postulates, in the cultivar Bluggoe (ABB) as well as only in Gros Michel (AAA) and Manzano (Apple, AAB).

Fusarium oxysporum is a multifaceted fungus of filamentous, anamorphic, morphologically undifferentiated fungal species which is saprophytic and pathogenic to plants, animals in addition to humans (O'Donnell and Cigelnick, 1999). The dispersal and reproduction of the fungus are done by producing microconidia, macroconidia, and chlamydospores structures. The microconidia and macroconidia are formed in orange structures termed as sporodochia (Fourie et al., 2011).

Diseases Cycle:

Panama wilt of banana is a "polycyclic" disease (Ploetz, 2015) and is disseminated through many ways but infected rhizomes are mostly proficient in the spread of the disease (Stover, 1962). The life cycle of *Fusarium oxysporum* has the phases, that is, saprophtytic and parasitic phases. The fungus survives saprophytically in the soil as chalmydospores, which remains latent and inert until they are enthused by the exudates that are produced from the extending banana roots to germinate (Stover, 1962; Beckman and Roberts, 1995). Surface waters are also infested with *F. oxysporum* f. sp. *cubense*, and infected water used for irrigation has been accountable for the pathogen's rapid dissemination along river basins and in banana fields. Furthermore, *F. oxysporum* f. sp. *cubense* is disseminated on contaminated tools, farm equipment, clothes, and footwear in the fields (Wardlaw, 1961). The pathogens move inter- and intra-cellularly through the root parenchyma tissue where they invade the vessels, blocking the xylem vessel which eventually leads to the demise of the crop. Notably, *F. oxysporum* f. sp. *cubense* infects roots of both the susceptible and resistant cultivars of banana, but infection of vascularized portions of the rhizome is most pronounced in susceptible genotypes (Beckman, 1990). The disease symptom usually becomes more apparent at the time of flowering. The roots of the banana plant are infected by the fungus, which colonizes the vascular system of the rhizome and pseudostem, as well as the typical wilting symptoms are expressed mostly after 5–6 months of planting and the symptoms are seen both externally and internally (Wardlaw, 1961; Stover, 1962).

Epidemiology:

The disease is more severe usually during the humid and moist months of the year as it helps in predisposing the disease to a greater extent but some aspects have a preponderant effect in the development of the disease. Moore et al. (1995) described that the fungus can subsist up to 30 years in the field when left unchecked. *F. oxysporum* thrives well between 9 °C and 38 °C under in vitro conditions, with a favorable extent of growth amid 23 °C –27 °C (Pérez et al., 2003) in the infected plant rubble as chlamydospores or survives in the roots of alternative hosts. The favorable soil condition for fusarium wilt disease is soil with a light texture than heavy clay texture soils (Stover, 1962). Thangavelu et al. (2001) reported that the incidence of the wilt disease from loose soil to heavy clay soil ranges from 4.80 to 8.45 pH and EC of 0.12–1.10 d sm^{-1}. High phosphorus present in the soil helps to lessen the disease incidence (Woltz and Jones, 1981).

Integrated Diseases Management:

Cultural Method:

Phyto-Sanitation:

- Disease-free tissue culture suckers should be employed to prevent the initiation of the disease in areas which are free from diseases (Moore et al., 1999).
- Infected plants that exhibit symptoms must be destroyed along with the other host species which may serve as inoculums for the spread of the disease (Dita et al., 2013).

Crop Rotation:

- Panama wilt was found to be decreased by 48% after banana was grown as an intercrop with sugarcane and left fallow (Sequeira, 1962).

- Banana crop used in crop rotation with Chinese leek (*Allium tuberosum*), was found to lessen TR4 incidence and the severity index in Cavendish and Guangfen (AAA) cultivars with 58% and 62%, respectively (Huang et al., 2012).

Physical Method:

- Ramakrishnan and Damodaran (1956) articulated that the pathogen survivability period was reduced up to 2 months when liming was done in soil.
- The incidence of fusarium wilt was reduced when soil solarization was done and the fusarium population was reduced when the soil temperature was maintained at 52.35 °C by Hermanto et al. (2012).

Biological Method:

- The biological agents such as *T. viride* and *P. fluorescens* were found to be efficient in managing the wilt incidence (Raghuchander et al., 1997).
- Thangavelu (2002) described that the application of *T. harzianum* Th-10, as dried banana leaf formulation @10 g/plant includes 4×1031 cfu/g in basal and top dressing on 2, 4, and 6 months after planting in cv. Rasthali documented the maximum reduction of disease incidence (51.16%) which was followed by *Bacillus subtilis* or *P. fluorescens* (41.17%) as talc based formulation in both glass house and field situations.
- Under field conditions, Saravanan et al. (2003) observed that either application of neem cake at 0.5 kg/plant along with sucker dipping in spore suspension of *P. fluorescens* for 15 min and soil application of *P. fluorescens* at 10 g/plant at 3, 5, and 7 months after planting in the basal region or basal application of neem cake at 0.5 kg/plant along with soil application of *P. fluorescens* at 10 g/plant at 3, 5, and 7 months after planting recorded the highest inhibition of fusarium wilt.
- In glass house trails, the strain WCS 417 of *P. fluorescens*, was found to minimize the disease incidence by 4%–87% in banana var. Cavendish (Nel et al., 2006).
- Pushpavati et al. (2015) examined that sucker treatment before planting with biocontrol agents *T. viride* and *P. fluorescens* and soil drenching with same biocontrol agents twice at 30 and 180 DAP as booster application, effectively reduced the fusarium disease incidence and intensity, thereby increasing the yield.

Chemical Method:

- A trial conducted at NRC for banana fields also indicated that dipping suckers in carbendazim (0.1%) at the time of planting for half an hour along with soil drenching with carbendazim (0.1%) near the pseudostem + pseudostem injected with carbendazim (0.1%) at 2, 4, and 6 months intervals after planting drastically reduced wilt severity (score 1.3) in contrast to the untreated control plants (score 4.0; Anonymous, 2009).

Integrated Approaches:

- Akila et al. (2011) reported that the application of botanical formulation and biological agents that is, *Datura metel*–Wanis 20 EC and (*P. fluorescens*, Pf1, and *B. subtilis*, TRC 540; Wanis 20 EC + Pf1 + TRC 54) in combination under greenhouse and field conditions minimized the disease incidence significantly by 64% and 75%, respectively.
- Zhang et al. (2011) evaluated that bio-organic fertilizers which contain amino acid fertilizer plus compost of pig manure along with the biological agents (*Paenibacillus polymyxa* SQR21, *Trichoderma harzianum* T37, and *B. subtilis* N11) minimized the incidence of wilt by 64%–82% in pot conditions.

6.2.2 SIGATOKA LEAF SPOT OF BANANA

Introduction and Economic Importance:

The Sigatoka leaf spot is one of the most severe diseases behind panama wilt in a banana plantation and is globally measured as an impending threat to banana production (Arzanlou et al., 2008). Sigatoka disease is linked with three species namely *Mycosphaerella fijiensis* (I S: *Pseudocercospora fijiensis)* causal organism of Black Sigatoka (Stewart et al., 1999), *Mycosphaerella musicola* (I S: *Pseudocercospora musicola)* causal organism of Yellow Sigatoka disease (Stover and Simmonda, 1987), and *Mycosphaerella eumusae* (I S: *Pseudocercospora eumusae*) causing Eumusae leaf spot (Carlier et al., 2000; Crous and Mourichon, 2002). These three species of fungi are hemibiotrophic and heterothallic (Mourichon and Zapater, 1990). These three species of Sigatoka are recognized worldwide as important constraints to production in banana crops (Carlier et al., 1996). Among the three genus species, Sigatoka leaf spot disease inhibited by *P. musicola* is regarded as a severe threat to banana production universally due to its epiphytotic nature (Mourichon and Fullerton, 1990; Selvarajan et al., 2000). Sigatoka disease is prevalent in all foremost banana growing areas, precisely in South East Asia, Pacific region, Latin America, and Africa (Jones, 2000). In India, Sigatoka disease has been accounted from several states such as Karnataka, Kerala, Tamil Nadu, Gujarat, Maharashtra, Andhra Pradesh, West Bengal, and the Northeast hill states of India (Selvarajan et al., 2001).

Sigatoka leaf spot has been documented to cause serious restrain to banana production worldwide (Blomme et al., 2011). In many of the producing areas, the disease is reported to cause losses from the early death of leaves, reduction in the growth and development of plants, and decrease in the quality of fruit (Castelan et al., 2013). Banana yield losses range from 20% to 50% due to Sigatoka leaf spot (Stover, 1983; Crous and Mourichon, 2002) and more than 50% of economic loss is also reported across the world (Burt et al., 1997). Sigatoka disease of banana continues to remain a major constraint in the production of banana globally (Raut and Ranade, 2004).

Symptoms:

Black and Yellow Sigatoka causes similar symptoms which mainly depend on the type of cultivar infected, stage of development of the disease, and the season of the year (winter or summer; Johanson, 1993).

1. **Black Sigatoka Disease:** The first characteristic symptom is the manifestation of dark-brown to black leaf streaks measuring roughly 1–2 mm in length on the lower surface of the banana leaf. The streaks enlarge into dark-brown which later advance to black streaks and then coalesce to form larger spots with a dark-brown colored to black center. The spots may often be enclosed by yellow halos (Stover and Simmonds, 1987).

2. **Yellow Sigatoka Disease:** The initial symptom is a light-green, narrow fleck on the upper leaf surface measuring approximately 1 mm in length. These flecks extend into a streak which runs analogous to the leaf veins and elongates, enlarges laterally to turn elliptical in shape which later turns rusty red in color. When the leaf turns turgid, a water-soaked halo is formed around the lesion. The infected part turns brown with the formation of a young spot. On younger leaves of the plants, the individual spot tends to be larger in size and more spherical in shape. The matured spots normally measure about 4–12

mm in length. Symptom development from specks through streaks to spots has been divided into various stages by different authors (Leach, 1946; Brun, 1963).

Causal Organism:

1. **Black Sigatoka Disease:** It was first illustrated from a material collected from the Sigatoka valley in Fiji island (Leach, 1964). *M. musicola* (I S: *P. musicola*) causing disease in banana was termed as "black leaf streak" by Rhodes in the year 1964. Leach (1964) depicted that the spread of this new banana disease is risky as it poses a serious threat due to the profuse quantity of ascospores present in the air produced by the fungus which may lead to the circulation of the disease more rapidly than Sigatoka leaf spot (Jones, 2003).

2. **Yellow Sigatoka Disease:** The fungus *M. musicola* (I S: *P. musicola)*, an ascomycete causing Yellow Sigatoka disease (Irish et al., 2013), was first depicted as a pathogen of banana plant in Java in the year 1902 (Zimmerman, 1902). The first and foremost disease epidemic of Yellow Sigatoka was articulated in Fiji in the year 1913 (Massee, 1914). It is a restricting aspect in the production of banana in India and other banana-growing regions worldwide (Nwauzoma et al., 2011).

Diseases Cycle:

Both *M. fijiensis* and *M. musicola* are alike in their disease cycle with only a trivial difference, as *M. fijiensis* produces significantly very less conidia and for a shorter phase of time than *M. musicola*, the key dispersing agents for this pathogen are ascospores (Stover, 1980). The ascospores are deposited by wind currents which generally lands on the terminal end of the leaf ensuing in a distinguishing leaf tip infection (Stover, 1972). The streaks usually tend to appear first near the leaf apex and then along the leaf margin, which is indicative of the infection made by ascospores (Meredith, 1970). Both the conidia and ascospores are vital for the dispersal of *M. musicola* (Stover, 1971) however, for both the pathogens ascospores are required in the progress of the pathogen to cover longer distances rather than the conidia. Conidia are firstly formed in lesions and then the infection spreads to other leaves on the same plant or to neighboring plants. The germination of the conidia occurs in water and the leaf is infiltrated through the stomata. Both the conidia and ascospores can germinate within a matter of 2–3 h, but stomata are not usually infiltrated until after 48–72 h of humidity at or near saturation, and at temperatures above 20 °C. After infection occurs, hyphae start to appear from the stomata and either grow into conidiophores or develop across the surface and infect the bordering stomata.

The disease cycle is slow in the case of Yellow Sigatoka and faster for Black Sigatoka, spotting related with *M. fijiensis* infections starts to appear 8–10 days faster than that related with *M. musicola* infections. The maturation time of ascospore for *M. fijiensis* is shorter at 2 weeks in contrast with *M. musicola* which occurs at 4 weeks (Stover, 1980). The production of perithecia and the successive discharge of ascospores persist for several months. Even in the case of severely necrotic conditions, ascospore discharge can persist for more than 2 months, this is also the case where the leaf has been detached and sited on the ground (Carlier et al., 2000). Ascospore ejection remains high for 3 weeks even after detachment of the leaf from the plant and then minimizes briskly over the next 6 weeks until the 10th week when the leaves themselves had collapsed (Gauhl, 1994).

Epidemiology:

1. **Black Sigatoka Disease:** Although the two pathogens may arise on the identical host, Black Sigatoka has been described to be more virulent than yellow Sigatoka (Ploetz, 1999). Highest germination of conidia and ascospores takes place in water and declines as the relative humidity (RH) decreases. Conidia and ascospore germination below 95% and 98% RH, respectively have not been found (Jacome et al., 1991). The ascospore germ tube development temperature of *M. fijiensis* at minimum, optimum, and maximum is 12 °C, 27 °C, and 36 °C, respectively and development does not take place at 11 °C and 38 °C (Porras and Pérez, 1997). Maximum disease development most likely takes place approximately at 25 °C –27 °C under moist conditions. Soils which are deficient in potassium or unbalanced N/K fertilizer relationship are more prone to higher attacks by the pathogen (Holderness et al., 1998; Pérez et al., 2002).

2. **Yellow Sigatoka Disease:** The windborne ascospores are the initial source of infection. From the same lesions, ascospores are produced from which conidia are discharged and are forcibly evicted from pseudothecia, due to high RH, and even in dry climates, but mostly due to greater leaf wetness periods (Simmonds, 1966). Thus, the density of conidia in the air is connected to the intensity always related with inconsistent temperature and erratic RH (Guyot and Cuille, 1958). The disease was reported to be more serious in orchards, which are

at the harvesting stage rather than those at the vegetative stage and predominantly more in ratoon crops in contrast to freshly planted crops (Thammiah, 2003). In the case of developing countries, the pathogen is spread through the infected suckers and leaves in long-distances, which are frequently used as packaging materials (Ploetz and Mourichon, 1999).

Integrated Diseases Management:

In order to reduce the usage of fungicide, numerous disease management strategies have been employed which include phyto-sanitation, disease fore-casting methods, use of biological agents for the commercially cultivated banana plantations (Ngongo, 2002).

Cultural Method:

- The efficient pruning of Black Sigatoka disease infected leaves and the exclusion of rubbles of leaf lamina with necrotic spots is vital to reduce *M. fijiensis* and other pathogens that may grow saprophytically (Stover, 1972; Pérez-Vicente, 1996).
- Mobambo and Naku (1993) reported that the severity of Black Siga-toka is less on fertile soil as compared to the poor soil since fertile soils contain higher soil organic matter content which incites not only root ramification, but ensues in better water and nutrient uptake and more vigorous plants.
- Elimination of early leaf is recommended as it shuns conidia produc-tion and secondary infection. This mainly consists of the removal of 15% of the length of the leaf tip from the third open leaf downwards in the banana plant (Martínez-Acosta et al., 2006).

Biological Method:

- Shanthiyaa (2014) observed that foliar spray with strains of Bacillus spp. (BB30 + BB15 + BB13) as the distinctive treatment prevails tremendously useful in decreasing the diseases, that is, Yellow Sigatoka.

Chemical Method:

- Knowles (1914) first supported to utilize Bordeaux mixture to control Sigatoka.

- Thammaiah et al. (2008) showed that spraying of propiconazole twice @0.05% effectively controlled the leaf spot disease of Sigatoka.

Patel (2009) reported that monthly intervals after 6 months of planting with any one of the four sprays of the fungicides, that is, carbendazim @1.0 g/L or tridemorph @0.7 ml/L or propiconazole @1.0 mL/L or thiophanate methyl @1.0 g/L g provided efficient control.

Hegde and Mesta (2014) reported that banana plants sprayed thrice with hexaconazole @0.1% reduced the disease incidence significantly up to an extent of 13.74%.

Integrated Approaches:

- Nayana and Radhakrishnan (2015) reported that the best treatment for the management of Sigatoka disease was found by spraying Magnesium (2 g/L), Zinc (3 gm/L), Boron (2 g/L) along with tebuconazole @0.1% which showed maximum disease suppression, minimum disease intensity reduction, and highest yield.

6.2.3   ANTHRACNOSE/FRUIT ROT DISEASE

Introduction and Economic Importance:

Anthracnose disease is one of the most severe diseases of banana caused by *Colletotrichum* sp. Anthracnose rot is regarded as a serious disease in banana fruit crops which is found in all banana-producing countries (Amani, 2008; Amani et al., 2006).

Symptoms:

Symptoms of *Colletotrichum* often become evident during storage and marketing processes. The infection on the banana usually begins during the development of the fruit but remains inactive until the fruit ripens (Prusky and Plumbley, 1992). On green banana fruit, it shows brown to black diamond-shaped lesions. On severely infected banana fruit, orange or salmon-colored rings may arise. On fruits that become yellow, brown spots primarily emerge which afterward become sunken and enclosed with orange spore masses. A tip rot may be developed and can lead to rotting of the entire fruit but the pulp is typically not affected except that the fruit remains overripe (Chillet et al., 2007). *C. musae* also commonly appear on the fruit peel after ripening having anthracnose lesions (Griffee and Burden, 1974).

Causal Organism:

Colletotrichum musae (Berk. and Curtis) Arx or *Gloeosporium musarum* (Cooke and Massee) causes fruit rot of anthracnose (Jones, 2000). Anthracnose is a significant postharvest rot disease of banana, caused by *C. musae*, which influences the fruit quality and marketability (Su et al., 2011; Ara et al., 2012).

Diseases Cycle:

Inoculum remains on dried leaves and they serve as primary inoculum. Secondary spread is through air-borne conidia. The fungus can enter the pores of green fruits. The latent infection of mature fruits may take place through lenticels. The fungus apparently infects the fruit while it is green and develops in flesh during ripening. The latent infection is carried from the field to storage. Healthy fruits develop infection after coming in contact with the diseased ones. The latent infection does not begin to spread until it reaches eating maturity.

Epidemiology:

Environmental conditions such as temperature and humidity are extremely important to the epidemics of anthracnose. The process of infection occurs throughout the flowering and fruit development. Symptoms may come from lesions of the infections occurring in green fruit, which remain quiescent until the fruit ripens (Chakravarty, 1957). In the old leaves and crop remains, spores are produced and are released by water, and dispersal is done with the help of wind or insects (Fitzell and Peak, 1984).

Integrated Disease Management:

Cultural Methods:

- According to Berangan (2012) fruit dipping in warm water at 50 °C for 20 min was discovered to be more efficient in suppressing disease development.

Biological Methods:

- According to the in vivo study by Khleekorn (2015), a selected antagonist known as *Enterobacter* sp. applied on a banana prior to harvest each week proved to show the most efficiency in managing anthracnose, with 87.6% inhibition.

- Jagana et al. (2017) showed that Neemgold and Discheck were found to be most efficient in the in vitro in addition to in vivo studies. Reduction in percent disease (92.11%) was found to diminish in most fruits which were treated with Neemgold, Discheck, and nimbicidin at 2% concentration which were found to be at par with same botanicals at 1.0% (89.47%) in addition to Neemgold at 0.5% (86.84%).

Chemical Methods:

- Wharton and Dieguez–Uribeondo (2004) reported that for the control of anthracnose diseases protectant (i.e., copper oxychloride, metallic dithiocarbamates, and chlorothalonil), systemic fungicides (i.e., benzimidazoles, triazole) and eradicant fungicides with multisite actions have been widely used either at pre or post harvest stages.

Integrated Approaches:

The successful management by postharvest dip treatment of bacterial antagonist (*Burkholderia spinosa*) in controlling anthracnose of banana when integrated with Tween 20 and hot water treatment has been reported by De costa et al. (2008).

6.3   BACTERIAL DISEASES

6.3.1   BACTERIAL SOFT ROT

Introduction and Economic Importance:

In the present scenario, banana soft rot is slowly becoming a severe menace causing huge loss to most banana growers all over the world. Up to 80%–90% and 93% in Gautemala was reported by Hildreth (1962). Soft rot disease was confirmed in India by Edward et al. (1973). It was also observed by Khan and Nagaraj in the year 1998, from several banana-growing regions of Karnataka in India. In Karnataka state, the districts of Bangalore and Kolar, the disease incidence varied from 30% to 35% (Nagaraj et al., 2012). The disease was located in major banana growing areas of north Karnataka ranging from 4.25% to 65.28% in Bijapur, Bagalkot, Belgaum, and Dharwad districts (Vijayalaxmi et al., 2014).

Symptoms:

In the initial stages, the infected rhizomes showed dark brown or yellow water-soaked lesions with dark peripheral rings profusely in the cortex region and then all throughout the rhizomes directing to cavity formation. When the infection reaches the advanced, pseudostem splitting was found to be the most prominent symptom (Thammaiah et al., 2005).

Rhizome and pseudostem rotting followed by marginal necrosis or scorching of leaves finally leading to the toppling of affected plants were found to be caused by the disease (Nagaraj et al., 2012). Sometimes the fruit-bearing trees topple before maturity or before bunch harvesting due to splitting and internal rotting of rhizome, pesudostem, and peduncle (Cetinkaya-Yildiz and Aysan, 2007).

Causal Organism:

Erwinia carotovora causes soft rot which was noticed first time in Israel during 1965 (Zutra and Volcani Zafrira, 1971). In India, Wardlaw (1950) gave the first report on the nature of bacterial head rot or rhizome rot in banana from Allahabad in Uttar Pradesh. In India, it was observed to be caused by *E. carotovora* subsp. *carotovora* (Edward et al., 1973 and Khan and Nagaraj, 1998).

Integrated Disease Management:

Chemical Methods:

- Thammaiah et al. (2006) evaluated different chemicals against bacterial soft rot of banana in lab conditions and stated that streptocycline @500 ppm and COC @2000 ppm recorded maximum inhibition.
- Nagaraj et al. (2002) conducted two trials in the field to assess the efficacy of various bactericides and antibiotics against bacterial soft rot. Three times drenching with streptomycin sulfate individually or in mixture with copper sulfate completely suppressed the tip over disease (100%) and enhanced the yield by 143.37%.
- Patel et al. (2011) revealed that paring + 0.15% Acephate dip + 1% B.M., and suckers dip in mancozeb 0.3% for about 10 min utes were found effective and economical in reducing the rhizome rot and also increased fruit yield.

Biological Methods:

- Cruz-Quiroz et al. (2011) evaluated the antagonistic capacity of freshly isolated *Pseudonomas fluorescens* strains alongside three important phytopathogenic bacteria (*Clavibacter michiganensis, Xanthomonas axonopodis,* and *E. carotovora*). Their product demonstrated that cell-free extracts showed a limited antagonist capacity in contrast to those extracts with cells, which demonstrated an excellent ability to reduce the growth of *C. michiganensis, X. axonopodis,* and *E. carotovora,* showing the intracellular nature of the bioactive metabolites linked to the bacterial growth inhibition.
- Kavita and Satish (2011) used nine different medicinal plants namely, *Acacia nilotica* (leaf), *Acorus calamus* (rhizome), *Carum copticum* (seeds), *Emblica officinalis* (leaf), *Eupatorium odaratum* (leaf), *Hyptis suaveolens* (leaf), *Millingtonia hortensis* (leaf), *Ocimium gratissium* (leaf), and *Pedalium murex* to evaluate against bacterial soft rot. The leaves and fruits were examined for antibacterial activity against important phytopathogenic bacteria for instance, *Xanthomonas campestris* pv. *vesicatoria, Xanthomonas axonapodis* pv. *malvacearum, Xanthomonas oryzae* pv. *oryzae,* and *E. carotovora* (MTCC 1428). Among the different solvent extracts investigated, methanol and ethanol extract of *E. officinalis, A. nilotica,* and *C. copticum* documented important inhibitory activity against all the pathogens tested followed by *P. murex, H. suaveolens, M. hortenesis,* and *E. odaratum.*

Integrated Approaches:

- The highest and lowest recording of bacterial soft rot disease was found to be done by soaking and spraying in the leaf using copper oxychloride 50WP at 3 g/L + streptomycin sulfate 0.5 g/L at 15 days interval, beginning from 15 days after planting and application of bleaching powder 25 g/plant/month 2 inches away from pseudostem around the collar region upto 4 months with an incidence of disease of about 7.67% during the year 2014 and 9.28% during the year 2015. (Kenganal et al., 2017).

6.3.2   MOKO DISEASE/BACTERIAL WILT DISEASE

Introduction and Economic Importance:

Schaumburg is regarded for his earliest citation to a bacterial wilt disease on banana fruit crop during his travels in British Guyana from the year

1840–1844 (Sequeira, 1998). So, the disease is first recorded in Guyana in 1840 in Moko plantain.

Disease caused an epidemic in the country of Trinidad, mainly of the susceptible cultivar "Moko," the common name of the disease was accepted. This was later depicted and published by Rorer after 65 years (1911). Until the early 1950s, commercial plantations stayed free from the Moko disease, but after 1950 three consecutive bacterial wilt epidemics have seized through Central and South America, where it is now deemed to be endemic (Sequeira, 1998; Buddenhagen, 2009). In some parts of the countries of Latin America and the Caribbean, the Moko disease, caused by *R. solanacearum* is regarded as a menacing disease to bananas and plantains crops, along with Black Sigatoka (*M. fijiensis*; Lehmann-Danzinger, 1987; Sequeira, 1998). Moko disease/bacterial wilt disease first reported from West Bengal in 1968 from India.

Symptoms:

The typical signs of Moko disease appear with wilting of the leaves where the underground infected banana crop displays swift yellowing, and once the pathogen has steadily colonized the pseudostem, it leads to vascular discoloration in the pseudostem, leaf sheaths, early fruit ripening or seize fruit development, fruit blackening, and dry rot of fruit pulp occurs (Thwaites et al., 2000; Denny, 2006). In most cases under the field conditions, it is difficult to differentiate Moko disease from Panama wilt based on the characteristics cited by Strover (1972). Moko disease can be authenticated by observing brown discoloration in the rhizome followed by using the method of positive ooze test. The negative response in the ooze test points out the chances of Panama wilt infection.

Causal Organism: *Ralstonia solanacearum* (*Pseudomonas* or *Burkholderia*)

Ralstonia solanacearum is described as a heterogeneous species, as it is revealed by its wide host range, physiological, and cultural properties, in addition to its phylogeny and pathogenic specialty (Hayward, 1991). Moko disease which is caused by *R. solanacearum* is presently found in all continents placed between the tropics of Cancer and Capricorn, causing serious disease on more than 200 crop species in over 50 families of different flora (Hayward, 1994; Belalcazar et al., 2004). *R. solanacearum* is regarded globally as one of the most destructive phytopathogenic bacteria because of its wide geographic distribution, broad host range, and its lethality (Mansfield et al., 2012).

Pathogen:

Moko disease is caused by race 2 of *R. solanacearum* which infects *Musa* and *Heliconia*. Cells of the bacterium are rod shaped and motile by 1–4 polar flagella. Pathogen is Gram-negative and lophotrichus bacteria. The bacterium multiplication occurs by bacterial fission.

Diseases Cycle:

The bacteria survive through infected rhizomes and also in the soil for 6 months to 2 years. Rhizome is the primary source of inoculum. While secondary infection through irrigation water and also through suckers used for planting. The spread is through the use of infected rhizomes, cutting machetes at the time of planting, and through insects which carry bacteria from oozing suckers and male flowers and bracts to healthy inflorescence and other parts of the plant. Entry into the host is mainly through injuries such as those caused during various cultural operations and attacks of insects and nematodes. The bacteria multiply rapidly in the xylem. Auxin balance of the plant is disturbed. IAA is synthesized by the bacterium and by the host and accumulates due to inhibition of the auxin degrading system.

Epidemiology:

Soil temp 28–32 °C, RH 87%–92%, PH slightly acidic to neutral, clay loam and sandy loam soil and susceptible variety.

Mode of Transduction:

According to the recent researches conducted, it has shown that numerous strains of bacteria can subsist over 25 years while constantly evolving and adapting in very different places. They are found to inhabit native flora, many diverse hosts, and organic matter in the soils (Buddenhagen, 1986). The inherent diversity of the pathogens in this cluster of microorganisms is also known as the *R. solanacearum* species complex (Fagan and Prior, 2005). In areas like the Caribbean and Latin America, Moko disease is locally transmitted from one plant to another by employing machete and other cutting implements on dessert bananas and "Bluggoe." Root to root transmission, movement of contaminated soil, and flood water are also some plausible reasons for the spread of the disease. Wardlaw (1972) has reported that insect transmission has occurred over distances of 90 km in Colombia and Venezuela. Over longer distances, it is spread by insects, particularly on "Bluggoe." Moko bacterial wilt has been moved across national boundaries on the infected planting material (Lehmann-Danzinger, 1987). Moko disease was spotted at Bombay airport in 1990 in a consignment of Heliconia spp. from Hawaii (Reddy and Nikale, 1992) in a post-entry quarantine nursery in Cairns, Australia (Hyde et al., 1992). A plausible hypothesis is that insect or mechanical transmission was involved. Aerial, transoceanic spread of Moko bacterial wilt by insects has not been cited, but insects may be carried along with the fruit.

Integrated Disease Management:

Host Plant Resistance:

The conventional breeding to manage Moko disease suffers due to the sterility of most of the edible cultivars, long generation period, and narrow genetic variability (Tripathi et al., 2004). However, Silva et al. (2000) artificially inoculated 31 different diploid (AA) genotypes (21 natural germplasm and 10 hybrids) with the Moko disease grown under greenhouse in Brazil (the hybrids F2P2, 1741-01, 1319-01, and SH3362 and "Babi Yadefana") where a variety from New Guinea, was reported to show good resistance to the Moko pathogen.

Cultural Methods:

Debudding is done after the last banana hand is formed, with the help of a forked wooden stick which was found to be more efficient in managing bacterial wilts of Musa spp. that resulted in larger and more uniformly full fruits (Blomme et al., 2005). Sanitation and early debudding reduced infection from an initial incidence of 88% to 6%, respectively after 12 months.

Biological methods:

In Uganda in vitro analysis, it has revealed promising levels of suppression by different bacterial isolates that is, *Burkholderia* spp., *Herbaspirillum* spp., and *Enterobacter* spp. isolated from the banana tissues accumulated in different parts of Uganda (Were, 2016).

Chemical Methods:

In Philippines, drenching of soil with formalin in the region of Ralstonia that infects Cavendish variety banana has reportedly resulted in a lower bacterial population (Pava et al., 2003).

6.4   VIRUS DISEASES

6.4.1   *BANANA BUNCHY TOP/CURLY TOP/CABBAGE TOP*

Introduction and Economic Importance:

Banana bunchy top disease (BBTD) is one of the most significant diseases of banana, causing severe crop losses in many banana-growing regions (Rybicki, 2015). BBTD is caused by BBTV and it is first reported in Fiji in the year 1891 on Cavendish varieties (Magee, 1927). Around 1940, it was introduced into India from Sri Lanka through a cyclone.

It is one of the most main constraints on the production of banana in 33 countries which includes Australia, Asia continent, and the Pacific islands but does not occur in the South and Central America (Kumar et al., 2011). BBTD has been cited from 13 countries in Africa, Eritrea, and Egypt (Blomme et al., 2013; IITA, 2009). A yield loss of about 90% was cited in severely infected plants by BBTD of susceptible cultivars, for instance, Poyo and AAA-Cavendish in a screening trial performed in the Rusizi valley of Burundi (Niyongere et al. 2012). In India, since the year 1970s BBTD has destroyed huge plantations of cv. Virupakshi (AAB), an elite cultivar also branded as "Hill Banana" in the lower Pulney hills of Tamil Nadu. In the past, BBTD has been the single cause for a severe reduction in hill banana cultivation from 18,000 ha to 2000 ha area (Kesavamoorthy, 1980). A survey conducted during the year 2006–2007 in the northern zones (Lucknow, Kanpur, Bahraich, and Etawah districts of the state in Uttar Pradesh) of banana cultivation on BBTD demonstrated a disease incidence of about 20%–30% (Vishnoi et al., 2009). BBTV was recorded as one of the 100 worst invasive species globally and the International Plant Protection Convention embraced it as a pathogen to be subjected to meticulous quarantine measures (IPPC, 2010).

Symptoms:

The primary symptoms are dark streaks down the secondary veins on the under surface of the lower segment of the leaf-blade, the leaf stalk, or the lower section of the midrib (Magee, 1927). This symptom is occasionally called as "Morse code streaking" because the streaks are uneven and look like a series of "dots" and "dashes." Another symptom can be observed as dark green hooked lines in the usually pale section of the leaf next to the midrib which is recognized as J-hooking (Ferreira et al., 1997; Nelson, 2004). The infected plants hardly produce a bunch, but even if they do, the fruit stalk is small and indistinct (Nelson, 2004). On plants which were infected very late, the simple symptoms found maybe a few dark green streaks on the tips of the flower bracts (Thomas et al., 1994). In case of field symptoms, it is expressed at 25–85 days after the first infection with BBTV (Hooks et al, 2008). Bunchy top symptoms are manifested approximately 25 days after the transmission, although this depends on the temperature and age of plants (Allen, 1987).

Causal Organism: Bunchy top virus or Banana virus-1 or Musa virus-1

Mode of Transmission:

BBTD is transmitted by an aphid vector that is, *Pentalonia nigronervosa* and is spread in the vegetative planting material, but it is not transmitted by mechanical inoculation (Magee, 1927). BBTV is a minute, isometric, nonenveloped, multipartite single-stranded DNA virus measuring about 18–19 mm in diameter (Xie and Hu, 1995) belonging to the family of Nanovirideae type member of the virus genus called Babuvirus (Association of Applied Biologists, 2013; Burnt et al. 1996).

The type of transmission that occurs in BBTV is mainly circulative, nonpropagative type spread by aphids. For individual aphids, the transmission efficiency has been cited between 46% and 67% (Hu et al., 1996). The transmission by the aphid is a persistent, nonpropagative method, through an acquisition feeding period of no less than 4 hours and inoculation feeding period of minimum 15 min (Hu *et al.,* 1996). The life span of an aphid varies from 19 to 26 days and during this period it will produce up to 20 offspring in optimal conditions of 24 °C–28 °C (Yasmin et al., 1999). The winged aphid transmits the virus to a healthy banana plant by feeding on it for between 15 min and 2 h (Ferreira et al., 1997). Kumar et al. (2011) have reported the spread of disease that occurs through the exchange of infected BBTV planting materials transfer over long gaps. There is no record of transovarial transfer to aphid progeny (Hu et al., 1996). Aphid numbers were found to decline throughout the drought periods (Wardlaw, 1961).

Integrated Disease Management:

Cultural Methods:

- Allen (1987) concluded that the exclusion of healthy plants contained by 5 m of an infected plant did not reduce the risk to neighboring farms and did not affect the number of new infections.

BBTD can be efficiently managed by means of disease-free planting material and by eliminating the diseased plants. Diseased plants ought to be uprooted and sprayed with power kerosene or insecticide to destroy all viruliferous aphids (Thomas et al., 1994).

Physical Methods:

Yellow traps filled with soapy water are generally used to collect and monitor winged aphids (McCartney and Fitt, 1985).

Regulatory Methods:

- Ploetz et al. (2003) observed that regulatory measures are to be adhered strictly to control the disease.

A new, striving programme of elimination of the disease is being instigated to replace the plantations where the disease occurs on a regular basis by the use of BBTV-tested, tissue-cultured, planting material of banana (Thomas et al., 1994).

Chemical Methods:

- Mallikuran (2007) reported that aphids can be controlled effectively by spraying insecticides like metasystox or dimecron or parathion @0.1%–0.5%. Besides that, injection of monocrotphos solution diluted with water at 1:4 at 30 day interval twice or thrice at 2–3 months was efficient in managing the disease.
- Vector control with systemic insecticides, namely, phosphomidon @1 mL/L or methyl demeton @2 mL/L.

KEYWORDS

- **banana**
- **disease**
- **etiology**
- **symptomatology**
- **management**

REFERENCES

Akila, R., Rajendran, L., Harish, S., Saveetha, K., Raguchander, T., and Samiyappan, R. (2011). Combined application of botanical formulations and biocontrol agents for the management of *Fusarium oxysporum* f. sp. *cubense* (Foc) causing Fusarium wilt in banana. *Biological Control*, **57**: 175–183.

Allen, R. N. (1987). Further studies on epidemiological factors influencing control of banana bunchy top disease, and evaluation of control measures by computer simulation. *Australian Journal of Agricultural Research*, **38**: 373–382.

Amani, M. (2008). Identification of Fungal Pathogens on Banana Trees (*Musa acuminata* L.) in Iran. 4 International Symposium on Tropical & Subtropical Fruits. Bogor, West Java, Indonesia.

Amani, M., Ershad, J., and Zare, R. (2006). Isolation and Identification of Fungi, the Causal Agents of Banana Fruit Cigar End Rot in Iran. 17th Iranian Plant Protection Congress.

Anonymous. (2009). Annual report. National Research Centre for Banana, Tiruchirapalli, India.

Ara, I., Rizwana, H., Al-Othman, M. R., and Bakir, M. A. (2012). Studies of actinomycetes for biological control of *Colletotrichum musae* pathogen during post harvest anthracnose of banana. *African Journal of Microbiology Research*, **6**: 3879–3886.

Arzanlou, M., Groenewald, J. Z., Fullerton, R. A., Abeln, E. C. A., Carlier, J., Zapater, M. F., Buddenhagen, I. W., Viljoen, A., and Crous, P. W., (2008). Multiple gene genealogies and phenotypic characters differentiate several novel species of Mycosphaerella and related anamorphs on banana. *Persoonia*, **20**: 19–37.

Ashby, S. F. (1913). Banana disease in Jamaica. *Science*, **31**: 754–755.

Association of Applied Biologists. (2013). Descriptions of plant viruses. Babuvirus. http://www.dpvweb.net/notes/showgenus.php,genus=Babuvirus [accessed 2015 July 02].

Bancroft, J. (1876). Report of the board appointed to enquire into the cause of disease affecting livestock and plants. In: Votes and Proceedings 1877, Vol 3, Queensland, pp. 1011–1038.

Basu, S. K. (1911). Report on the banana disease of Chinsurah. *Quarterly Journal of Department of Agriculture Bengal*, **4**: 196–198.

Beckman, C. H. (1987). The Nature of Wilt Disease of Plants. American Phytopathological Society Press, St. Paul, Minnesota, USA, p. 175.

Beckman, C. H. (1990). Host responses to the pathogen. In: Ploetz RC, ed. Fusarium Wilt of Banana. St Paul, MN, USA: APS Press, pp. 93–105.

Beckman, C. H. and Roberts, E. M. (1995). On the nature and genetic basis for resistance and tolerance to fungal wilt diseases of plants. *Advances in Botanical Research*, **21**: 35–77.

Belalcazar, S. C., Rosales, F. E., and Pocasangre, L. E. (2004). "El Moko del banano y el plátano y el rol de las plantas hospederas en su epidemiología," in Proceedings of the XVI International ACORBAT Meeting. September 26–October 1, (M. Orozco-Santos, J. Orozco-Romero, M. Robles-Gonzalez, J. Velazquez-Monreal, V. Medina-Urrutia, and J. A. Hernandez-Bautista, eds.), Oaxaca: Artturi, 16–35.

Bhuvanendra, K. H., Udaya, S. A. C., Chandra, N. S., Ramachandra, K. K., Shetty, H. S., and Prakash, H. S. (2010). Biochemical characterization of *Fusarium oxysporum* f. sp. cubenseisolates from India. *African Journal of Biotechnology*, **9(4)**: 523–530.

Blomme, G., Eden-Green, S., Mustaffa, M., Nwauzoma, B., and Thangavelu, R. (2011). Major diseases of banana. In: Banana Breeding Progress and Challenges (Michael Pillay and Abdou Tenkouano, eds.), CRC Press, Taylor Francis Group, Boca Raton, FL, USA. pp. 85–119.

Blomme, G., Mukasa, H., Ssekiwoko, F., and Eden-Green, S. (2005). On-farm assessment of banana bacterial wilt control options. *African Crop Science Conference Proc*eedings, **7**: 317–320.

Blomme, G., Ploetz, R., Jones, D., De Langhe, E., Price, N., Gold, C., Geering, A., Viljoen, A., Karamura, D., Pillay, M., Tinzaara, W., Teycheney, P. Y., Lepoint, P., Karamura, E., and Buddenhagen, I. (2013). A historical overview of the appearance and spread of *Musa* pests and pathogens on the African continent: highlighting the importance of clean *Musa* planting materials and quarantine measures. *Annals of Applied Biology*, **162**: 4–26.

Brandes, E. W. (1919). Banana wilt. *Phytopathology*, **9**: 339–389.

Brun, J. (1963). La Cercosporiose du bananier en Guine'e. Etude de la phase ascospore'e de *Mycospaerella musicola* Leach. The'se Doctorat e's science, Orsay, Paris.

Brunt, A. A., Crabtree, K., Dallwitz, M. J., Gibbs, A. J., Watson, L., and Zurcher, E. J. (eds.) (1996). Plant Viruses Online: Descriptions and Lists from the VIDE Database. Version: 20 August. http://pvo.bio-mirror.cn/descr056.htm [accessed 2015 July 4].

Buddenhagen, I. W. (1986). Bacterial wilt revisited. In: Persley GJ, ed. Bacterial wilt disease in Asia and the South Pacific—Proc International Workshop held at PCARRD, Los Baños, the Philippines, 8–10 Oct 1985. pp. 126–143. ACIAR Proceedings 13. Australian Centre for International Agricultural Research, Bruce, ACT, Australia.

Buddenhagen, I. W. (2009). Blood bacterial wilt of banana: history, field biology and solution. *Acta Horticulturae*, **828**: 57–68.

Burt, J. A., Rutter, J., and Gonzalez, H. (1997). Short distance wind dispersal of the fungal pathogens causing Sigatoka diseases in banana and plantain. *Plant Pathology*, **40(4)**: 451–458.

Carlier, J., Fouré, E., Gauhl, F., Jones, D. R., Lepoivre, P., Mourichon, X., Pasberg-Gauhl, C., and Romero, R. A. (2000). Black Leaf Streak. In 'Diseases of Banana, Abaca and Enset.' (D. R. Jones, ed.), CABI Publishing: Wallingford, pp. 37–79.

Carlier, J., Lebrun, M. H., Zapater, M. F., Dubois, C., and Mourichon, X. (1996). Genetic structure of the global population of banana black leaf streak fungus, *Mycosphaerella fijiensis*. *Molecular Ecology*, **5**: 499–510.

Carlier, J., Zapater, M. F., Lapeyre, F., Jones, D. R., and Mourichon, X. (2000). Septoria leaf spot of banana: A newly discovered disease caused (anamorph *Septoria eumusae*) by *Mycosphaerella eumusae*. *Phytopathology*, **90**: 884–890.

Castelan, F. P., Abadie, C., Hubert, O., Chilin-Charles, Y., De Bellaire, L. L., and Chillet, M. (2013). Relation between the severity of Sigatoka disease and banana quality characterized by pomological traits and fruit green life. *Crop Protection*, **50**: 61–65.

Cetinkaya-Yildiz, R. and Aysan, Y. (2007). A new seed-borne pathogen on tomato: Erwinia carotovora subsp. carotovora and some seed treatments. *Acta Horticulturae*, **729**: 63–68.

Chakravaty, T. (1957). Anthracnose of banana (*Gloeosporium musarum* Cke. & Massee) with special reference to latent infection in storage. *Transactions British Mycological Society*, **40**: 337–345.

Crous, P. W. and Mourichon, X. (2002). *Mycosphaerella eumusae* and its anamorph *Pseudocercospora eumusae* spp. nov.: Causal agent of eumusae leaf spot disease of banana. *Sydowia*, **54**: 35–43.

Cruz-Quiroz, R. L., Catalina, C., Hernandez, M., Rodriguez, R., Hernandez, D., and Aguilar, C. N., (2011). Antagonist capacity of newly isolated strains of *Pseudomonas Fluorescens* against three important Phytopathogenic bacteria. *American Journal of Agricutural and Biological Sciences,* **6**(2): 267–272.

De costa, D. M., Zahra A. R. F., Kalpage M. D., and Rajapakshe, E. M. G. (2008). Effectiveness and molecular characterization of Burkholderia sinosa a prospective biocontrol agent for controlling postharvest disease in banana. *Biological Control,* **47**: 257–267.

Denham, T. P., Haberle, S. G., Lentfer, C., Fullagar, R., Field, J., Therin, M., Porch, N., and Winsborough, B. (2003). Origins of agriculture at kuk swamp in the highlands of New guinea. *Science,* **301**: 189–193.

Denham, T. P., Haberle. S., and Lentfer, C. (2004). New evidence and interpretations for early agriculture in Highland New Guinea. *Antiquity,* **78**: 839–857.

Denny, T. P. (2006). "Plant pathogenic ralstonia species," in Plant-Associated Bacteria, (S. S. Gnanamanickam, ed.), Dordrecht: Springer, pp. 573–644.

Dita, M. A., Garming, H., Bergh, I. V., Staver, C., and Lescot, T. (2013). Banana in Latin America and the Caribbean: current State, challenges and perspectives. Proc. Int. ISHS-ProMusa Symp. on Bananas and Plantains: Towards Sustainable Global Production and Improved Uses Eds.: I. Van den Bergh et al. *Acta Horticulturae, 986,* ISHS 2013: 365–380.

Edward, J. C., Tripathi, S. C., and Singh, K. P. (1973). Observations on a "Tip-over" disease of banana in Allahabad. *Current Science, 42*: 696–697.

FAO (2013) Climatic Database.

FAO statistics. http://faostat.fao.org. Final area and production estimates for horticulture crop 2014 2015.http://nhb.goc.in/. As accessed on 24th April 2016.

Fegan, M. and Prior, P. (2005). How complex is the *Ralstonia solanacearum* species complex? (C. Allen; P. Prior; A. C. Hayward, eds.) Bacterial wilt disease and the *Ralstonia solanacearum* complex. American Phytopathological Society (APS) Press, St. Paul, MN, USA, pp. 449–461.

Ferreira, S. A., Trujillo, E. E., and Ogata, D. Y.(1997). Banana bunchy top virus. College of Agriculture and Human Resources, University of Hawaii at Manoa. PD-12. 4p. http://www.issg.org/database/species/reference_files/BBTV/CoExSe.pdf [accessed 2015 July 02].

Fitzell, R. D. and Peak, C. M. (1984). The epidemiology of anthracnose disease of mango: inoculum sources, spore production and dispersal. *Annals of Applied Biology,* **104**: 53–59.

Fourie, G., Steenkamp, E. T., Ploetz, R. C., Gordon, T. R., and Viljoen, A. (2011). Current status of the taxonomic position of *Fusarium oxysporum* formae speciale *cubense* within the *Fusarium oxysporum* complex. *Infection, Genetics and Evolution,* **11**: 533–542.

Gauhl, F. (1994). Epidemiology and ecology of Black Sigatoka (*Mycosphaerella fijiensis* Morelet) on Plantain and Banana (Musa spp.) in Costa Rica, Central America.' (INIBAP: Montpellier, France).

Griffee, P. J. and Burden, O. J. (1974). Incidence and control of *Colletotrichum musae* on bananas in the Windward Islands. *Annals of Applied Biology,* **77**: 11–16.

Guyot, H. and Cuille, J. (1958). Essai deprevision d_ataques de Cercospora em Guadalupe. *Fruits,* **13**: 85–94.

Hayward, A. C. (1991). Biology and epidemiology of bacterial wilt caused by Pseudomonas solanacearum. *Annual Review of Phytopathology,* **29**: 65–87.

Hayward, A. C. (1994). "The hosts of *Pseudomonas solanacearum,*" in Bacterial wilt: The Disease and its Causative Agent, *Pseudomonas solanacearum,* (A. C. Hayward and G. L. Hartman, eds.), Wallingford: CAB International, pp. 9–24.

Hedge, G. M. and Mesta, R. K. (2014). Integrated management of sigatoka leafspot of banana. Mesta College of Forestry, Sirsi, University of Agricultural Sciences, Dharwad. *The Bioscan,* **9(1)**: 359–362, (Supplement on Plant Pathology).

Hermanto C., Eliza, I., Djatnika, D. E., Mujiman, and Subhana (2012). Pre-planting treatments for management of banana Fusarium wilt. Asian Research Publishing Network (Arpn). *Journal of Agricultural and Biological Science,* 7(4), April 2012.

Higgins, J. E. (1904). The banana in Hawaii. Hawaii, US. Hawaii Agricultural Experiment Station, University of Hawaii. p. 51. (Bulletin No. 7).

Hildreth, R. C. (1962) Studies on bacterial rhizome rot of bananas. *Review of Plant Pathology,* **42**: 79.

Holderness, M., Tushemereirwe, W. K., and Gold, C. S. (1998). Cultural controls and habitat management in the integrated management of banana leaf diseases. In: (Frison, E. A., Gold, C. S., Karamura, E. B., Sikora, R. A. (Eds.) Proceedings of a Workshop Mobilizing IPM for Sustainable Banana Production in Africa. Nelspruit, South Africa. 23–28 November. pp. 149–163.

Hooks, C. R. R., Wright, M. G., Kabasawa, D. S., Manandhar, R., and Almeida, R. P. P. (2008). Effect of *Banana bunchy top virus* infection on morphology and growth characteristics of banana. *Annals of Applied Biology,* **153**: 1–9.

Hu, J. S., Wang, M., Sether, D., Xie, W. and Leonhardt, K. W. (1996). Use of polymerase chain reaction (PCR) to study transmission of banana bunchy top virus by the banana aphid (*Pentalonia nigronervosa*). *Annals of Applied Biology,* **128**: 55–64.

Hu, J. S., Xu, M. Q., Wu, Z. C., and Wang, M. (1993). Detection of *Banana bunchy top virus* in Hawaii. *Plant Disease,* **77**: 952–952.

Huang, Y. H., Wang, R. C., Li, C. H., Zuo, C. W., Wei, Y. R., Zhang, L., and Yi, G. J. (2012). Control of *Fusarium* wilt in banana with Chinese leek. *European Journal of Plant Pathology,* **134**: 87–95.

IITA (2009). Two diseases could wipe out African bananas, experts set up control efforts. Available online https://old.iita.org/cma/details/news_details.aspx. articleid=2735&zoneid=81.

IPPC (2010). Occurrence of Banana Bunchy Top Disease on banana and plantain in Cameroon. IPPC Official Pest Report, No. CMR-01/1, No. CMR-01/1. Rome, Italy: FAO. https://www.ippc.int/.

Irish, B. M., Goenaga, R., Rios, C., Chavarria–Carvaja, J., and Ploetz, R. (2013). Evaluation of banana hybrids for tolerance to black leaf streak (*Mycosphaerella fijiensis* Morelet) in Puerto Rico. *Crop Protection*, **54**: 229–238.

Jacome, L. H., Schuh, W., and Stevenson, R. E. (1991). Effect of temperature and relative humidity on germination and germ tube development of *Mycosphaerella fijiensis* var. difformis. *Phytopathology,* 81(12): 1480–1485.

Johanson, A. (1993). Molecular methods for the identification and detection of the *Mycosphaerella* species that cause Sigatoka leaf spots of banana and plantains. PhD Thesis, University of Reading, England.

Jones, D. R. (2003). The Distribution and Importance of the Mycosphaerella Leaf Spot Diseases of Banana. (L. Jacome, P. Lepoivre, D. Marin, R. Ortiz, R. Romero, and J. -V. Escalant, eds.), Mycosphaerella Leaf Spot Diseases of Bananas: Present Status and Outlook. Proceedings of the Workshop on Mycosphaerella Leaf Spot Diseases held in San JosT, Costa Rica on 20–23 May 2002. Montpellier, France: INIBAP pp. 25–42.

Jones, D. R. (2000). Diseases of Banana. In: Abacá and Enset. (eds DR Jones). CABI Publishing, Wallingford, United Kingdom. pp. 544.

Kavita, H. U and Satish, S., (2011). Eco- friendly management of plant pathogens by some medicinal plant extracts. Journal of *Agriculture Technology*, **7(2)**: 449–461.

Kenganal, M., Nimbaragi, Y. A., and Guruprasd, G. S. (2017). Management of soft rot of banana caused by *Erwinia carotovora* sub sp. *Carotovora*. *International Journal of Plant Protection*, **10(2)**: 381–385.

Kesavamoorthy, R. C.(1980) Radical changes in ecosystem in the Pulney hills. Proceedings of the 13th National Seminar on Banana production Technology. Coimbatore: TNAU. pp. 23–28.

Khan, A. N. A. and Nagaraj, M. S. (1998). Occurrence of a new bacterial disease on banana in Karnataka, paper presented in the Symposium on Integrated disease management and crop loss assessment held on December 10–12. Indian Phytopathological Society (southern chapter), UAS, Bangalore, pp. 73.

Khleekorn, S., McGovern, R. J., and Wongrueng, S. (2015). Control of the banana anthracnose pathogen using antagonistic microorganisms. *International Journal of Agricultural Technology*, **11(4)**: 965–973.

Knowles, C. H. (1914). Disease in bananas. Pamphlet of the Department of Agriculture, Fiji, 8.

Kumar, P. L., Hanna, R., Alabi, O. J., Soko, M. M., Oben, T. T., Vangu, G. H. P., and Naidu, R. A. (2011). *Banana bunchy top virus* in sub-Saharan Africa: Investigations on virus distribution and diversity. *Virus Research*, **159**: 171–182.

Leach, R. (1964). Report on investigations into the cause and control of the new banana disease in Fiji, black leaf streak. Council Papers, Fiji, 38, Suva.

Lehmann-Danzinger, H. (1987). "The distribution of Moko disease in Central and South America and its control on plantains and bananas," in Proceedings of the CTA Seminar: Improving citrus and Banana Production in the Caribbean through Phyto-Sanitation, St Lucia, 130–152.

MacHardy, W. E. and Beckman, C. H. (1981). Vascular wilt Fusaria: Infections and Pathogenesis. Pages 365–390 in: Fusarium: Diseases, Biology and Taxonomy (Nelson, P. E., Toussoun, T. A. and Cook, R. J. eds.). The Pennysylvania State University Press, University Park and London.

Magee, C. J. (1927). Investigation on the bunchy top disease of the banana. Commonwealth of Australia Council for Scientific and Industrial Research Bulletin No. 30. 64 p. + 22 plates.

Mallikarjun, Y., Kenganal, Hemavati, R., and Byadgi, A. (2007). Management of bunchy top disease in banana. Sci-tech and Agriculture, *The Hindu*.

Mansfield, J., Genin, S., Magori, S., Citovsky, V., Sriariyanum, M., and Ronald, P. (2012). Top 10 plant pathogenic bacteria in molecular plant pathology. *Molecular Plant Pathology*, **13**: 614–629.

Martínez-Acosta, A. M., Castañeda-Sánchez, D. A., Bornacelly-Horta, H., and Merchan, V. (2006). Early cut-off practice to the integrated management of black Sigatoka in bananas. Proceedings of the XVII International Meeting of ACORBAT. Joinville, Brazil, October 15–20. pp. 690–697.

Massee, G. (1914). Fungi exotici. XVIII. Kew Bull., 159.

Meredith, D. S. (1970). Banana Leaf Spot Disease (Sigatoka) caused by *Mycosphaerella musicola* Leach. Phytopathological Papers, No. 11. Kew, UK: Commonwealth Mycological Institute.

Mirshekari, B. A., Ding, P., Kadir J., and Mohd Ghazali, H. (2012). Effect of hot water dip treatment on postharvest anthracnose of banana var. *African Journal of Agricultural Research,* **7(1)**: 6–10.

Mobambo, K. N. and Naku, M. (1993). Situation de la cercosporiose noire desbananiers et plantains (Musa spp) sous différents systémes de culture á Yangambi, Itaut—Zarie. Tropicultura, **11**:7–10.

Molina, A. B. and Valmayor, R. V. (1999). Banana production system in South East Asia. In : *Bananas and Food security.* (C. Pica, E. Fource, E. A. Frision, eds.), INIBAP, Montpellier, France. pp. 423–436.

Moore, N. Y., Bentley, S., Pegg, K. G., and Jones, D. R. (1995). *Fusarium* wilt of banana. *Musa* disease fact sheet. INBAP, Montpellier, France, **5**:4.

Moore, N. Y., Pegg, K. G., Bentley, S., and Smith, L. J. (1999). Fusarium wilt of banana : global problems and perspectives. (A. B. Molina, N. H. N. Masdek, K. W. Liew, eds.), Banana Fusarium Wilt Management: Towards Sustainable Cultivation. Proceedings of the International.

Mourichon, X. and Fullerton, R. A. (1990). Geographical distribution of two species *Mycosphaerella musicola* Leach (*Cercospora musae*) and *Mycosphaerella fijiensis* Morelet (*Cercospora fijiensis*), respective agents of Sigatoka disease and Black leaf streak disease in bananas and plantains. *Fruits*, **45**: 213–218.

Mourichon, X. and Zapater, M. F. (1990). Obtention in vitro du stade *Mycosphaerella fijiensis*, forme parfaite de *Cercospora fijiensis. Fruits*, **45(6)**: 553–557.

Nagaraj, M. S., Khan, A. N. A., Ravikumar, M. R., and Amarnanjundeswara, H. (2002). Management of tip over disease of banana, Global Conference on Banana and Plantain, Bangalore, India Oct., pp. 28–31.

Nagaraj, M. S., Umashankar, N., Palanna, K. B., and Khan, A. N. A. (2012). Etiology and management of tip-over disease of banana by using biological agents. *International Journal of Advanced Biological Research*, **2(3)**: 483–486.

Nayana, K. and Radhakrishnan, N. V. (2015). Integrated disease management of sigatoka leaf spot of banana caused by *Mycosphaerella musicola*. Krishi Kosh, MSc Thesis published by college of Agriculture, Vellayani.

Nel, B., Steinberg, C., Labuschagne, N., and Viljoen., A. (2006). The potential of non-pathogenic *Fusarium oxysporum* and other biological control organisms for suppressing fusarium wilt of banana. *Plant Pathology*, **55**: 217–223.

Nelson, S. (2004). Banana bunchy top: Detailed signs and symptoms. College of Tropical Agriculture and Human Resources, University of Hawaii at Manoa. http://www.ctahr. hawaii.edu/bbtd/downloads/bbtv-details.pdf [accessed 2015 July 02].

Ngongo, P. M. K. (2002). Integrated crop management strategies for plantain production and control of black leaf streak (black Sigatoka) disease in the Democratic Republic of Congo, Disease control of black Sigatoka. Info Musa, **11(1)**: 3–6.

NHB (2015). *Indian Horticulture Database, 2015.* National Horticulture Board, Gurgaon, Haryana, India.

Niyongere, C., Losenge, T., Ateka, E. M., Nkezabahizi, D., Blomme, G., and Lepoint, P. (2012) Occurrence and distribution of banana bunchy top disease in the Great Lakes region of Africa. Tree and Forestry. *Science and Biotechnology*, **6**: 102–107.

Nwauzoma, A. B., Uma, S., Saraswathi, M. S., and Mustaffa, M. (2011). Developing markers for Sigatoka leaf spot disease (*Mycosphaerella leach*) resistance in banana (musa spp.). *Afr. Journal of Biotechnology,* **10(13)**: 6213–6219.

O'Donnell, K. and Cigelnik, E. (1999). A DNA sequence-based phylogenetic structure for the Fusarium oxysporum species complex. *Phytoparasitica*, **27**: 69.

Patel P. (2009). Chemical control of Sigatoka leaf spot (*Mycosphaerella musicola*) of banana. *International Journal of Plant Protection*, **2(1)**: 98–100.

Patel, P. R., Sharma, H., and Shukla, A.,(2011). Efficacy of chemicals against rhizome rot of banana. *Karnataka* Journal of *Agriculture Science*, **24(5)**: 712–713.

Pava, H. M., Franje, N. F., and Timario, T. J. (2003). Banana pilot demonstration studies for bukidnon: table salt and early debudding to control 'Bugtok' disease of cooking banana cultivars 'Saba' and 'Cardaba' Philippines. *Journal of Crop Sci*ence, **28**: 31–43.

Pawar, B. R., Landge, V. V., Deshmukh, D. S., and Yeware, P. P. (2010). Economics of banana production in drip irrigated and flood irrigated gardens. *IJCBM*, **3(1): 88–91**.

Perez, L., Alvarez, J. M., and Pérez, M. (2002). Economic impact and management of black leaf streak disease in Cuba. In: Mycosphaerella leaf spot diseases of bananas: present status and outlook. Proceedings of the 2nd. International Workshop on Mycosphaerella Leaf Spots Diseases held at San José Costa Rica. (L. Jacome, P. Lepoivre, D. Marin, R. Ortiz, R. Romero, and J. V. Escalant, eds.), pp. 71–84

Pérez-Vicente, L. (1996). Manual para el manejo integrado de Sigatoka negra (*Mycosphaerella fijiensis* Morelet) y Sigatoka amarilla (*Mycosphaerella musicola* Leach ex Mulder) en banano y plátano. Proyecto FAO - Ministerio de Agricultura. TCP/CUB/4454.

Pérez-Vicente, L. (2004). Fusarium wilt (Panama disease) of bananas: an updating review of the current knowledge on the disease and its causal agent. In. Memorias de XV Reunion Internacional de ACORBAT (Oaxaca, MX). pp. 1–14.

Pérez-Vicente, L., Batlle-Viera, A., and Chacón-Benazet, J. (2003). *Fusarium oxysporum* f. sp. *cubense* en Cuba: biología de las poblaciones, reacción de los clones híbridos de la FHIA y biocontrol. En: Memorias del Taller "Manejo convencional y alternativo de la Sigatoka negra, nemátodos y otras plagas asociadas al cultivo de Musáceas en los trópicos. (Guayaquil, EC). Rivas, G.; Rosales, F. (eds.). pp. 141–155.

Ploetz, R. C. (1999). The most important disease of a most important fruit. APSnet Educ Cent Marzo. 2004.

Ploetz, R. C. (2005). Panama disease, an old enemy rears its ugly head: Parts 1 and 2. In: *Plant Health Progress*, APSnet: Online doi:10.1094/PHP-2005-1221-01-RV.

Ploetz, R. C. and Pegg, K. G. (1997). *Fusarium* wilt of banana and Wallace's line: Was the disease originally restricted to his Indo-Malayan region. *Australasian Plant Pathology*, **26**: 239–249.

Ploetz, R. C., Thomas, J. E., and Slabaugh, W. R. (2003). Diseases of banana and plantain. Pages 73–134 *in* R. C. Ploetz, Ed. Diseases of Tropical Fruit Crops. CABI, London. 527p. (See pages 122–125 on BBTV.)

Ploetz, R. C. (2015). Management of Fusarium wilt of banana: A review with special reference to tropical race 4. *Crop Protection* **73**: 7–15.

Ploetz, R. C. and Mourichon, X. (1999). First report of black Sigatoka in Florida. *Plant Disease*, **83**: 300.

Ploetz, R. C. and Churchill, A. C. L. (2011). Fusarium wilt: the banana disease that refuses to go away. In: Van den Bergh, I., Smith, M., Swennen, R. and Hermanto, C. (Eds.). Proceedings of International ISHS—ProMusa Symposium on Global Perspectives on Asian Challenges, Guangzhou, China, 14–18/09/2009. Acta Horticulturae 897. ISHS, Leuven, Belgium. pp. 519–526.

Porras, A. and Ptrez L. (1997). The role of temperature in the growth of the germ tubes of ascospores of *Mycosphaerella* spp., responsible for leaf spot diseases of banana. *Infomusa*, 6(2): 27–32.

Prusky, D., Plumbley, R. A. (1992). Quiescent infections of *Colletotrichum* in tropical and subtropical fruits. (J. A. Bailey, M. J. Jeger, eds.) *Colletotrichum*: *Biology, Pathology and Control*. Wallingford, UK: CAB International. pp. 289–307.

Pushpavathi, Y., Dash, S. N., Mishra, M. K., and Triveni, V. (2015). Management of Fusarium wilt of banana under coastal Odisha conditions. *International Journal of Farm Science*, **5(4):** 241–247.

Raguchander, T., Jayashree, K., and Samiyappan, R. (1997). Management of *Fusarium* wilt of banana using antagonistic microorganisms. *Journal of Biological Control*, **11:** 101–105.

Ramakrishnan, T. S. and Damodaran, S. (1956). Observation on the wilt disease of banana. *Proceedings of Indian Academy of Science*, **43:** 213–22.

Raut, S. P. and Ranade, S. (2004). Diseases of banana and their management. In: Diseases of fruits and vegetables (S. A. M. H Naqvi, ed). Kluwer Academic Publishers, Netherlands, pp. 37–52.

Rhodes, P. L. (1964). A new banana disease in Fiji. *Commonwealth Phytopathological News*, **10**: 38–41.

Robinson, J. C. (1996). Banana and plantains. CAB International. Wallingford. UK. pp. 238.

Rorer, J. B. (1911). A bacterial disease of bananas and plantains. *Phytopathology*, **1**: 45–49.

Rybicki E. P. (2015). A top ten list for economically important plant viruses. *Archives of Virology*, **160**: 17–20.

Samson, J. A. (1986). Bananas and plantains. Pages 139–189 in: Tropical fruits. 2nd ed. Tropical Agriculture Series, Longman, London, UK.

Saranavan, T., Muthusamy, M., and Marimuthu, T. (2003). Development of integrated approach to manage the fusarial wilt of banana. *Crop Protection*, **22**: 1117–1123.

Selvarajan, R., Uma, S. and Sathiamoorthy, S. (2000). Etiology and survey of banana leaf spot diseases in India. *Adv Banana Plant R&D Asia Pac*, **10**: 94–102.

Sequeira, L. (1962). Influence of organic amendments on survival of *Fusarium oxysporum* f. sp. cubense in soil. *Phytopathology*, **52**: 976–982.

Sequeira, L. (1998). "Bacterial wilt: the missing element in international banana improvement programs," in Bacterial Wilt Disease: Molecular and Ecological Aspects, (P. Prior, C. Allen, and J. Elphinstone, eds.) (Berlin: Springer-Verlag), 6–14.

Shanthiyaa, V. (2014). Biological management of yellow sigatoka disease of banana caused by *mycosphaerella musicola* (leach) by using liquid bioformulation of bacillus spp. Thesis submitted in part fulfillment of the requirements for the degree of doctor of philosophy (agriculture) in plant pathology to the Tamil Nadu Agricultural University, Coimbatore 641 003.

Silva, S. de O., Veras, S. de O., Gasparotto, L., Matos, A. P. de, Cordeiro, Z. M., and Boher, B. (2000). Evaluation of *Musa* spp. For resistance to moko disease (*Ralstonia solanacearum*, Raza 2). *Infomusa*, **9**: 19–20.

Simmonds, N. W. (1966). Bananas. Longman, Green & Co, London.

Smith, E. F. (1910). A Cuban banana disease. *Science*, **31**: 754–755.

State wise horticulture status: Indian Horticulture Database Data period: 2010–11, National Horticulture Mission, 2011.

Stewart, E. L., Liu, Z., Crous, P. W., and Szabo, L. J. (1999). Phylogenetic relationships among some cercosporoid anamorphs of *Mycosphaerella*-based on rDNA sequence analysis. *Mycological Research,* **11**: 1491–1499.

Stover, R. H. (1980). Sigatoka leaf spots of bananas and plantains. *Plant Disease,* 64(8): 750–756.

Stover, R. H. (1962). Fusarium wilt (Panama disease) of bananas and other Musa species. Kew, UK. *Commonwealth Mycological Institute.* pp. 177.

Stover, R. H. (1971). A proposed international scale for estimating intensity of banana leaf spot (*Mycosphaerella musicola* Leach), *Tropical Agriculture,* **48**: 185–196.

Stover, R. H. (1983). The effect of temperature on ascospore germ-tube growth of *Mycosphaerella musicola* and *Mycosphaerella fijiensis* var. *difformis. Fruits,* **38**: 625–628.

Stover, R. H. (1972). Banana, plantain and abaca diseases. Commonwealth Mycological Institute (CMI), Kew, Surrey, UK, p. 316.

Stover, R. H. (1972). Intercontinental spread of banana leaf spot (*Mycosphaerella musicola* Leach). *Tropical Agriculture (Trinidad),* **29**: 327–338.

Stover, R. H., Simmonds, N. H. (1987). Bananas, 3rd edit. Longman Scientific and Technical, Harlow, Essex. UK, p. 468.

Su, Y. Y., Noireung, P., Liu F., Hyde, K . D., Moslem, M. A., Bahkali, A. H., Abd-Elsalam, K. A., and Cai, L. (2011). Epitypification of *Colletotrichum musae,* the causative agent of banana anthracnose. *Mycoscience,* **52**: 376–382.

Thammaiah, N., Kalmadi, V. C., Shirol, A. M., Gangadharappa, P. M., and M. S. (2006). Incidence of bacterial rhizome rot of banana in northern Karnataka and *in-vitro* evaluation of chemicals, antibiotics and plant extracts against *Erwinia chrysanthemi.* Abstracts published In: *Nat. Semi. Int. Prod. Post Harvest Manag. Trop. Fruits.* p. 58.

Thammaiah, N., Kanamadi, V. C., and Shirol, A. M.(2008). Management of sigatoka leaf spot disease (*Mycosphaerella musicola*) in banana at different locations in Belgaum district of Karnataka, India. *International Journal of Agriculture Science,* **4(1)**: 57–58.

Thammaih, N. (2003). Studies on the epidemiology and management of sigatoka leafspot of banana. PhD thesis, submitted to the University of Agricultural Sciences, Dharwad. pp. 1–231.

Thangavelu R. and Mustaffa, M. M. (2010). First report on the occurrence of a virulent strain of Fusarium wilt pathogen (Race-1) infecting Cavendish (AAA) group of bananas in India. *Plant Disease,* **94**: 1379.

Thangavelu, R. (2002). Characterization of *Fsarium oxysporum* schlecht. f. sp. *cubense* (e.f. smith) snyd. & hans. and Molecular Approaches for the Management of Fusarium wilt of Banana. PhD thesis. Tamil Nadu Agricultural University, Coimbatore, Tamil Nadu, India, pp. 254.

Thangavelu, R., Palaniswami, A., Ramakrishnan, G., Sabitha, D., Muthukrishnan, S., and Velazhahan, R. (2001). Involvement of Fusaric acid detoxification by *Pseudomonas fluorescens* strain Pf10 in the biological control of Fusarium wilt of banana caused by *Fusarium oxysporum* f. sp. *cubense. Journal of Plant Disease and Protection,* **108**: 433–445.

Thangavelu, R., Sundararaju, P., Sathiyamoorthy, S., Reghucharnder, T., Velazhahan, R., Nakkeeran, S., and Palaniswamy, A. (1999). Status of *Fusarium* wilt of banana in India. In: *Proc. Inter. Workshop on Banana Fusarium Wilt Disease,* pp. 58, 18–20 October, Mysore.

Thomas, J. E., Iskra-Carvana, M. L., and Jones, D. R. (1994). Banana bunchy top disease, Musa Disease Fact Sheet No. 4. Montpellier, France: INIBAP.

Thurston, H. D. and Galindo, J. J. (1989). "Moko del banano y el plátano," in Enfermedades de Cultivos en el Trópico, (Turrialba: CATIE), 125–133.

Thwaites, R., Eden-Green, S. J., and Black, R. (2000). "Diseases caused by bacteria," in Diseases of Banana, Abacá and Enset, (D. R. Jones, ed.) (Wallingford: CAB International), 213–239.

Tripathi, L., Tripathi, J. N., and Tushmereirwe, W. K. (2004). Strategies for resistance to bacterial wilt disease of bananas through genetic engineering. *African Journal of Biotechnology*, **3**: 688–692.

Vijayalaxmi, S. T., Rajput, R., Roopa, R. S., and Jahagirdar, S. (2014). Studies on host-range of tip-over disease of banana caused by Erwinia carotovora subsp. carotovora. *Internat. Journal of Plant Protection,* **7(1)**: 270–271.

Vishnoi, R., Shri, K. R., and Vivek, P. (2009). Molecular characterization of an Indian isolate of banana bunchy top virus based on six genomic components. *Virus Genes,* **38**: 334–344.

Wardlaw, C. W. (1950). Banana diseases VIII, Notes on the various diseases occurring Trinidad. *Tropical Agriculture*, **11**: 143–149.

Wardlaw, C. W. (1961). Banana Diseases Including Plantains and Abaca. London, UK: Longmans, Green and Co. Ltd.

Wardlaw, C. W. (1972). Banana Diseases Including Plantain and Abaca. Harlow: Longman, 878.

Wardlaw, C. W. (1961). Banana Diseases Including Plantain and Abaca. John Wiley & Sons, Inc., New York.

Were, E. (2016). Endophytic Bacteria Associated with Banana and Their Potential for Controlling Banana Xanthomonas wilt. A thesis for the award of the degree of Masters in Molecular Biology and Biotechnology, Makerere University, Kampala.

Wharton, P. S. and Dieguez-Uribeondo, J. (2004). The biology of Colletotrichum acutatum. *Anales del Jardin Botanico de Madrid,* **61**: 3–22.

Woltz, S. S. and Jones, J. P. (1981). Nutritional requirements of *Fusarium oxysporum*. Basis for a disease control system. (P. E Nelson, T. A. Tousson, R. J. Cook, eds.), Fusarium: Diseases Biology and Taxonomy. The Pennsylvania State University Press. Univ. Park. Pennsylvania, pp. 340–349.

Workshop on Banana Fusarium Wilt Disease. Kuala Lumpur, Malaysia: INIBAP, 11–30.

Xie, W. S. and Hu, J. S. (1995). Molecular cloning, sequence analysis, and detection of banana bunchy top virus in Hawaii. *Phytopathology,* **85**: 339–347.

Yin, X. M., Jin, Z. Q., Xu, B. Y., Ma, W. H., Fu, Y. G., and Wang, J. B. (2011). Characterization of early events in banana root infected with the GFP-tagged *Fusarium oxysporum* f. sp. cubense. *Acta Horticulturae*, **(897)**: 371–376.

Zhang, N., Wu, K., He, X., Li, S., Zhang, Z., Shen, B., Yang, X., Zhang, R., Huang, Q., and Shen, Q. (2011). A new bioorganic fertilizer can effectively control banana wilt by strong colonization with *Bacillus subtilis* N11. *Plant Soil*, **344**: 87–97.

Zimmerman, A. (1902). Ueber einige an tropischer kulturpflanzen beobachtete pilze. II. Zentralblatt für Bakteriologie, Parasitenkunde Infektionskrankheiten und Hygiene, Abt., **28**: 219.

Zutra, D. and Volcani Zafrira (1971). Bacterial soft rot and wilt of banana plants. *Review of Plant Pathology,* **50**: 279.

CHAPTER 7

BER (*ZIZIPHUS MAURITIANA* LAMK.): KEY DISEASES AND THEIR MANAGEMENT

S. K. MAHESHWARI*, HARE KRISHNA, and SHAKTI KHAJURIA

ICAR-Central Institute for Arid Horticulture, Beechwal, Bikaner 334 006, Rajasthan, India

*Corresponding author. E-mail: maheshwariskciah@gmail.com

ABSTRACT

Ber (*Ziziphus mauritiana* Lamk) is one of the most important fruit crops grown in arid and semiarid regions of India and worldwide. Out of more than 90 cultivars, only 11 are commonly cultivated commercially in different agro-climatic regions of India. In view of the gaining popularity, area under this fruit is being increased gradually day by day. Ber fruits are natural antioxidants for people in arid regions. The fruits are quite nutritious, rich in vitamin C, second only to aonla and guava and much higher than citrus and apple. Its fruit products are getting commercial angle in different countries, particularly for their health benefiting properties. The crop is gaining popularity among the growers because of its adaptability to adverse climatic condition and good return. However, the avoidable loss is more due to insect pest and diseases. It is affected by many serious diseases like powdery mildew, sooty mold, leaf spots, and rust in ber fruit crops have been identified in arid and semi- arid regions.

7.1 INTRODUCTION

Arid region in India spread over 38.70 million ha covers about 12% of the total area of the country. The arid region is having nutrient deficient sandy soil along with fragile edaphic and climatic conditions. Growing annual crops during the most congenial part of the year is the exclusive source of livelihood for the people of the arid region. These crops often encounter abiotic stresses like drought, hot and desiccating winds, nutrient deficiency during their growth period leading to poor productivity. These regions have tremendous potential for the cultivation of perennial fruit trees. The arid regions have been identified for successful fruit production zones. The tracts of hot arid region are seen in the states of Rajasthan (19.6 m ha), followed by Gujarat (6.2 m ha). *Ber* also called as desert apple, jujube, *Badari* (Sanskrit), *Kul* or *Boroi*, Beri, Indian plum hails from the genus *Ziziphus* of the family Rhamnaceae and order Rhamnales. It is a tropical and subtropical fruit indigenous to the northern hemisphere. Fifty genera and more than 600 species fall under the Rhamnaceae family (Pareek, 2001). Among *Ziziphus* species, *Z. jujube* Mill (Chinese date or Chinese jujube), *Z. mauritiana* Lamk. (*ber* or Indian jujube), and *Z. spina-christi* (L.) wild (Christ's thorn) are the most economically important.

Ber (*Ziziphus mauritiana* Lamk) is considered as one of the most important fruit crops grown under arid and semiarid environments of India (Pareek, 2001) and worldwide. *Ber* is hailed as one of the key tree species with respect to distribution and economic viability in arid regions (Abbas, 1997). Mostly, *ber* cultivation is common in water-deficient and low rainfall areas of Indian states of Maharashtra, Rajasthan, Haryana, Gujarat, Bihar, Madhya Pradesh, Uttar Pradesh, and Andhra Pradesh. In India, it occupies a large area (50,000 ha) and is a well-admired dry land fruit crop. In Punjab, it was being cultivated on 2894 ha having annual production 43,410 MT (Anonymous, 2006a, b). Out of more than 90 cultivars, only 11 are commonly cultivated commercially in different agroclimatic regions of the country (Morton, 2005). Owing to its growing popularity, the area under this fruit is witnessing a gradual increase. *Ber* fruits are a rich source of natural antioxidants for the inhabitants of arid regions. The fruits are quite nutritious, rich in vitamin C, second only to aonla and guava and much higher than citrus and apple (Khera and Singh, 1976). *Ber*-based food products are acquiring commercial importance in several countries, especially, for their health benefiting attributes (Vithlani and Patel, 2010). The crop is gaining popularity among the growers because of its adaptability to adverse climatic conditions and good return.

However, the avoidable loss is more due to insect-pest and diseases (Singh, 2008). It is affected by many serious diseases like powdery mildew, sooty mold, leaf spots (*Alternaria, Septoria, Cercospora, Pestalotiopsis, Cladosporium,* etc.), and rust owing to fungal infections and witches broom mediated by pleuropneumonia-like organisms. Many fungal diseases such as powdery mildew, black leaf spot, fruit rot, *Cercospora* leaf spot, and *Alternaria* leaf spot in *ber* fruit crop have been acknowledged in arid and semiarid regions.

7.2 POWDERY MILDEW

Introduction/Economic Importance

Powdery mildew of *ber* is caused by a fungal pathogen. It was first reported from Allahabad (Uttar Pradesh). Kumar et al. (1978) noted the incidence of this disease in the Indian arid zone. This disease is one of the dreaded diseases of *ber* and causes fruit yield losses in semiarid and subtropical regions but it is not serious in Bikaner and nearby regions due to prevailing unfavorable climatic conditions. It is a major problem in growing successful healthy *ber* orchards and fruits resulting in heavy yield losses (Rawal, 1988; Thind et al., 2002).

Symptoms:

Initial symptoms appear on young fruits at the pea stage of development. White specks may appear on immature fruit and later may cover the entire fruit. Mildew produces symptoms as a whitish powdery mass on all the aerial plant parts causing premature drop of flower buds. The infected fruits become misshapen and corky. Fruits finally drop. In severe conditions, floral parts, whole fruits, tender branches, and leaves would appear with powdery mass of fungal conidia. The whole tree would appear unhealthy with less fruit setting and malformed fruits. Such *ber* orchards are devastated completely. Infected fruits fail to develop, show discoloration, and become mummified. Young shoots and spurs are also attacked. They fail to grow and are vulnerable to the die-back pathogen. Severe dropping can be seen under the canopy area of the tree.

FIGURE 7.1　Powdery mildew of *Ber*.

Casual Organism: *Oidium erysiphoides* f. sp. *ziziphi* Yen and Wang

Taxonomic Position

Division—Eumycota
　　Subdivision—Ascomucotina
　　　　Class—Plectomycetes
　　　　　　Order—Erysiphales
　　　　　　　　Family—Erysiphaceae
　　　　　　　　　　Genus—*Oidium*
　　　　　　　　　　　　Species—*erysiphoides* f. sp. *ziziphi*

Disease Cycle:

Limited works were done on perpetuation of the fungus. It can survive on bud wood of ber plants during the nonflowering period. The pathogen may survive as active mycelium on *ber* leaves and as dormant mycelium in bark tissues and fruits in cliestothecia form. Conidia are wind-borne. The secondary disease spread is by air-borne spores.

Epidemiology:

This study was done by Prakash and Jhooty (1987). In North India, this disease is seen from October to January. Maximum temperature of 21–23 °C,

59%–88% relative humidity, and 9.6 h sunshine per day was found congenial for the spread of the disease. Cloudy, humid with mild temperature together with some rainy days favor the occurrence of this disease in moderate to a severe form in this crop.

Integrated Disease Management:

Successful management is achieved through integrated strategies alone. Following combined applications of control can be useful for disease management.

Cultural Control:

- Disposal of infected leaves, twigs, and collateral hosts from orchard.
- Ber orchards should not have wild species (*Ziziphus nummularia*) growing in vicinity.
- Summer ploughing should be practiced.

Chemical Control:

- Karathane (0.1%) proved the best control followed by calixin @0.1%.
- Maximum disease control has been found by the combined application of *Psendomonas fluorescens* (1%) and Karathane @0.05% (Nallathambi et al., 2003).
- Three fungicides namely, bayleton 25 WP (0.05%), karathane 48 EC (0.01%), and sulfex 80 WP (0.25%) were applied during flowering, after fruit set, and the first week of November against powdery mildew. The studies showed the superiority of bayleton in managing the disease effectively. Bayleton significantly checked the disease, reduced the fruit drop, and increased the fruit yield as compared to karathane and sulfex (Thind and Kaur, 2006).
- Three sprays of karathane (0.1%) followed by difenoconazole (0.05%) at 15 days interval proved 51.0% and 40.3% disease control against powdery mildew of *ber* (Anonymous, 2016a, b).

Biological Control:

- Spraying of *Trichoderma* spp. (CIAH-240) @5% also reduced the disease.
- *Psendomonas fluorescens* strain CIAH-196 (1%) was also moderately effective against this disease (Nallathambi, et al., 2006).

Host Resistance:

- Plantation of resistant varieties is an excellent means for reducing the disease. Use of resistant/tolerant cultivars like Sanur-2, Chinese, Jogia, and Vikas must be grown.

7.3   BLACK LEAF SPOT

Introduction/Economic Importance:

For the first time, this disease was reported from Haryana by Gupta and Madan (1977). It is common in Southern and Northern parts of the country. Verma et al. (1995) also reported this disease from Punjab. Disease incidence was noted during the survey programme of Rajasthan and Uttar Pradesh from 1 to 8.0 and 4.75% to 38.75%, respectively (Anonymous, 2010).

Symptoms:

Sooty tuft black spots, which may vary from circular to irregular in shapes, develop on the ventral surface of plant leaves. At an advance stage, the larger part of the lower surface of leaves is also covered with spots while the upper leaf surface shows brownish discoloration. Leaves and twigs are dried rapidly under severe infection.

Casual Organism: *Isariopsis indica* var. *ziziphi*

Epidemiology

Most prevalent temperature was found during January, when the disease was severe.

Cloudy weather with moderate temperature during October–November is favorable for disease initiation.

Integrated Disease Management:

Cultural Control:

- Eradication of infected leaves and twigs from orchard.
- To keep the *ber* orchard free from all weeds.
- Proper pruning to be done in the orchard.

Chemical Control:

- Fungicidal sprays at 15 days interval (mancozeb or copper oxychloride @0.2%) (Verma and Cheema, 1988).
- Two sprays of propiconazole (tilt), followed by difenoconazole (score) @0.1% was noted to be useful against this disease and also increased fruit yield (Anonymous, 2016).

Host Resistance:

- Use of resistant varieties such as ZG-3, Seo, Bahadurgarhi, and Safeda Rohtak.

7.4 ALTERNARIA FRUIT ROT

Introduction/Economic Importance

The rot was reported from different locations of Rajasthan, particularly in Chomu (Jaipur) and nearby areas. It is also an economically important disease.

Symptoms:

Disease incidence can be noted on immature, young, and mature fruits. It starts from the pedicel and in the bottom portion of fruits. Ripened fruits are almost free from this rot. Light brown to dark brown spots are produced on *ber* fruits. Mature fruits rot under severe conditions by brown blotching. Such fruits ultimately fall down resulting in huge losses. Some fruits have concentric rings loaded with fungal spores. All the young fruits also fall down in severe cases and *ber* plants appear with moderate fruiting.

Casual Organism: *Alternaria alternata*

Taxonomic Position

Subdivision—Deuteromycotina
   Class—Hyphomycetes
      Subclass—Sporomycetidae
         Order—Moniliales
            Family—Dematiaceae
               Genus—*Alternaria*
                  Species—*alternata*

FIGURE 7.2   *Alternaria* fruit rot in *Ber*.

Disease Cycle:

The fungus continues to remain alive as mycelium in infected crop debris. The infection begins from the fruits containing numerous conidia. These conidia are disseminated by wind, rain splashes, water and cause secondary disease. The conidia germinate in the presence of moisture at a temperature nearly 25–30 °C giving rise to germ tubes, which enter the host tissue.

Epidemiology:

Availability of high moisture, followed by warm weather conditions is most congenial for disease development. Its development is favored between 20 °C and 30 °C with an optimum temperature at 25 °C. High relative humidity coupled with frequent rainy days seems to be more crucial than temperature alone in disease development.

Integrated Disease Management:

Cultural Control:

- Removal and destruction of infected leaves and twigs from orchard.
- Keeping the *ber* orchard free from all weeds.
- Pruning should be done properly in the orchard.

Chemical Control:

- Fungicidal sprays at 15 days interval (mancozeb @0.2%) was given to reduce the disease.
- Mancozeb and copper oxychloride (0.2%) are also useful for suppressing the fungal growth.
- Neem oil effectively suppressed the severity of fruit rot of *ber* (caused by *Alternaria alternata*; Kumar et al., 2013).

Biological Control:

- This disease can be controlled by *Trichoderma harzianum* and *Bacillus* spp. (Mercer and Papadopolous, 1990) and Fluorescent Pseudomonads (Prasad and Kulshrestha, 1999).

7.5 *CERCOSPORA* LEAF SPOT

OCCURRENCE AND DISTRIBUTION

It is a very frequent disease of *ber* and is predominantly observed in different *ber*-growing regions of the country, especially, Northern parts of India.

Symptoms:

Circular to oval discolored spots (up to 4 mm diameter) develop on leaves which later turn to brown color and encircled by dark brown margins.

Causal Organism: *Cercospora ziziphi*

Taxonomic Position

Division—Eumycota
  Subdivision—Deuteromycotina
    Class—Hyphomycetes
      Order—Moniliales
        Family—Dematiaceae
          Genus—*Cercospora*
            Species—*ziziphi*

Disease Cycle:

Active pathogen overwinters as mycelium and spores in crop residue in the soil. Conidia produced from mycelium are wind-borne and are spread by water, rain splashes, and insects and cause the secondary infection.

Epidemiology:

Moist weather is congenial for fungal sporulation. Conidial germination and penetration takes place at temperature ranging from 20 °C to 30 °C and RH (90%).

Integrated Disease Management:

Cultural Control:

- Removal and destruction of infected leaves, twigs, and weeds from orchard.
- Proper pruning in the orchard.

Chemical Control:

- Spraying of mancozeb (0.3%) in *ber* trees was useful for reducing the disease (Bhardwaj and Sharma, 1999).
- Systemic fungicides (benlate and bavistin @0.1%) are also effective against this disease.

7.6   *ALTERNARIA* LEAF SPOT

Introduction/Economic Importance

In Uttar Pradesh, *ber* orchards are found in Varanasi, Faizabad, Agra, Raebareli, and Aligarh districts. It is caused by *Alternaria alternata* and has now attained the status of one of the most common diseases of *ber* in Uttar Pradesh. It generally appears during December to January month and become severe during the month of February to March.

Symptoms:

Infection was manifested as dark grayish-black colonies (of the fungus) on leaves. Infection was quite prominent during foggy days in *ber* foliage. At later stage, infected surface turns gray by fungal growth.

Casual Organism: *Alternaria alternata*

Epidemiology:

The disease was first recorded during 1st week of December (49th meteo-rological week) on cv. Gola. The conducive weather parameters such as temperature maximum (23.2 °C) and minimum (7.9 °C), relative humidity morning (97.8%) and evening (59.5%), sunshine hrs/day (1.9), and rainfall (0.0 mm) are required for its incidence. Maximum PDI (58.0) were recorded when temperature maximum and minimum, RH morning and evening, sunshine hrs and rainfall were 27.0 °C, 11.6 °C, 81.7%, 51.2%, 7.0, and 0.0 mm, respectively. The optimum conditions determined for lesion development were incubation at 25–30 °C with 98%–100% RH. During February, when temperature ranges from 20 °C to 25 °C, severe foliar infections have been noticed in some pockets of South Bengal (Maiti et al., 2006).

Integrated Disease Management:

Cultural Control:

- Removal and destruction of infected leaves, twigs, and also weeds from orchard.
- Pruning should be done properly in the orchard.

Chemical Control:

- For its efficient management, spray difenconazole (0.1%) followed by propiconazole @0.1% and copper oxychloride @0.2%. Neem oil @3.0% has been proven least effective (Singh et al., 2016).

KEYWORDS

- **Ber**
- **disease**
- **etiology**
- **symptomatology**
- **management**

REFERENCES

Abbas, M. F. (1997). Jujube. *In:* Post-harvest physiology and storage of tropical and subtropical fruits. (Ed. S. K. Mitra), CAB International. pp. 405–415.

Anonymous (2006). District wise area and production of different fruits in Punjab state.

Anonymous (2010). Disease management. *In:* Annual Report, All India Co-ordinated Research Project on Arid Zone Fruits. pp. 135–144.

Anonymous (2016a). Cost effective control of ber powdery mildew. *In*: Annual Report, ICAR-All India Co-ordinated Research Project on Arid Zone Fruits. pp. 106–107.

Anonymous (2016b). Management of black leaf spot of ber. *In*: Annual Report, ICAR-All India Co-ordinated Research Project on Arid Zone Fruits. pp. 104–105.

Bhardwaj, S. S. and Sharma, L. M. (1999). Diseases of minor fruits. *In*: Diseases of Horticultural Crops-Fruits (Eds. Verma, L. R. and Sharma, R. C), Indus Publishing Company, New Delhi, pp. 540–562.

Gupta, P. C. and Madan, R. L. (1977). Diseases of fruits from Haryana—a new leaf spot disease of ber. *Curr. Sci.*, 46: 237–238.

Khera, A. P. and Singh, J. P. (1976). Chemical composition of some ber cultivars (*Zizyphus mauritiana*). *Haryana J. Hort. Sci.,* 5: 21–24.

Kumar, A., Bhansali, R. P., and Arya, H. C. (1978). A note on the occurrence of powdery mildew on ber (*Ziziphus* spp.) in Indian arid zone. *Ann. Arid Zone,* 17 (3): 323–325.

Kumar, S., Godara, S. L., Gangopadhyay, S., Kumar, R., and Singh, S. (2013). Effect of climatic factors on fruit rots (*Alternaria alternata* and *Aspergillus niger*) of ber (*Zizyphus mauritiana* Lamk.) and their management systems. *African J. Microbiol. Res.*, 7: 5742–5748.

Maiti, C. K., Sen, S. Acharya, R., and Acharya, K. (2006). First report of *Alternaria alternata* causing leaf spot on Stevia rebaudiana. *New Dis. Reprt.*, 14: 22.

Mercer, R. C. and Papadopolous, S. (1990). Biological control of seed borne diseases of linseed. Procd. on biocontrol of pests and diseases. 20–21, Sept., at Agriculture and Food Science Centre, Bolfast.

Morton, J. F. (2005). *Indian jujube. J. Exp. Bot.,* 56: 3082–3092.

Nallathambi, P., Umamaheswari, C., Joshi, H. K., and Dhandar, D. G. (2003). Management of ber (*Ziziphus mauritiana Lamk.*) powdery mildew using Fluorescent Pseudomonads. *In:* Proc. of 6th International Workshop on plant growth Rhizobacteria held during 5–10 October, 2003 at IISR, Calicut, Kerala, pp. 184–187.

Nallathambi, P., Umamaheswari, C., Nagaraja, A., and Dhandar, D. G. (2006). Pomegranate diseases and management. Technical Bulletin, CIAH, Bikaner, p. 30.

Pareek, O. P. (2001). Ber: International Centre for Underutilized crops, Southampton, UK, pp. 248–266.

Prakash, V. and Jhooty, J. S. (1987). Epidemiology of powdery mildew of *Ziziphus mauritiana* caused by *Microsphaera alphitoides* f. sp. *ziziphi. Indian Phytopathol.* 40 (4): 491–494.

Prakash, R. D. and Kulshreshtha, D. D. (1999). Bacterial antagonists of *Alternaria helianthi* of sunflower. *J. Mycol. Pl. Pathol.*, 29 (1): 127–128.

Rawal, R. D. (1988). Assessment of yield losses in ber fruit due to powdery mildew. *Plant Dis. Res.*, 3: 138.

Singh, M. P. (2008). Managing menace of insect-pests on ber. *Indian Hort.*, 53 (1): 31–32.

Singh, S., Kumar, S., Kumar, S., and Tiwari, P. K. (2016). Epidemiology and management of Alternaria leaf spot (*Alternaria alternata*) of ber (*Zizyphus mauritiana* Lamk.). *Res. Environ. Life Sci.,* 9 (7): 845–848

Thind, S. K., Kaur, N., and Aulakh, P. S. (2002). Studies on diseases of miscellaneous fruits. Ann. Progress Report, RFRS, Abohar, 2001–2002, pp. 108–13.

Thind, S. K. and Kaur, N. (2006). Management of ber powdery mildew with fungicides. *Indian J. Hort.*, 63 (3): 267–269.

Verma, K. S. and Cheema, S. S. (1988). Chemical control trials against mouldy leaf spot of ber caused by *Isariopsis indica* var. *ziziphi. Plant Dis. Res.*, 3 (1): 32–36.

Verma, K. S., Kumar, S., and Kumar, S. (1995). Survival and dispersal of *Isariopsis indica* var. *ziziphi* causing mouldy leaf spot of ber. *Indian Phytopathol.*, 48 (1): 45–48.

Vithlani V. A. and Patel H. V. (2010). Production of functional vinegar from Indian jujube (*Zizyphus mauritiana*) and its antioxidant properties. *J. Food Tech.*, 8 (3): 143–149.

CHAPTER 8

MAJOR DISEASES OF CITRUS AND THEIR MANAGEMENT

N. M. GOHEL[1,*], B. K. PRAJAPATI[2], and J. N. SRIVASTAVA[3]

[1]B. A. College of Agriculture, Anand Agricultural University, Anand 388 110, Gujarat, India

[2]Directorate of Research, S. D. Agricultural University, Sardarkrushianagr 385 506, Gujarat, India

[3]Department of Plant Pathology, Bihar Agricultural University, Sabour 813210, Bhagalpur, Bihar, India

*Corresponding author. E-mail: nareshgohel@aau.in

ABSTRACT

Citrus fruits are among the most consumed fruits in the world. Citrus fruits are very common and the most important fruit crops in India, and are grown in 9.85 lakh hectares with a total production of 114.19 lakh tons. Low productivity in citrus is caused by many factors such as, plant pathogens, insect-pests, nutritional imbalances, and physiological disorders. However, the world production of citrus fruits is faced with several constraints which hinder its development. Furthermore, the spread of pests such as mealybugs, mites, and Mediterranean fruit fly has impacted negatively on citrus quality and yield. Therefore, the main objectives of this review were to identify the key harmful organisms to the citrus crops and to highlight the appropriate methods to manage them.

8.1 INTRODUCTION

The citrus fruits include lemons, orange limes, tangerines, pomelo, and grapefruit. Not only the citrus fruits in this varied group have delicious and refreshing all star fruits in this varied group, they all include compound flavonoids. Compound flavonoids have anticancer effects.

Citrus fruits are very common and the most important fruit crops in India, and are grown in 9.85 lakh hectares with a total production of 114.19 lakh tons (Anonymous, 2017a). Low productivity in citrus is due to many factors such as pathogens, insect-pests, nutritional imbalances, physiological disorders, and improper cultural practices. Citrus fruits growing countries are Brazil, Cuba, Israel, Japan, Morocco, Spain, South Africa, Turkey, USA, and India in the world. India ranks sixth in the worldwide production of citrus fruits. After mango and banana production, citrus fruits occupy third place in production in India. Citrus fruits originated in Southeast Asia's tropical and subtropical areas, essentially India and China. Northeast India is the native site of numerous citrus plants. Citrus fruits hit with a high content of vitamin C and a refreshing drink. India grows and produces several commercially important citrus fruits, such as orange (mandarin or santra), sweet orange (mosambi, malta, or satgudi), lime, and lemon (Anonymous, 2012).

Citrus plants are very liable to infestation many diseases, these are as given below:

Diseases

(1) Gummosis: *Phytophthora parasitica, P. palmivora, P. citrophthora*
(2) Scab/Verucosis: *Elsinoe fawcetti Sphaceloma fawcettii (anamorph)*
(3) Canker: *Xanthomonas campestris* pv *citri*
(4) Tristeza or fast decay: *Citrus tristeza virus* (CTV)
(5) Huanglongbing/Citrus greening: *Candidatus Liberibacter asiaticus*

8.2 GUMMOSIS: *PHYTOPHTHORA* SPP.

Introduction/Economic Importance:

There are six *Phytophthora* spp. capable of causing disease in citrus fruit trees, but the greatest reduction in fruit yield may be attributed to the following two spp.: (1) *Phytophthora parasitica* Dast. (mostly in tropical and subtropical regions), (2) *Phytophthora citrophthora* (R.E. Srn. and E.H. Srn.) Leonian (mostly in temperate climates; Timmer and Menge, 1988). *Phytophthora* spp. has a back history as a major constraint in citrus fruit production throughout the world. Gummosis or Phytophthora disease occurred and reported from Australia in 1860 and later in Florida in 1952 (Broadbent, 1977). Considerable losses to the citrus industry in South Africa by root pathogens including

Phytophthora spp./Gummosis in every year (Kotze, 1984). The loss due to *Phytophthora* spp. was estimated by Themann and Werres (1995) in citrus nurseries alone up to 80%. Gummose is a common disease which occurs as frequently in Punjab and Assam, India. Lemons are more vulnerable than lemons and whole grapefruits. In Southern Indian regions, it is very popular in sweet orange.

Symptoms:

Symptoms appeared when large patches of water soaked nearly ground level on the basal portions of the stem. Bark of such parts of stem/trunk showed symptoms in lengthwise vertical strips such as dries, shrinks, and cracks and shreds. Stem/trunk bark profuse later stage gum exudation (Anonymous, 2018a).

FIGURE 8.1   Citrus gummosis disease.

Casual Organism: *Phytophthora parasitica, P. palmivora, P. citrophthora, P. hibernalis, P. syringae, P. cactorum*

Pathogen:

Phytophthora spp. have various types of inoculum: mycelia, chlamydo-spores, sporangia, zoospores, and oospores.

Scientific Classification (Kirk et al., 2008)

Kingdom: Fungi
Phylum: Oomycota
Class: Oomycetes
Order: Pythiales
Family: Pythiaceae
Genus: *Phytophthora*

Disease Cycle and Epidemiology:

Species of *phytophthora* can spread in many ways, including soil movement with nursery stocks, irrigation water, and infected root sections. Irrigated citrus often suffers as runoff water can carry the pathogen into channels, streams, or rivers. Water from those sources can then contaminate areas that were previously uninfected. On farm machinery, the fungus may be brought into the soil. Seeds taken from contaminated fruits are rarely contagious (Graham and Menge, 1999).

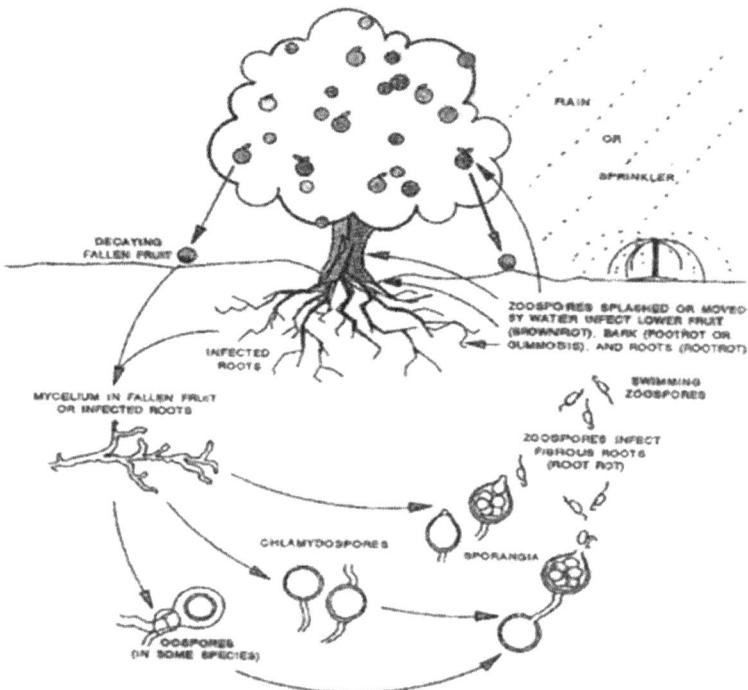

FIGURE 8.2   The disease cycle of gummosis.

Favorable Conditions:

High soil moisture increases infection, mainly due to increased sporangia formation and improved conditions for zoospore release, motility, and movement to the site of infection. In soils with restricted drainage and soil pH between 6.0 and 6.5, the occurrence of the disease is typically more extreme. Stress from either excess moisture or low moisture can also make certain hosts more vulnerable to infection (Erwin and Ribeiro, 1996).

Management:

Preventive Measures:

- Selecting of field for planting of citrus fruits with well developed drainage system for water.
- Citrus plant must be planted with little higher than the ground level.
- Avoiding excess irrigations.
- Range of planting material (30–45 cm or above) with large budded grafts.
- Prevent mechanical damage to the crown roots of citrus trees or stem base during cultural activities.
- To propagate popular/commercial varieties, use resistant sour orange or trifoliate orange rootstock.
- Citrus trees paint root at almost ground level with Bordeaux paste at least once a year or with $ZnSO_4 + CuSO_4 + lime$ (5:1:4) at a height of around 60 cm above ground level.
- Apply *Trichoderma viride* multiplied on neem cake.

Curative Measures:

- Injuries to coronary roots or stem base should be avoided during cultural operations.
- Scrape/chisel the sick part out.
- Protect the Bordeaux paste cut surface.
- Two sprays, drilled with Aliette (2.5 g/L) or Ridomil MZ-72 (2.50 g/L) covering the entire plant canopy and basin of the affected plant at a 40-day interval from the onset of the monsoon provided considerable control (Anonymous, 2018b).

8.3   SCAB/VERUCOSIS: *ELSINOE FAWCETTI* (ANAMORPH: *SPHACELOMA FAWCETTII* JENK)

Introduction/Economic Importance:

Disease from citrus scab is common in regions. In areas with rainfall to provide favorable temperature and high humidity, the disease is prevalent. Apart from that, disease that occurs on a new flush and fruit setting coincides with warm and humid weather. The disease often occurs when planting citrus in warm, low-lying soils and dense or shaded conditions. Serious outbreaks of citrus scab disease only occur in areas where susceptible species or cultivars of citrus fruits are produced. Yield losses often rely on seasonal and local weather variations. In which areas where annual rainfall is limited (<1300 mm) the disease is not an issue. Citrus-developing zones with a dry climate state, whether normal, rare, or immaterial (Gopal et al., 2014).

FIGURE 8.3   Citrus scab disease (a) Citrus scab symptoms on sour orange leaves; (b) Citrus scab symptoms on sour orange fruit.

Symptoms:

Symptoms occur on citrus leaves and berries, as wart lesions/pustules. The lesions on the leaf are initially semitranslucent dots that become tiny nipple-like projections with pale yellow cream. Lesions can go up by as much as 3 mm in diameter and grow a scab-like appearance. Older scab lesions are usually colored, warty, deeply cracked, from cinnamon to honey, and can separate as they mature. Young shoots and fruit that develop similar leaf symptoms but are not seen as much as they do as young shoots. Pustules are the location where the fungal spores are produced. The lesions on grapefruit appear to be flattened scabby sheets, and sweet orange. Tissue age and time of infection also affect the lifting and size of the lesions. The lesions on young tissues appear to grow, and those on mature tissues are more flatter.

Because of variations in cultivars and infected tissue age it is difficult to differentiate between citrus scab and citrus cancer based on symptoms alone (Gopal et al., 2014).

Casual Organism: *Elsinoe fawcetti* (Anamorph: *Sphaceloma fawcettii* Jenk)

Pathogen:

Citrus scab disease is caused by the *Elsinoe fawcetii* Bitancourt and Jenkins. *E. Fawcetii* and that contains a number of pseudothecial locules with round asci. Growing ascus contains eight filamentous ascospores in sizes of 5–6 × 10–12 μm, hyalin and elongated circular. *Elsinoe* spp. produce two conidium types, such as hyaline conidia and spindle conidia (Gopal et al., 2014).

Scientific Classification (Kirk et al., 2008)

Kingdom:	Fungi
Phylum:	Ascomycota
Class:	Dothidiomycetes
Order:	Myriangiales
Family:	Elsinoaceae
Genus:	*Elsinoe*

Disease Cycle and Epidemiology:

Over the course of the year, young fruits and leaves produce spores on the surface of scab lesions. Rough lemon rootstock is also very susceptible to scabbing and can act as a source of fungal inoculum. Fungal spores are distributed by splashes of rain, overhead irrigation, and spray operations. Dew may also cause the spores to be released from the lesions but due to the inhibited splashing action, only localized dispersal may occur (Whiteside, 1975).

Favorable Conditions:

Citrus scab disease can be especially serious on flushes from summer development. Summer wet cycles linked to rain showers and dew are highly favorable for spore germination and infection (Anon., 2018a).

Management:

• Follow the crop sanitation practices.

- Establishment of citrus nurseries in appropriate drainage soil, in dry areas or in greenhouses and appropriate treatment may help to produce pathogen-free rootstocks and bud wood.
- Protective fungicides (copper, ferbam, thiram, difenoconazole, and chlorothalonil) or systemic fungicides, such as benomyl, carbendazim, may be applied prior to flushing and after petal fall. Certified citrus planting material should be planted in a newly established orchard (Anonymous, 2018c).

8.4 CANKER: *XANTHOMONAS CAMPESTRIS* PV. *CITRI*

Introduction/Economic Importance:

Among all citrus diseases, Citrus canker is the most devastating disease. The disease affects all of the important crops of citrus fruits in that region. The disease causes tremendous losses to the citrus, and along with cultivars and the prevailing climate, the severity of the disease varies. The prevalence of the disease as endemic in India, Japan, and other countries of Southeast Asia but except in all other continents producing citrus except Europe. In arid citrus growing areas, canker usually does not occur. However, the occurrence of the disease worldwide in many areas is a continuous threat to citrus fruit production, particularly in canker-free zones. Citrus canker disease in India appears to be a serious problem wherever acid lime (*C. aurantifolia*) is grown on a large and commercial scale and has become a major permanent problem for citrus growers. In 1942, Luthra and Sattar first reported Citrus canker disease from Punjab (Luthra and Sattar, 1942; Bedi, 1961). In 1954, Ramakrishan further recorded the occurrence of citrus canker disease in Tamil Nadu (Ramakrishnan, 1954). Canker has also been reported from the Punjab at kinnow, mandarin nursery (Anonymous, 2000).

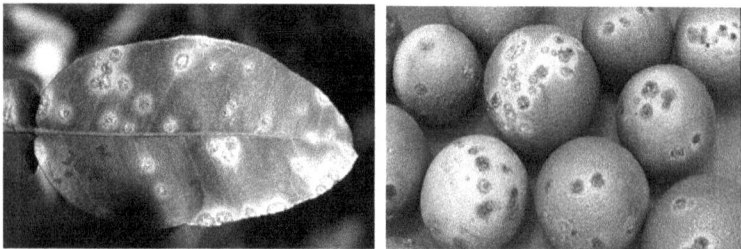

FIGURE 8.4   Citrus canker disease (a) Citrus canker symptoms on sour leaves; (b) Citrus canker symptoms on citrus fruits.

Symptoms:

Leaf Lesions:

Canker lesions are evident on the leaves' underside, and then on the upper surface, about 7–10 days after infection. With the elevated margin and sunken middle, the pustules are corky. The yellow halo around it is a common symptom of the disease on the leaves.

Fruit and Stem Lesions:

In fruit and roots, citrus canker lesions range up to 1 mm in size and are close to those on leaves. Crop deficiency results in the premature fall of the crop. Usually, the fruit's internal quality is not affected but individual lesions penetrate the rind deeply enough to expose the fruit's interior to secondary microorganism infection. Stem lesions allow the bacteria to live in the long term.

Casual Organism: *Xanthomonas campestris* pv. *citri*

Pathogen:

The causal organism is the bacterial pathogen *Xanthomonas citri*, now called *X. campestris* pv. *citri* (Hasse) Dowson. Single polar flagellum (monotrichous). It lacks endospore formation. It is a Gram-negative, aerobic form surrounded by a capsule (Anonymous, 2017b).

Scientific Classification (Kirk et al., 2008)

Kingdom:	Bacteria
Phylum:	Proteobacteria
Class:	Gammaproteobacteria
Order:	Xanthomonadales
Family:	Xanthomonadaceae
Genus:	*Xanthomonas*

Disease Cycle and Epidemiology

Bacteria grow on leaves, stems, and fruit in lesions. The bacteria ooze out when free moisture is present on the lesions. Rain splash is the principal dispersal agent and wind helps to penetrate bacteria through natural openings or wounds created by thorns, pruning, and insects (leaf miner). Bacterial death increases when exposed to direct sunlight. These bacteria can live in infected tissues of plants that have been kept dry and free from the soil for years.

Favorable Conditions:

Wind-driven rain plays a major role in the dispersal of *Xanthomonas* spp. Sprinkled rain and wind, and the volume of *Xanthomonas* spp., should easily disperse the bacteria. Decreases with wind-blown rain dispersal after the main event. Besides that, bacteria also prefer warm weather. Citrus infection is increasingly extreme in regions with high precipitation and high mean temperature (Bock et al., 2005).

Management:

- Complete eradication of infected trees is advised.
- Remove contaminated branches from pruning scissors and then spray the trees with 1% of the Bordeaux mixture at regular intervals. Falling infected leaves and twigs should be gathered and burned.
- Spray 1 g streptocycline and 30 g copper oxychloride in 10 L of water at intervals of 15 days in nurseries and at intervals of fortnight in orchards during rainy season.

8.5   TRISTEZA: CTV

Introduction/Economic Importance:

Tristeza is common in India in Andhra Pradesh, Tamil Nadu, Karnataka, Madhya Pradesh, Maharashtra, Punjab, Bihar, West Bengal, Sikkim, and others. Side effects indicate snappy or ceaseless declines in tree species are especially common and extreme on trees spread over hard orange root stocks. To represent the tragic appearance of the ailing citrus trees, the name "Tristeza" has been proposed.

Symptoms:

The initial stages of infected trees appear chlorotic and sickly. Gradually the leaves fall and result in the appearance of die-back. Pitting stem seen on infected trees. Honeycombing, a fine pitting of the inner side of the bark in the portion of the trunk under the bud union, is the principal symptom of Tristeza in sweet orange. Tristeza contaminated citrus trees on rootstocks of sour orange cause the phloem at the graft union to become necrosed. Ailing trees appear to bloom heavily. Stunted are the trees with stem pitting and set fewer fruits.

Casual Organism: CTV

Flexible filamentous virion with +ve ssRNA belonging to Closterovirus group.

Survival and Spread:

Long distance dissemination occurs through the movement of infected citrus planting material, or through the movement of infested plant material with infected aphids.

Management:

- Implement strict quarantine legislation.
- Use certified and healthy planting materials. Destroy all diseased trees whenever and when the disease happens.
- Daily insecticidal sprays such as dimethoate or 1 ml/l methyl-*o*-demetone to avoid further spread of the disease.

8.6 GREENING OR HUANGLONGBIN (HLB)

Introduction/Economic Importance:

In sweet oranges, kinnow, lemons, and other members of the citrus family, citrus greening/huanglongbing disease also occurred. Citrus-greening disease is a major citrus disease in India. Several Asian and African countries confirmed the disease (Garnier and Bove, 1996). The disease was documented for the presence of citrus greening bacterium disease in different parts of India in different citrus species.

Symptoms:

- Stunting of leaves, scant foliage, twig dying back, bad harvest of mostly greenish, useless fruits. The twigs are upright and grow smaller leaves.
- Young leaves appear normal but soon assume an upright position, become leathery, and develop prominent veins and a slender green color.

- Small, curved fruit columella. Low in juice, and solids soluble, high in acidity.

Casual Organism: *Candidatus Liberobacter asiaticus* (Fastidious Phloem limited Bacterium), obligate Gram-negative bacterium.

Mode of Spread:

The disease is transmitted by Diaphorina citri, an infected wood bud and citrus psylla. The illness is also transmitted through the dodder from the citrus to the Periwinkle (*Catharanthus roseus*).

Management:

- Control psyllids with insecticides.
- Use healthy bud wood for propagation.

KEYWORDS

- **citrus**
- **disease**
- **etiology**
- **symptomatology**
- **management**

REFERENCES

Anonymous (2000). Proceedings of the group discussion of the All India Coordinated Research project and ICAR ad hoc schemes on tropical fruits. 5–8 January 2000, Rahuri. Tech. Doc. No. 72, p. 31.

Anonymous (2012). http://citrusindia.blogspot.in/.

Anonymous (2017a). http:/agricoop.gov.in/sites/default/files/201718%281st%20Adv.%20 Est%29.pdf.

Anonymous (2017b). http://www.biologydiscussion.com/plants/plant-diseases/citrus-canker -disease-with-diagram/64321.

Anonymous (2018a). http://vikaspedia.in/agriculture/crop-production/integrated-pest-managment/ipm-for-fruit-crops/ipm-strategies-for-citrus/diseases-and-symptoms #section-4.

Anonymous (2018b). http://www.ikisan.in/mh-citrus-disease-management.html#Gummosis.

Anonymous (2018c). https://www.plantwise.org/KnowledgeBank/Datasheet.aspx?dsid=20777.

Bedi, K.S. (1961). Some important observations on the citrus canker in Punjab. Punjab Hort. J., 2: 89–91.

Bock, C. H., Parker, P. E., and Gottwald, T. R. (2005). Effect of simulated wind-driven rain on duration and distance of dispersal of *Xanthomonas axonopodis* pv. *Citri* from canker-infected citrus tree. *Plant Dis.* 89 (1): 71–80.

Graham, J. H. and Menge, J. A. (1999). Root diseases. In: "Citrus health management". (eds. Timmer, L.W. and Duncan, L.W.) American Phytopathological Society, St. Paul, MN. pp. 126–135.

Gopal, K., Govindarajulu, B., Ramana, K. T. V., Kumar, C. S. K., Gopi, V., Sankar, T. G., Lakshmi, L. M., and Sarada, G. (2014). Citrus Scab (*Elsinoe fawcettii*): a review. *Res. Rev.* 3 (3). 49–58.

Luthra, J. C. and A. Sattar, (1942). Citrus canker and its control in Punjab. Punjab Fruit J., 6 (1): 179–182.

Ramakrishnan, T. S. (1954). Common diseases of citrus in Madras state. Government of Madras Publication.

Whiteside, J. O. (1975). Biological characteristics of *Elsinoe fawcettii* pertaining to the epidemiology of sour orange scab. *Physiopathology,* 65: 1170–1177.

CHAPTER 9

INTEGRATED MANAGEMENT OF CITRUS CANKER DISEASE (*XANTHOMONAS AXONOPODIS* PV. *CITRI*)

AMAR BAHADUR[1,*] and J. N. SRIVASTAVA[2]

[1]College of Agriculture, Tripura, Lembucherra, Agartala, India

[2]Department of Plant Pathology, Bihar Agricultural University, Bhagalpur, Bihar, India

[*]Corresponding author. E-mail: amarpatel44@rediffmail.com

ABSTRACT

Citrus canker is one of the most important diseases in citrus caused by a bacterium *Xanthomonas citri* (syn. *Xanthomonas campestris* pv. *citri*). The importance of this disease can be ascertained with its dominance in almost all the citrus growing countries worldwide especially in countries like Asia, South America, Oceania, Africa, as well as in the United States. The disease causes tremendous yield losses in citrus cultivars as well as citrus relatives. The pathogen produces necrotic lesions almost in all aerial parts of plant and initiates severe infections on premature fruit that results in dropping of the fruits. Naturally, bacterial infection is caused by wounds or by mechanical damages and is disseminated by wind and rainfall splashes which assist infection much faster. Citrus leaf miner is also one of the vital aspects causing infestation of citrus plants and indirectly raises the disease to many folds. The disease first occurred in an endemic form in India, Japan, and other Southeast Asian countries and then spread to other citrus growing continents except Europe. The aim of the chapter is to focus and review on *Xanthomonas citri* causing economic losses in India and many other countries by producing poor quality fruits, reduction in yield, and management practices for controlling canker diseases.

9.1 INTRODUCTION

Citrus canker is a very poor citrus disease that exists in citrus-growing regions around the world. Citrus canker probably originated in Southeast Asia, and its geographic range continues to expand. It is the critical policy that many countries are obliged to avoid implementing. Citrus canker occurs in pacific Asia, Ocean Islands, South America, and the United States in over 30 countries. Citrus canker was first portrayed in 1915 in the Gulf States (U.S.). However, specimens of herbariums collected in India in 1827 were well-known as canker lesions.

The disease emerged in South Africa and Australia earlier too. This was removed at nursery and orchard in these countries as well as in the Gulf States during inspections, quarantines, and trees burning. Subsequent epidemics have occurred in Australia, Argentina, Brazil, Oman, Saudi Arabia, Reunion Island, and Uruguay. Extensive outbreak was discovered in Metropolitan Miami, Florida in 1995 and outbreak in September 1995 was about 14 square miles, and spread through December 1998 to over 202 square miles. Host species, citrus, and citrus-related cultivars include orange, sour orange, grapefruit, pummelo, mandarin, lemon, lime, tangerine, tangelo, rough lemon, and orange trifoliate.

For several citrus-growing countries, the bacterium is under quarantine disease and controlled by an international phytosanitary program. The disease is associated with different pathotypes. Citrus canker is common in tropical and subtropical climates, and can occur in drier climates as well. Asian cancer (Canker A) is the disease's most acute form, affecting nearly all varieties of citrus growing and the disease's most economically significant.

Symptoms:

Citrus canker disease occurs in severe forms whereas during rainfall and warm temperatures, periods of shoot emergence and early fruit development occur. Citrus canker is mainly leaf-spotting and fruit rind-blemishing disease, causing defoliation shoot dieback and fruit drop when conditions are highly favorable for infection. Lesions start at the Leaf canker as pinpoint spots and reach diameters of up to 2 to 10 mm. The extent of the lesions depends on the age of the host tissue during infection and of the citrus cultivar. Lesions on the underside of the leaves are visible about 7 to 10 days after infection, and then on the upper surface. The raised young lesions can be seen on both leaf surfaces, especially on the lower leaf surface. The pustules gradually grow into corky, raised margins, and sunken centers.

FIGURE 9.1   Symptoms on leaf, stem, and fruits.

The characteristic symptom of the disease occurs on the leaves, since yellow halo covers the lesions. The diagnostic symptom of citrus canker disease is the water-soaked margin formed around the necrotic tissue and can be easily detected by transmitting light. Fruit and stem canker lesions reach a depth of 1 mm, and are similar to those on trees. Owing to the predisposition of the crust, fruit lesions may vary in size over a longer period of time than for leaves. If the fruit remains on the tree until such fruit ripening has reduced marketability, the infected fruit may cause premature drop in yield. In general, the internal quality of the fruit is not affected, but individual lesions, stems, lesions, which are viable for several seasons, penetrate deeply into the crust. The bacteria can sustain lengthy survival of stem lesions. Asian leafminer (*Phyllocnistis citrella*) can infest leaves, stems, and fruits, and increases the number of individual lesions that quickly coalesce and form large irregularly shaped lesions and feed on the epidermis just below the leaf cuticula. Throughout the cuticula, numerous cracks occur that cover leafminer galleries that supply bacteria that penetrate directly into the highly susceptible parenchyma and spongy mesophyll infectious palisade. Citrus foliar extensive wounds consist of all leafminer-feeding galleries and do not form callus for up to 10 to 12 days, usually callus farm for 1 to 2 days, increasing the susceptibility of the gallery to infection. Leafminer infestations occur creating thousands of possible courts of infection on individual trees. If bacterial dispersal occurs in the presence of the leafminer, this greatly exacerbates the risk for infection across the entire spreading range.

Causal Organism:

Bacterium *Axonopodis Xanthomonas pv. Citri (Xac)* causes disease which causes Asian citrus canker. The Bacterium *Axonopodis Xanthomonas pv. Citri (Xac)* is a gram-negative, rod-shaped, and single polar flagellum. Media colonies of bacteria are normally yellow due to the production of pigments called *xanthomonadin*. When glucose is applied to the culture medium, exopolysaccharide slime develops that causes colonies to become mucoid. It is possible to prepare on semi-selective medium the addition of an antibiotic (kasugamycin) which inhibits many contaminants but not xanthomonads. The optimal elevated temperature ranges between 28 °C and 30 °C.

Identification of bacterial isolates into pathovars by serology, bacteriophage typing, fatty acid profiles, polymerase chain reaction (PCR), and DNA analysis are useful techniques unavailable, strains of *X. axonopodis* pv. *citri* distinguished from other pathovars by susceptible and resistant citrus hosts. On detached leaves, bioassays can be carried out. Bioassays can be performed on detached leaves. Lesions may become visible on separated leaves 2 to 4 days after inoculation, and different isolates may be checked for comparison on the same leaf. Under favorable conditions, the bacteria enter the host through natural openings such as stomata and wounds caused by spines, wind, and leaf miners. After entering the host cells, the bacteria multiply in the intercellular space, dissolve the middle lamella, and settle in the cortex zone. The canker pustules develop throughout the affected areas, and the bacteria exude in the form of gum material. They are easily distributed by wind and rain. Young tissues of plants get more easily infected. Fast humidity, temperature between 20 °C and 35 °C on the host surface, and 20 minutes of humidity to help the disease (Peltier and Frederich, 1926; Ramakrishnan, 1954). Citrus leaf-miners *(Phyllocnistis cilre/fa* and *Throscoryssa citri)* are reported to help in the spread and infection of bacteria (Mundkur, 1961).

According to Naik (1949), sweet orange, acid limes, and grapefruit are highly susceptible to citrus canker while Nepali is highly resistant to oblong, round, and seedless lemons. Mundkur (1961) noted that infections found on sweet orange and pummolo are highly susceptible to citrus cancer except for Jamberi, sour orange, and Kaghzi lime. Kishore and Chand (1975) recorded that total phenols more in resistant C. *reticulata* than in susceptible C. *aurantifolia*. Citrus cancer is extremely susceptible to grapefruit, Mexican limes, and orange trifoliate; sour orange, lemon, and sweet orange are moderately susceptible; and mandarins are moderately resistant. Cultivars that mature early are more susceptible than cultivars of the mid-season which are more susceptible than cultivars of the late season.

The Asiatic canker (canker A) caused by *X. axonopodis* pv. *citri* are most prevalent and severe form of the disease. This group of *X. axonopodis* pv. *citri* strains that causes the Asiatic citrus canker. Minor genetic variation of this canker strain was found in canker A strains in Florida and other world-wide citrus-growing regions. Citrus cancrosis B caused by *X. axonopodis* pv. *aurantifolii* is a disease of lemons in Argentina, Paraguay, and Uruguay, but Mexican lime, sour orange are also susceptible. Group of strains of Cancrosis B can be easily differentiated from strains of the canker A. Citrus cancrosis C, caused by *X. axonopodis* pv. *aurantifolii*, isolated from Mexican lime in Brazil. Many known host for this bacterium is Sour orange. Citrus canker community of this type cannot easily be differentiated from group B strains of cancrosis. Diagnostic tests may use PCR samples to check this type of cancer strain including monoclonal antibodies, fatty-acid profiling, and genetics (Schubert et al., 2001). Isolates of Cancrosis B can be differentiated serologically from Canker A but not from Cancrosis C.

Citrus canker disease is caused by different types of pathovars and variants of the bacterium *X. axonopodis*. Symptoms are usually distinguished from one another by these types based on host selection, cultural, and physiological characteristics, sensitivity to bacteriophages (Civerolo, 1984), serology (Alvarez et al., 1991), DNA-DNA homology (Egel et al., 1991), plasmid fingerprints (Pruvost et al., 1992) by various RFLP and PCR analyses (Hartung and Civerolo, 1989; Hartung, 1992., Hartung et al., 1993; Hartung et al., 1996; Vernière et al., 1998, Cubero and Graham, 2002). Such strains are distinct genetically and pathologically, based on DNA assays. PCR methods have established a rapid and reliable detection of isolated bacteria in leaves and fruit culture and lesion extracts (Cubero et al., 2001; Hartung et al., 1993). The primers used for diagnosis of citrus canker are based on plasmid containing the *pth A* gene, primary virulence part in all citrus canker strains (Yang and Gabriel, 1995; Hartung et al., 1996). *PthA*-based primers are available to detect all citrus cancer strains in Florida (Cubero and Graham, 2002), but A-strain-specific monoclonal antibodies have failed to identify A strain variant Aw (Wellington strain), newly discovered in Palm Beach County, Florida and able to cause cancer symptoms on a small range of citrus hosts including Mexican lime and alemow (C. *macrophylla)* (Taka-hishi and Doke, 1984). The result is a universal primary for Xac detection and recognition for unique sequences in the bacterium's chromosome. PCR primers are dependent sequence variations of 16S and 23S ribosomal DNAs in intergenic spacer regions specifically designed to differentiate strains and their variants. This set of primers promptly differentiates all A strains from the B and C strains of *X. axonopodis* pv. *aurantifolia*. Rep-PCR with BOX and ERIC primers is used within the same pathotype for different forms and strains of citrus cankers (Louws et al.; 1999 Cubero and Graham, 2002). This methodology can assess the diversity of *Xanthomonas* strains in Florida that cause citrus canker and worldwide collection strains and their possible geographic origin (Cubero and Graham, 2002).

Disease Cycle:

Citrus canker bacterium is propagated in leaf, stem, and fruit lesions. Once free moisture is present on the lesions, other plants ooze out and spread for new growth. Rain water obtained with lesion on foliage produces bacteria (Goto, 1962; Stall et al., 1980). Wounds are made of thorns, Asian leafminer, and blowing sand on plant pieces (Timmer, 2000). Serpentine mines under the leaf cuticle are caused by the Asian citrus leafminer larvae (Heppner, 1993) and cause profuse injury to new growth and increase infection with citrus cankers (Cook, 1988., Rodrigues et al., 1998., Sohi and Sandhu, 1968; Sinha et al., 1972). Inoculums associated with water blockage have demonstrated how one to two bacterial cells induce infection and lesion formation through stomatal openings (Gottwald and Graham, 1992; Graham et al., 1992a, 1992b). Windblown inoculums are detected up to 32 m from infected trees in Argentina (Stall et al., 1982). Although, in Florida, confirmation for longer dispersals (7 miles) when severe rainstorms and tropical storms (Gottwald et al., 1992; Gottwald et al., 2001). Multiplication of bacteria occurs due to the continuous increase in lesions and the amount of bacteria generated per lesion depends on the susceptibility of the host. The bacterium lives in the margins of the leaves and fruit lesions until it decomposes. Bacterial death is increased by direct exposure to sunlight. The exposed bacteria only live in the soil for a few days and in the plant for a few months. The bacteria will reside in polluted tissues which have kept them dry and soil free for years (Goto, 1992).

All the pieces of citrus above ground are susceptible to *X. axonopodis* pv. *citri* and optimum resistance during the half stage of growth expansion (Gottwald and Graham, 1992; Stall et al., 1980). During rainy weather, bacterial cells emanate from open lesions to provide inoculum for further disease growth (Timmer et al., 1996). At optimal conditions, the initial symptoms on the leaves appear low, somewhat elevated blister-like lesions about 7 days after inoculation. Favorable host-infection temperature varies from 20 °C to 30 °C (Koizumi, 1985). Symptoms can take more than 60 days to show up below the optimum conditions of infection and incubation (Gottwald and Graham, 1992; Loucks, 1934). At first, we turn white and orange-to-brown light; the water-soaked margin appears, often surrounded by a chlorotic halo. The lesion's middle height rise and is spongy/corky. Such elevated stomatal infection lesions are usually evident on both sides of a leaf. Defoliation occurs by intensifying the disease on a plant (Goto and Yaguchi, 1979; Gottwald et al., 1988, 1989).

Symptoms of citrus canker on twigs and fruits are similar elevated corky lesions surrounded by water-soaked margins (Timmer et al., 1991). No chlorosis surrounds twig lesions but can occur on fruit lesions. Twig lesions on young shoots effect by *X. axonopodis* pv. *citri,* where citrus canker is endemic. Twig dieback and early falling fruit are major later-stage impacts of the disease. Complete *X. axonopodis* pv. *citri*, is most theoretically capable of infecting green tissues when they are at half the growth stages of development. As the leaves, twigs, and fruit mature, a thickened cuticule develops, and they start to harden physiologically and become more resistant to infection (Stall and Seymour 1983). From one growing season to the next, bacteria simply survive, often lesions are produced during the growing season. In host cells near the lesion, bacteria can remain viable although the bacterial may drop significantly (Timmer et al., 2000, 1996; Vernière et al., 1998). Bacteria that have been viable for many years in stem lesions and can survive under natural conditions on non-host plant material for several weeks (Pereira et al., 1976). The bacterial population declines to an undetectable amount within 1 to 2 months due to antagonism and competition with saprophytic microorganisms (Hartung, 1992; Leite, 1990). The proliferation of extracellular slime layer polysaccharides encapsulates the survival bacterial cells (Goto and Hyodo, 1985).

Transmission of the disease by Asian citrus leafminer, feeding on the epidermal cell layer forms galleries under the cuticle. The cuticula cracks can lead to mesophyll tissue contact with bacterial infection and cause massive lesions. The combination of bacteria and leafminer leads to even extremely resistant cultivars developing major field infections. Asian citrus leafminer in Florida, causing wounding by feeding the larval, is now vulnerable to all citrus cultivars. Recently, Kalita et al. (1997) reported that goat weed (*Ageratum conyzoides* L.) in India, common in citrus orchards in India serves as a host of *X. axonopodis* pv. *citri.*

Epidemiology:

Citrus canker disease transmitted by *Xanthomonas axonopodis* pv. *citri* (*Xac*) in the Nurseries of Citrus and commercial plantations. Spreading of the citrus nursery is mostly through splash dispersal (Serizawa and Inoue, 1974; Gottwald et al., 1989). The development of multiple foci that gradually coalesce and form wider areas of the disease are irregularly formed, making it difficult to narrate and measure the illness. Disease gradients are associated with citrus canker in nurseries and disease-induced defoliation in nursery plants and subsequent infection of newly developed leaves (Gottwald et al., 1989); Citrus canker-infected trees, dispersal of bacteria with splash, decreased the secondary foci coalesced (Gottwald et al., 1989). When the leaves, stems, and fruit mature, they become resistant to infection, unless they are injured. Infections often occur on the leaves and stems within the first 6 weeks after development begin. The most dangerous time for fruit skin infection is first 90 days after the petals have fallen. After this time when infection occurs, the pustules form small and inconspicuous. The fruits are vulnerable to longer periods compared to leaves and get more than one dispersal infection and different-age lesions on the same fruit. Fruits act as an indicator for infection with the citrus canker.

Citrus canker spread within trees and adjacent trees for short distances by wind and rain. Spread over longer distances results from severe weather events like storms, hurricanes, and tornadoes. Hurricanes and storms significantly increase contamination with the citrus canker and disperse the bacteria over several miles. The movement of diseased propagating material, such as bud wood, rootstock, seedlings, and budded trees, frequently spread over long distances. Nursery workers can move bacteria by hands, clothing, and equipment from one nursery to another. Equipment for pruning and spraying also spread the disease throughout the orchards. Wooden boxes containing diseased fruit and leaves have also been mixed up in long distance spread to disease-free orchards. Outbreak first occurred in the Florida citrus orchards in the United States in 1995. Extreme rainstorm occurred in January 1996 in tornadoes that traveled from southwest to northeast across this polluted area (Gottwald et al., 1997a, 1997b, 2001). These storms are very common at the Florida front boundary and diagonally across Florida with current winds along the front boundaries pushing moisture and winds to the northeast as well. It is spread alongside meteorologically irregular human movement suspected of being continuous migration of citrus canker north along the east coast of the Florida suburban region. The citrus leafminer was restricted primarily to Southeast Asia when distributed. The leafminer spread to most of the world's major citrus growing region, new to Florida in 1993 (Heppner, 1993), as first recorded back in 1996 in Brazil. In the Miami area Asian citrus leafminer infestation and citrus canker are identified immediately. With the flush of citrus the leafminer infests new leaves and new stems. Immature fruit and grapefruit in particular, is also sporadically attacked. Leafminer larvae form feeding galleries of young leaves and other parts of plant tissue in the epidermal cell layer, and then break cuticles (Achor et al., 1996; Graham et al., 1996). Leafminer's feeding practices promote infections by removing the cuticle, which exposes the mesophyll to direct bacterial contamination when splash-dispersed or wind-blown-rain-dispersed on the surface of the leaf and larvae infected with bacteria by feeding galleries. This results in multiple infections of mesophyll within galleries (Graham et al., 1996). Many leaf-miner-induced lesions develop; break through the epidermis, and coalesce to form huge infections covering large areas of the leaf lamina and significantly increase the area of infection. The bacterial inoculums also become contaminated where no leafminer is present (Graham et al., 1996). Huge quantities of inoculum leave from induced lesions, facilitating the bacteria's spread by rain splash and spreading the outbreak faster.

Citrus leaf minor wounds are very susceptible to bacterial infection. Leaf minor wounds are normal wounds, and delay in plant healing reaction, causing disease in lower doses of bacteria inoculums. In Florida the Asian citrus leafminer spread citrus canker faster. An epidemiological research carried out in the Western Hemisphere formerly of citrus canker to the presence of leafminer (Danos et al., 1982, 1984; Gottwald, et al., 1992a, 1992b) since the wounds serve as infection courts and the amount of inoculum produced in lesion increased significantly (Bergamin-Filho et al., 2000).

Integrated Disease Management:

Citrus canker cannot eradicate by quarantine measures to prevent the introduction and establishment of *Xanthomonas axonopodis* pv. *citri* (*Xac*). Only kindergartens can be as in zones free of citrus canker. Nursery production free of diseases, marketing of fresh fruit and nursery stocks, orchards omitting canker by management practices (Leite, 1990).

To maintain pathogen exclusion, Orchard staff carry disinfection of clothing, shoes, orchard machinery, and harvesting equipment. Bacteria on their bodies, shoes, boots, hand tools, picking bags, ladders etc. may be borne by humans. Apparatus such as tractors, sprayers, and hedgers are infected and parts of plants are often mistakenly transported. In areas where citrus canker is resident, sprayed with bactericidal compounds, the construction of decontamination stations for workers, vehicles, and machinery is important.

The most traditional method of suppression is the elimination of inoculums once inserted into a region by removing and killing infected and bare trees. Throughout Florida, trees have been rooted and burnt across urban areas, and legislation mandates all citrus trees within 579 m of infected trees removed in residential and commercial circumstances. Where illness is a major problem, it is necessary to incorporate appropriate cultural practices such as sanitation, windbreaks, and control of leafminers with frequent copper spray applications.

Citrus fruit is susceptible to canker during the first 90 days after petal dropping, during this time maintaining a protective coating of a copper content on the fruit surface. Two to three sprays required for this purpose, and depending on susceptibility to rainfall and cultivar. Windbreaks can dramatically reduce the spread and severity of disease, and increase the efficacy of copper spray. For young trees and some cultivars which have flushes of vegetative growth, leafminer control is vital.

Cultural activities of pruning and defoliation of diseased plants in summer and autumn are important events for citrus canker management (Kuhara, 1978; Leite, 1990; Stall and Seymour, 1983). Windbreaks are the disease's very efficient measure of control of susceptible citrus cultivars (Gottwald and Timmer, 1995; Leite, 1990). In combination with copper sprays, windbreaks can minimize the incidence of disease on leaves and fruits to unrecognizable citrus cultivar rates (Leite, 1990). The pruning and defoliation of diseased plants in combination with copper sprays has also been important. The pruning of the citrus trees is achieved during the dry season. Preventive sprays of copper-based bactericides treat citrus canker (Koizumi, 1977; Kuhara, 1978, Leite, 1990; Leite et al., 1987; Stall and Seymour, 1983) and reduce inoculums build up on new leaf flushes and fruit surfaces. Susceptible citrus cultivars spray for managing citrus leafminer in the young stages to prevent damage to the emerging leaf flushes predisposing to infection. A suitable level of resistance to the operation of Leaf miners enables successful disease management with the integrated system including windbreaks and chemical control (Leite and Mohan, 1984), where the disease is well known and serious, it may be recommended and profitable only for the more resistant forms of citrus such as Valencia oranges and mandarins. Though canker is prevalent, other cultural activities are employed to reduce disease severity. Hold away for work in polluted orchards when trees are damp and rainy, or decrease in dew or rain and wind, decreased by windbreaks in orchards and between rows. Wind speed reduction decreases and guides bacterial invasion of stomata, as well as the entry of vegetation and fruit injuries.

Copper sprays are necessary for an effective measurement of citrus canker with intermediate resistance levels (Leite, 1990), whereas the weather is highly conducive to the epidemic development of citrus canker, recommended by more than five sprays (Leite et al., 1987). Once *Xanthomonas campestris* pv. *citri* is introduced into the citrus-growing region, removal of inoculum and destruction of infected trees is the most widely accepted practice for further spreading (Stall et al., 1987). Eradication efforts were carried out in Florida between 1910 and 1933 and in the 1980s and 1990s, and trees in Florida and elsewhere were eradicated. Citrus canker is still not present in some countries where climate conditions are favorable for pathogen establishment, which are likely restrictions on the spreading material and fruit from the cancer zone. Reddy and Papa Rao (1960a, 1960b) Application of neem-cake foliage spray in citrus nurseries has been reported to reduce cancer infections. Sprinkling 1 kg of neem-cake in 20 liters of water every fortnight during the rainy season increases shooting and leaf production, and reduces cancer infection (Reddy and Papa Rao, 1960a, 1960b).

KEYWORDS

- **citrus canker**
- *Xanthomonas citri*
- symptomatology
- management

REFERENCES

Achor, D. S., Browning, H. W., and Albrigo, L. G. (1996). Anatomical and histological modification in citrus leaves caused by larval feeding of citrus leaf-miner (*Phyllocnistis citrella* Staint). Page 69 in: Proceedings of the International Citrus Leafminer, Orlando, Florida, April 23–25, 1996. University of Florida, Gainesville.

Alvarez, A. M., Benedict, A. A., Mizumoto, C. Y., Pollard, L. W., and Civerolo, E. L. (1991). Analysis of *Xanthomonas campestris* pv. *citri* and *X. c. citrumelo* with monoclonal antibodies. *Phytopathology.*, **81**: 857–865.

Bergamin-Filho, A., Amorim, L., Laranjeira, F., and Gottwald, T. R. (2000). Epidemiology of citrus canker in Brazil with and without the Asian citrus leafminer. (Abstr.) In: Proceedings of the International Citrus Canker Research Workshop, Ft. Pierce FL, June 20–22, 2000.

Online. Division of Plant Industry, Florida Department of Agriculture and Consumer Services.

Civerolo, E. L. (1984). Bacterial canker disease of citrus. *J. Rio Grande Valley Hort. Assoc.* **37**: 127–146.

Cook, A. A. (1988). Association of citrus canker pustules with leafminer tunnels in North Yemen. *Plant Dis.*, **72**: 546.

Cubero, J., and Graham, J. H. (2002). Genetic relationship among worldwide strains of *Xanthomonas* causing canker in citrus species and design of new primers for their identification by PCR. *Appl. Environ. Microbiol.*, **68**: 1257–1264.

Cubero, J., Graham, J. H., and Gottwald, T. R. (2001). Quantitative PCR method for diagnosis of citrus bacterial canker. *Appl. Environ. Microbiol.*, **67**: 2849–2852.

Danos, E., Berger, R. D., and Stall, R. E. (1984). Temporal and spatial spread of citrus canker within groves. *Phytopathology.*, **74**: 904–908.

Danos, E., Bonazzola, R., Berger, R. D., Stall, R. E., and Miller, J. W. (1982). Progress of citrus canker on some species and combinations in Argentina. *Proc. Fl. State Hort. Soc.*, **94**: 15–18.

Dopson, R. N. (1964). The eradication of citrus canker. *Pl. Dis. Rept.*, **48**: 30–31.

Egel, D. S., Graham, J. H., and Stall, R. E. 1991. Genomic relatedness of *Xanthomonas campestris* strains causing diseases of citrus. *Appl. Environ. Microbiol.*, **57**: 2724–2730.

Goto, M. (1962). Studies on citrus canker. *I. Bull. Fac. Agri. Shizuoka University*, **12**: 3–72. (In Japanese, English summary)

Goto, M. (1992). Citrus canker. Pages 250–269 in: Plant Diseases of International Importance. J. Kumar, H. S. Chaube, U. S. Singh, and A. N. Mukhopadhyay, eds. Prentice-Hall, Englewood Cliff, NJ, USA.

Goto, M., and Hyodo, H. (1985). Role of extracellular polysaccharides of *Xanthomonas campestris* pv. *citri* in the early stage of infection. *Ann. Phytopathol. Soc. Japan.*, **51**: 22–31.

Goto, M., Ohta, K., and Okabe, N. (1975). Studies on saprophytic survival of *Xanthomonas citri* (Hasse) Dowson: 1. Detection of the bacterium from a grass (*Zoysia japonica*). *Ann. Phytopath. Soc. Japan,* **41**: 9–14.

Goto, M., and Yaguchi, Y. (1979). Relationship between defoliation and disease severity in citrus canker (*Xanthomonas citri*). *Ann. Phytopathol. Soc. Japan.*, **45**: 689–694.

Gottwald, T. R., and Graham, J. H. (1992). A device for precise and nondisruptive stomatal inoculation of leaf tissue with bacterial pathogens. *Phytopathology.*, **82**: 930–935.

Gottwald, T. R., Graham, J. H., and Egel, D. S. (1992a). Analysis of foci of Asiatic citrus canker in a Florida citrus orchard. *Plant Dis.*, **76**: 389–396.

Gottwald, T. R., Graham, J. H., and Schubert, T. S. (1997a). An epidemiological analysis of the spread of citrus canker in urban Miami, Florida, and synergistic interaction with the Asian citrus leafminer. *Fruits*, **52**: 371–378.

Gottwald, T. R., Graham, J. H., and Schubert, T. S. (1997b). Citrus canker in urban Miami: An analysis of spread and prognosis for the future. *Citrus Industry*, **78**: 72–78.

Gottwald, T. R., Hughes, G., Graham, J. H., Sun, X., and Riley, T. (2001). The citrus canker epidemic in Florida: The scientific basis of regulatory eradication policy for an invasive species. *Phytopathology*, **91**: 30–34.

Gottwald, T. R., McGuire, R. G., and Garran, S. (1988). Asiatic citrus canker: spatial and temporal spread in simulated new planting situations in Argentina. *Phytopathology*, **78**: 739–745.

Gottwald, T. R., Reynolds, K. M., Campbell, C. L., and Timmer, L. W. (1992b). Spatial and spatiotemporal autocorrelation analysis of citrus canker epidemics in citrus nurseries and groves in Argentina. *Phytopathology*, **82**: 843–851.

Gottwald, T. R., and Timmer, L. W. (1995). The efficacy of windbreaks in reducing the spread of citrus canker caused by *Xanthomonas campestris* pv. *citri*. *Trop. Agric.*, **72**: 194–201.

Gottwald, T. R., Timmer, L. W., and McGuire, R. G. (1989). Analysis of disease progress of citrus canker in nurseries in Argentina. *Phytopathology*, **79**: 1276–1283.

Graham, J. H., Gottwald, T. R., Browning, H. S., and Achor, D. S. (1996). Citrus leafminer exacerbated the outbreak of Asiatic citrus canker in South Florida. Page 83 in: Proc. Int. Conf. Citrus Leafminer, held in Orlando, Florida, April 23–25, 1996. University of Florida, Gainesville.

Graham, J. H., Gottwald, T. R., Riley, T. D., and Achor, D. (1992a). Penetration through leaf stomata and strains of *Xanthomonas campestris* in citrus cultivars varying in susceptibility to bacterial diseases. *Phytopathology*, **82**: 1319–1325.

Graham, J. H., Gottwald, T. R. Riley, T. D., and Bruce. M. A. (1992b). Susceptibility of citrus fruit to bacterial spot and citrus canker. *Phytopathology.*, **82**: 452–457.

Hartung, J. S. (1992). Plasmid-based hybridisation probes for detection and identification of *Xanthomonas campestris* pv. *citri*. *Plant Dis.*, **76**: 889–893.

Hartung J. S., and Civerolo, E. L. (1989). Restriction fragment length polymorphims distinguish *Xanthomonas campestris* strains isolated from Florida citrus nurseries from *Xanthomonas campestris* pv. *citri*. Pages 503–508 in: Proc. 7th Int. Conf. Plant Path. Bacteria, Akademiai Kiado, Budapest, Hungary.

Hartung, J. S., Daniel, J. F., and Pruvost, O. P. (1993). Detection of *Xanthomonas campestris* pv. *citri* by the polymerase chain reaction. *Appl. Environ. Microbiol.*, **59**: 1143–1148.

Hartung, J. S., Pruvost, O. P., Villemot, I., and Alvarez, A. (1996). Rapid and sensitive colorimetric detection of *Xanthomonas axonopodis* pv. *citri* by immunocapture and nested-polymerase chain reaction. *Phytopathology*, **86**: 95–101.

Heppner, J. P. (1993). Citrus leafminer, *Phyllocnistis citrella*, in Florida. *Tropical Lepidoptera*, **4**: 49–64.

Hogg, D. R. (1985). Citrus canker in Argentina: A case history. Pages 8–10 in: Citrus Canker: An International Perspective. L. W. Timmer, ed. Citrus Research & Education Center, University of Florida, Lake Alfred.

Kalita, P., Bora, L. C., and Bhagabati, K. N. (1997). Goat weed: a host of citrus canker (*Xanthomonas campestris* pv. *citri*). *J. Mycol. Pl. Pathol.*, **27**: 96–97.

Kishore, V., and Chand, J. N. (1975). Resistance of citrus to citrus canker caused by *Xanthomonas* citri-Analysis of phenols and sugars. *Indian Phytopath.*, **28(1)**: 46–50.

Koizumi, M. (1977). Relation of temperature to the development of citrus canker lesions in the spring. *Proc. Int. Soc. Citriculture*, **3**: 924–928.

Koizumi, M. (1985). Citrus canker: The world situation. In *Citrus Canker: An International Perspective* (L. W. Timmer, ed.) Citrus Research & Education Center, University of Florida, Lake Alfred, pp. 2–7.

Kuhara, S. (1978). Present epidemic status and control of the citrus canker disease (*Xanthomonas citri* (Hasse) Dowson) in Japan. *Rev. Plant Prot. Res.*, **11**: 132–142.

Lee, H. A. (1920). Behavior of the citrus canker organism in the soil. *J. Agr. Res.* **19**: 189–206.

Leite, Jr., R. P. (1990). Citrus canker: prevention and control in the state of Parana. Fundacao IAPAR, Circular Instituto Agronomico do Paraná. No. 61.

Leite, Jr., R. P., and Mohan, S. K. (1984). Evaluation of citrus cultivars for resistance to canker caused by *Xanthomonas campestris* pv. *citri* (Hasse) Dye in the State of Paraná, Brazil. *Proc. Int. Soc. Citriculture*, **1**: 385–389.

Leite Jr., R. P., and Mohan, S. K. (1990). Integrated management of the citrus bacterial canker disease caused by *Xanthomonas campestris* pv. *citri* in the State of Paraná, Brazil. *Crop Protection*, **9**: 3–7.

Leite, Jr., R. P., Mohan, S. K., Pereira, A. L. G., and Campacci, C. A. (1987). Integrated control of citrus canker: Effect of genetic resistance and application of bactericides. *Fitopatologia-Brasileira*, **12**: 257–263.

Loucks, K. W. (1934). Citrus Canker and its Eradication in Florida. Unpublished manuscript in the archives of the Florida Department of Agriculture, Division of Plant Industry, Gainesville.

Louws, F. J., Rademaker, J. L. W., and de Bruijn, F. J. (1999). The three D's of PCR-based genomic analysis of phytobacteria: Diversity, detection, and disease diagnosis. Ann. Rev. *Phytopathology*, **37**: 81–125.

Mundkur, B. B. (1961). *Fungi and Plant Diseases*, p. 246. Macmillan and Co. Ltd., New York.

Naik, K. C. (1949). *South Indian Fruits and Their Culture*, p. 335. P. Varadachary and Co., Madras.

Peltier, G. L., and Frederich, W. J. (1926). Effects of weather on the world distribution and prevalence of citrus canker and citrus scab. *J. Agric. Res.*, **32**: 47–64.

Pereira, A. L., Watanabe, K., Zagato, A. G., and Cianciulli, P. L. (1976). Survival of *Xanthomonas citri* (Hasse) Dowson [the causal agent of citrus canker] on sourgrass (*Trichachne insularis* (L.) Nees) from eradicated orchards in the State of Sao Paulo, Brazil. *Biologico.*, **42**: 217–221.

Pereira, A. L., Watanabe, K., Zagatto, A. G., and Cianciulli, P. L. (1978). Survival of *Xanthomonas citri* (Hasse) Dowson, the causal agent of "citrus canker" in the rhizosphere of guineagrass (*Panicum maximum* Jacq.). *Biologico*, **44**: 135–138.

Pruvost, O., Hartung, J. S., Civerolo, E. L., Dubois, C., and Perrier, X. (1992). Plasmid DNA fingerprints distinguish pathotypes of *Xanthomonas campestris* pv. *citri*, the causal agent of citrus bacterial canker disease. *Phytopathology*, **82**: 485–490.

Ramakrishnan, T. S. (1954). *Common Diseases of Citrus in Madras State*, pp 112. Government of Madras Publication: Chennai, India.

Reddy, G. S., and Papa Rao, A. (1960a). Control of canker in citrus nurseries. *Andhra agric. J.*, **7(3)**: 11–13

Reddy, G. S., and Papa Rao, A. (1960b). Problem of sweet orange decline in Andhra Pradesh. *Andhra agric. J.*, **7(5)**: 175–188.

Rodrigues, J. C. V., Rossetti, V., Machado, M. A., Sobrinho, J. T., and de Lima-Nogueira, N. (1998). Citrus leafminer: A factor for increase of pests and citrus canker. *Laranja*, **19**: 49–60.

Schubert, T. S., Rizvi, S. A., Sun, X., Gottwald, T. R., Graham J. H., and Dixon, W. N. (2001). Meeting the challenge of eradicating citrus canker in Florida-Again. *Plant Dis.*, **85**: 340–356.

Serizawa, S., and Inoue, K. (1974). Studies on citrus canker, *Xanthomonas citri*: III. The influence of wind on the infection of citrus canker. Bull. Shizuoka Prefect. *Citrus Exp. Stn. Komagoe Shimizu City, Japan*, **11**: 54–67.

Sinha, M. K., Batra, R. C., and Uppal, D. K. (1972). Role of citrus leaf-miner (*Phyllocnistis citrella* Staintan (sic) on the prevalence and severity of citrus canker [*Xanthomonas citri* (Hasse) Dowson]. *Madras Agr. J.*, **59**: 240–245.

Sohi, G. S., and Sandhu, M. S. (1968). Relationship between citrus leafminer (*Phyllocnistis citrella* Stainton) injury and citrus canker [(*Xanthomonas citri* (Hasse) Dowson] incidence on citrus leaves. *J. Res. Punjab Agric. University (Ludhiana)*, **5**: 66–69.

Stall, R. E., Civerolo, E. L., Ducharme, E. P., Krass, C. J., Poe, S. R., Miller, J. W., and Schoulties, C. L. (1987). Management of citrus canker by eradication of *Xanthomonas campestris* pv. *citri*. Pages 900–905 in: Plant Pathogenic Bacteria, Current Plant Science and Biotechnology in Agriculture. E. L. Civerolo, A Collmer, R. E. Davis, and A. G. Gillaspie, eds. Martinus Nijhoff Publishers, Dordrecht, The Netherlands.

Stall, R. E., Marcó, G. M., and Canteros de Echenique, B. I. (1982). Importance of mesophyll in mature-leaf resistance to cancrosis of citrus. *Phytopathology.*, **72**: 1097–1100.

Stall, R. E., Miller, J. W., Marco, G. M., and Canteros de Echenique, B. I. (1980). Population dynamics of *Xanthomonas citri* causing cancrosis of citrus in Argentina. *Proc. Fla. Hort. Soc.*, **93**: 10–14.

Stall, R. E., and Seymour, C. P. (1983). Canker: a threat to citrus in the Gulf Coast states. *Plant Dis.*, **67**: 581–585.

Takahishi, T., and Doke, N. (1984). A role of extracellular polysaccharides of *Xanthomonas campestris* pv. *citri* in bacterial adhesion to citrus leaf tissues in preinfectious stage. *Ann. Phytopath. Soc. Japan*, **50**: 565–573.

Timmer, L. W. (2000). Inoculum production and epiphytic survival of *Xanthomonas campestris* pv. *citri*. (Abstr.) In: Proceedings of the International Citrus Canker Research Workshop, Ft. Pierce FL, June 20–22, 2000. Online. Division of Plant Industry, Florida Department of Agriculture and Consumer Services.

Timmer, L. W., Garnsey, S. M., and Graham, J. H., eds. (2000). Compendium of Citrus Diseases, 2nd edition. American Phytopathological Society, St. Paul, MN, USA.

Timmer, L. W., Gottwald, T. R., and Zitko, S. E. (1991). Bacterial exudation from lesions of Asiatic citrus canker and citrus bacterial spot. *Plant Dis.*, **75**: 192–195.

Timmer, L. W., Zitko, S. E., and Gottwald, T. R. (1996). Population dynamics of *Xanthomonas campestris* pv. *citri* on symptomatic and asymptomatic citrus leaves under various environmental conditions. *Proc. Int. Soc. Citriculture*, **1**: 448–451.

Vernière, C., Hartung, J. S., Pruvost, O. P., Civerolo, E. L., Alvarez, A. M., Maestri, P., and Luisetti, J. (1998). Characterization of phenotypically distinct strains of *Xanthomonas axonopodis* pv. *citri* from Southwest Asia. *Europ. J. Plant Pathol.*, **104**: 477–487.

Yang, Y., and Gabriel, D. W. (1995). *Xanthomonas* avirulence/pathogenicity gene family encodes functional plant nuclear targeting signals. *Mol. Plant-Microbe Interact.*, **8**: 627–631.

CHAPTER 10

STATUS AND STRATEGIES FOR MANAGING QUICK DECLINE/ TRISTEZA DISEASE OF CITRUS IN INDIA

J. N. SRIVASTAVA[1,*], A. K. SINGH[2], RAKESH KUMAR SHARMA[3], and PRASHANT BAKSHI[3]

[1]Department of Plant Pathology, Bihar Agricultural University, Bhagalpur, Bihar, India

[2]Division of Plant Pathology, Sher-e-Kashmir University of Agriculture Sciences and Technology, Jammu, Jammu and Kashmir, India

[3]Division of Fruit Science, Sher-e-Kashmir University of Agriculture Sciences and Technology, Jammu, Jammu and Kashmir, India

*Corresponding author. E-mail: j.n.srivastva1971@gmail.com

ABSTRACT

Citrus tristeza virus (CTV) (genus *Closterovirus*, family Closteroviridae) is the causal agent of devastating epidemics that changed the course of the citrus industry. Adapted to replicate in phloem cells of a few species within the family Rutaceae and to transmission by a few aphid species, CTV and citrus probably coevolved for centuries at the site of origin of citrus plants. CTV dispersal to other regions and its interaction with new scion varieties and rootstock combinations resulted in three distinct syndromes named tristeza, stem pitting, and seedling yellows. The first, inciting decline of varieties propagated on sour orange, has forced the rebuilding of many citrus industries using tristeza-tolerant rootstocks. The second, inducing stunting, stem pitting and low bearing of some varieties, causes economic losses in an increasing number of countries. The third is usually observed by biological indexing, but rarely in the field. Measures to control CTV damage include quarantine and budwood certification programs, elimination of infected trees, use of tristeza-tolerant rootstocks, or cross-protection with mild isolates, depending on CTV incidence and on the virus strains and host varieties predominant in each region. A deep understanding of the interactions between viral proteins and host and vector factors will be necessary to develop reliable and sound control measures. The aim of the chapter is to focus and review on CTV causing economic losses in India and many other countries of Citrus grower as well as Citrus industry and management practices for *Citrus tristeza*/Citrus decline diseases.

10.1   INTRODUCTION

Citrus fruits (Mandarin, Sweet orange, acid lime, lemon etc.) are most important fruit crops in India, covering about 3.49 lakh ha. with total production of 27.58 lakh tones annually but very low productivity. Citrus fruits are grown all over India and Andhra Pradesh, Maharashtra, Punjab, Madhya Pradesh, Gujarat, Karnataka, Utter Pradesh, Bihar, and Assam are the leading citrus growing states. During past decades, citrus cultivation has been suffering due to several diseases. Quick decline disease of citrus is most devastating disease in India. In India, citrus decline first reported in Madhya Pradesh in 18th century. This disease is also known as Decline disease or die-back disease or Tristeza disease by its symptom and causal organism. The infected trees show general decline. Citrus decline in India refers to a particular syndrome known as die-back. Die-back of Citrus was recorded from India in 1912. Citrus quick decline disease or die-back disease was generally attributed to various soil and nutrient factors by earlier workers but now, it was found and proved by scientist, Citrus quick decline disease or die-back disease in Citrus in India caused by virus. This virus is *Citrus tristeza virus* (CTV) and it is a viral species of the *Closterovirus* genus. The name Tristeza was suggested in 1942 to describe the sad appearance of the diseased citrus trees. The Tristeza was first observed in South Africa in 1899 but it was suspected in Indian orchards by Lal (1953). The Tristeza was confirmed by other worker (Vasudeva and Capoor, 1958; Reddy and Papa Rao, 1960; Reddy and Govinda Rao, 1961; Bakshi and Dillon, 1964; and Nariani et al., 1965).

The types of symptoms produced by CTV depend upon scion cultivars, rootstock, and strain of virus. Since most of the citrus crop is propagated vegetatively the pathogens are consistently transmitted through planting material thereby causing heavy economic losses to the growers.

Another aspect of this disease is the various types of citrus plants are not affected equally by the CTV. Studied have shown that seedlings of sweet orange and mandarin are not damaged by CTV. It is the budded or grafted plants which are most susceptible to CTV. Thus, tristeza appears in most serious form when sweet oranges, mandarins, acid lime etc., are grown on sour orange rootstocks. The disease damages the conducting tissue of plant with the result the translocation of food material is affected and the plants produce symptoms of malnutrition. It involves the defoliation of young shoots and dying back of twigs from the tip downwards. Sieve tubes of tristeza-affected trees become necrotic and degenerated below the bud union resulting in decline of affected trees (Ahlawat, 2005). CTV infects nearly all Citrus species and relatives and hybrids.

The infected trees show general decline, leaves loose dark green luster, become dull chlorotic, exhibit curling lengthwise, and lead to pre mature defoliation. Twigs show die-back symptoms. Therefore, this disease is known for reducing of life expectancy, fruit size, quality, and yield (Vasudeva and Capoor, 1958; Reddy and Papa Rao, 1960). Citrus tristeza disease is cause of death, millions of *Citrus* trees all over the world besides that, other millions have lost their production capability. During the First World War, tristeza destroyed about 30 million trees. Estimates indicate that tristeza destroyed about a million trees in India (Ahlawat, 1997).

10.2   HISTORY OF CTV

The name "trustees" was coined by Moreira (1942) in Brazil, a Portuguese word meaning "Sad disease." Tristeza was first observed in South Africa as early as 1899 as a disease of sweet orange, tangerine, and grapefruit budded on sour orange root stock. Probably around this time, CTV was introduced into India from South African countries (Capoor, 1966).

Experimental evidence for the presence of tristeza disease in India was first obtained during 1955–1956 (Vasudeva and Capoor, 1958) and later confirmed by Nagpal (1954b) in Bombay state (Nagpal, 1959b). Thereafter, this disease has been reported from various parts of the country by several workers (Vasudeva and Capoor, 1958; Vasudeva, 1959; Reddy and Papa Rao, 1960; Nariani et al., 1965, 1966, 1970; Capoor and Rao, 1967).

The first serious epidemic causing the decline of large new plantings of sweet orange trees on sour orange root stocks was recorded in 1930 in Argentina, where it was considered a root Tristeza. There after the CTV has been found in most citrus growing areas. Tristeza virus was isolated from Citrus quick decline disease or die-back affected trees from different regions of India. From Delhi, Nariani et al. (1965) observed CTV in Malta cv. Jafna and Mosambi from Northern and central India, sweet orange cvs. Jafna, Mosambi, grapes fruit cv. Marsh seeded; Mandarin hill; Eureka lemon; Sweet lime, sweet orange (S.O.) cv. Vanile on florida rough lemon, S.O. cv. Pineapple on rough lemons; S.O. cv. Red blood on Jatti khatti, S.O. cv. Malta on Jatti khatti and S.O. cv. Mosambi on rough lemon CTV was isolated by Nariani et al. (1966). In Bihar and West Bengal CTV was found on S.O. Mosambi on Jambheri, Lemon seedling, mandarin seedlings and kagzi lime seedlings (Nariani and Raychaudhuri, 1968). In Assam and Orissa, CTV was found on seedling of Khasi mandarins, S.O. cv. Valencia; New Zealand grape fruit; Nagpur mandarin of citrus Karna (Nariani et al., 1970).

10.3　CAUSAL VIRUS

The disease is caused by CTV and it is a viral species of the *Closterovirus* genus. CTV is a highly flexuous rod virus with dimensions of 2000 nm long and 10–12 nm in diameter and present in phloem of affected trees (Nariani and Bhagabati, 1977). The existence of three strain of CTV namely mild, severe and seedling yellow (SY) have been reported (Balaraman and Ramakrishnan, 1978). The virus has been found to contain low band of coat protein. The major coat protein (CP-1) having the molecular weight 21000 Daltons presence in minor amounts and the smaller coat protein (CP-2), it has also a molecular weight 21,000 Daltons. The ratio of CP-1:CP-2 in purified virion is 5:1. Both react with polyclonal antibodies prepared against unfixed purified CTV virions (Cheema and Kang, 2000). The genome size of CTV is a largest among RNA viruses reported. CTV virus genome contains a single-stranded RNA with an estimated size of 6.5×10^6 or 20 kb. The CTV genome appears to be plus-stranded RNA (Bar-Joseph et al., 1989).

10.4　HOST RANGE

The CTV virus infected most of the *Citrus* spp, varieties, hybrids and some closely citrus relatives are the hosts. This virus induces a range of symptoms and severity in different citrus species and scion stock combinations (Ahlawat, 1987).

The virus has been found naturally occurring on the following hosts: Sweet orange (S.O.) mosambi on Jamberi, S.O. Valencia on jatti Katti, S.O. jaffa on Jatti Katti, S.O. vanile on Florida rough lemon, S.O. pineapple on rough lemon, S.O. mosambi on rough lemon, S.O. malta on Karnakhatta, S.O. blood red on jatti khatti. S.O. malia on Jatti Khatti; Nagpur mandarin on citrus karna, seedling trees of khasi mandarin, lemons, kagzi lime, mandarin seedlings, Eureka lemon, sweet lime, New Zealand grape fruit; Coorg mandarin etc. (Capoor, 1963; Nariani et al., 1965, 1970). On Artificial inoculation of virus, seedlings of sweet oranges (*Citrus sinensis*) Sour orange (*Citrus aurantium*) mandarin (*Citrus reticulata*), lemon (*Citrus lemon*), rough lemon (*Citrus jambhri*), Eureka lemon (*C. limon*), sweet lime (*Citrus limmettoides*), West Indian and Kagzi lime (*Citrus aurantifolia*) grape fruit (*Citrus paradisi*), Calamondin (*Citrus madurensis*), *Citrus amblycarpa, Citrus pennivisiculata, Aeglopsis chevalieri*, etc., are infected. Budlings of calamondin on rough lemon root stocks were killed by SY tristeza. Whereas nucellar seedlings of Calamondin remain very much dwarfed due to SY tristeza infection (Capoor, 1965). Some scion-stock, other than the members of the *Rutacease* family CTV infects certain species of Passiflora.

Among all the members of the Rutaceae family some plants, such as poncirus trifoliate orange and some of its hybrids with sweet orange of grape fruits (*Citrus paradise*), *Severinia buxifolia*, and *Swinglea glutinosa* are resistant to CTV (Garnsey et al., 1987). However, trifoliate orange was found to be immune against the isolates of tristeza obtained from Darjeeling and Sikkim Hills lime plantations.

10.5   SYMPTOMS

The types of symptoms produced by CTV depend upon different species of scion cultivars, rootstock, and strain of virus. So symptoms vary according to the different species of citrus. The original symptoms of the disease have been described as "reaction of sweet orange on sour orange root stocks" in the form of sudden wilting and drying of all leaves. Similarly mandarins, grape fruits, acid limes also exhibit decline and sudden wilting and drying of the affected plants grown on sour orange rootstock. The infected tree show general decline. The chief symptoms of the disease on susceptible rootstock and scion combination are the partial or complete suppression of new flushes of growth and appearance of various types of leaf discoloration. The leaves dull or slightly bronzed at first but later on may take various shades of yellow, the yellowing being more intense in the mid rib or lateral veins. The leaves curl lengthwise and upward. The disease damages the conducting tissue of plant with the result the translocation of food material is affected and the plants produce symptoms of malnutrition. It involves the defoliation of young shoots and dying back of twigs from the tip downward, resulting in loss of vigor, general health, and decreased fruit production. Affected trees blossom heavily in earlier stages of attack. Lastly trees with heavy crop collapse (wilting) suddenly (Vasudeva and Capoor, 1958; Nagpal 1959a; Reddy and Papa Rao, 1960). The affected trees also exhibit severe root injury.

No all CTV isolates induce visible decline and many seedlings trees of mandarins, sweet oranges another citrus sp, have been infected for many years without showing symptoms. Vein clearing symptoms can be observed in several CTV infected hosts under proper condition. The most characteristic symptoms are observed on acid lime (*C. aurantifloia*) which is also used as an indicator or diagnostic plant for CTV detection (Capoor, 1961). When seedlings of this are budded from tissues of an infected tree, some of the newly forming lime leaves develop fleck-like vein clearing which can be readily seen by transmitted light. The infected leaves are reduced in size, often cupped and chlorotic. In several cases infected leaves drop down and defoliated twigs die-back from tip.

Most strains of CTV also cause stem pitting (SP) and root pits (RP) on this plant. Some severe CTV isolates will cause SP in grape fruits, some sweet oranges, and mandarins. SP and RP is apparently the most important CTV problem in many areas and is often associated with other symptoms that include tree decline, reduced fruit size, and poor performance.

10.5.1   STEM PITTING

"Stem pitting" is a term used to describe depressions of different sizes and shapes in the wood of plants. In citrus pits consists of depressions in the outer wood with corresponding pegs of projections on the inner face of barks. In case of wood pitting induced by CTV, there are variation in numbers and shapes of pits dependent upon the strain of virus used and the host variety affected. In acid limes, pitting start appearing on small, green hardened shoots. At times pitting can be seen even without removing the bark. The severity of the SP, the size, and the number of pitting vary according to the isolate. Pegs on the inner surface of the bark match the pits in the wood. Some isolates of the virus cause severe stunting and almost a complete cessation of growth. In such plants, pitting is so extensive that the entire stem surface assumes a rogues appearance. It appears that some strains of tristeza virus induce more pitting than others. In India, some strains of CTV were found causing severe SP on Kagzi lime, mandarin, and sweet orange also.

10.5.2 ROOT PITTING

Similarly, SP the RPs were also observed in different species of citrus in glasshouse as well as field grown plants caused by tristeza virus (Sharma, 1987a). RPs were very prominent in field grown stunted acid lime plants. In experimentally inoculated plants visible RPs developed in about 4-month time. Severity of RPs was maximum on the main roots adjacent to collar region and tender roots of acid lime plants. In several cases when pits were numerous, they coalesced to form longer depressions. In tender roots, severe resulted in brown discoloration giving an appearance of brown root rot. A significant correlation was observed between root pitting index and height of plants in acid lime. Severe RPs were also observed in Seville lemon, lime karna, Adjamir, and grape fruit (Sharma, 1987a).

10.5.3 CORKY VEIN

Some strains of CTV induce corky eruptions in the main and lateral veins, usually on the upper surface of the leaves. This symptom cause considerable disfigurement of the seedlings. The leaves curl downwards, become leathery and brittle. In severe isolates, internodes of the stem and shots are short and multiple buds appear at the nodes, which give a bushy appearance to the plant.

10.5.4 SEEDLING YELLOWS

SY was first reported from Australia (Fraser, 1952). SY causes specific yellowing and stunting reaction on inoculated seedlings of Eureka lemon, sour orange and citron. This reaction was described as SY. Indication of the presence of the SY strain of tristeza disease in India was first obtained in

1955–1956 in some of the sweet orange trees grown on rough lemon root stock in orchards in Karnataka, Andhra Pradesh, and Maharashtra. Affected 12-year-old trees showed severe symptoms of decline and when indexed on seedling of C. lemon produced typical SY symptoms in the form of chlorotic patches, followed by vein yellowing on subsequently developed leaves after about 4 moths. Further growth of the affected seedling completely stopped, their leaves gradually turned yellow and after about 13 months of deterioration they collapsed suddenly. On bud inoculation similar reaction was exhibited by sour orange seedlings (Capoor, 1965).

10.5.5   SOURCES OF SY VIRUS

In Karnataka, mandarin seedlings are tolerant to CTV. Some trees, however, exhibit decline-leaf rolling, yellowing, twig decline, heavy leaf fall, and severe SP. Inoculums taken from such trees induce typical SY reaction in seedling of Eureka Lemon, sour orange, and grape fruit. Acid limes develop severe decline, deep SP, thick bark, and corking of the veins. Budlings of Calamondin (*C. madurensis Loureira*) on rough lemon root stocks were killed by SY. Whereas nucellar seedling of *calamondin* remain very much dwarfed due to SY infection.

Sweet orange cv. Mosambi growing on rough lemon root stock exhibiting severe decline was found to contain SY. Inoculums from these trees produce on seedlings of Rangpur Lime-Patchy chlorotic mottling of leaves; on rough lemon retardation of growth; on trifoliate orange-faint vein clearing and chlorosis of younger leaves.

A SY produces seedling yellow reaction in Eureka lemon, grape fruit and sour orange seedlings. Indian SY does not induce SP disease of grape fruit as reported from Argentina, Brazil, South Africa, and Australia (Capoor, 1965).

10.6   TRANSMISSION

Tristeza virus is transmitted by various types of graft inoculations and bud wood but through back grafting the virus was transmitted more frequently and with higher percentage of transmission but also transmission of CTV by the aphid species namely *Toxoptera citricidus* (Kirk), *Toxoptera aurantii* (Fonse), *Aphis spiraecola* (Patch), *Aphis gossypii* (Glover), *Aphis craccivora* (Koch), *Dactynotus jaceae* L. and *Myzus persicae* (Sulz) (Guttwald et al., 1994). Recently Sharma (1987b) has reported *Aphis fabae*, *Aphis nerii*, and

Acrythosiphum pisum to be additional new vectors of CTV. Among all the aphid vectors tried *T. citricida* has been found to be the most efficient vector (Sharma, 1987b).

CTV is also transmitted by three species of Cuscuta (Dodder), that is, *Cuscuta reflexa*, *Cuscuta subinclusa* and *Callicarpa americana* (Garnsey and Muller, 1986).

Besides aphid transmission, the following modes of transmission of CTV are as given below:

1. Tristeza virus is readily transmitted by grafting and budding from citrus to citrus.
2. There is no report of successful transmission of tristeza virus complex by sap-inoculation technique.
3. Several isolates of CTV, including SY isolates, can be transmitted by stem cut inoculation (Garnsey, 1984).

10.6.1 APHID TRANSMISSION

The mode of CTV transmission by aphids is semi persistent (Bar-Joseph et al., 1979). Balaraman and Ramakrishnan (1979) reported that aphids, *Toxoptera citricida*, *T. auranti, and A. gossypii* were able to transmit tristeza virus strains. A minimum of at least 15 viruliferous aphids of *T. citricida* per plant were required for 100% transmission with 24 hours each for acquisition and incubation feeding periods, and more than 100 aphids per plant reduced the incubation period of the virus in the host by 15–20 days. Capoor and Rao (1967) claimed 100% transmission with three aphids per plant with 5-min acquisition and 3-min transmission feeding periods. A single aphid is able to transmit the disease although with a low transmission rate of 10% (Capoor and Rao, 1967; Manjunath, 1985).

The aphid vectors have a tendency to pick up one or more strains from the mixture in source plant but both of them at one time. Capoor and Rao (1967) studied the behavior of 6 aphid species, that is, *Toxoptera citricda, A. gossypii, and A. craccivora* can pick up and transmit both mild and severe strains while *M. persicae, D. jaceae,* and *T. aurantii* can pick up and transmit only the mild strain. Whereas Sharma (Personal communication) has reported that *A. craccivora, A. gossypii, A. fabae, A. nerii, Acrythosiphum pisum, M. persicae*, and *T. citricida* transmitted both mild and severe strains of CTV. Among the above aphids tested *T. citridus* was the most efficient vector.

10.6.2   DODDER TRANSMISSION

Weathers and Harjung (1964) reported transmission of tristeza virus of citrus by dodder (*C. subinclusa*). Nariani and Raychaudhuri (1970) transmitted the tristeza virus by the parasitic dodder (*Cusucta refexa* Roxb.) from CTV infected Kagzi lime plants to healthy Kagzi lime plants. Virus transmission was noticed after 6-month time. Stem cut inoculation transmission: Several isolates of CTV including SY isolates are transmitted by stem cut inoculations. Many citrus varieties, species and citrus relatives could be infected by slash cut mechanically inoculation with CTV (Muller and Garnsey, 1984).

10.7   MANAGEMENT STRATEGIES

Virus and virus-like diseases can be effectively managed by adopting an integrated approach, that is, use of virus-free planting material (Ahlawat and Srivastava, 1997), host resistance, sanitization, cultural practices, managing the of insect-vectors and regulatory measures (Ahlawat, 2005). The virus-free planting material can easily be produced by adopting a biotechnological tools and it is have opened new dimensions for production of planting materials on larger scale (Ahlawat, 2000).

The spread of the virus takes place through the use of infected bud wood used for propagation. Once a tree gets infected with CTV there is practically no possibility to eliminate the virus. Therefore, the fresh planting should be done with virus free scions on tolerant root stocks like rough lemon and Cleopatra mandarin (Cheema et al., 1985). Virus free plants can now be obtained easily by shoot tip grafting (STG) (Hoa et al., 2004; Vijayakumari and Singh, 1999).

Citrus is a perennial crop and only healthy planting materials will not serve the purpose. Therefore, citrus orchard would require regular monitoring and testing to prevent the infection of CTV. Infected plants need to be quickly replaced by healthy plants. The disease can also be managed by adopting following measures:

10.7.1   ERADICATION

Eradication is a popular term and in practice the complete eradication of CTV and its strains may be extremely difficult program, it is most important to eradicate and destroy all the citrus plants infected with CTV or SY to avoid further spread of the virus. Once the virus infection has established in certain areas, it would be highly difficult and impossible to control the disease and its spread. Since Tristeza and their strains are transmitted by several aphid vectors, the spread may be very fast, if inoculum is present. Though the eradication is a difficult task, nevertheless it can be done by eradication programs either by citrus growers or by governments.

10.7.2   USE OF RESISTANT ROOTS TOCKS

Early workers have observed wide difference in reaction to CTV among citrus varieties and stionic combination. Soon in many countries where CTV has destroyed the citrus crop, citrus industry was re-established by using new root stocks which has shown resistance to tristeza decline.

However, many CTV resistant/tolerant root stocks were susceptible other citrus viruses and especially to citrus exocortis viroids. These root stocks were also found susceptible to citrus blight, a spreading disease of unknown etiology which is causing considerable damage to citrus crops in many countries of the world.

Many citrus species are quite tolerant when grown on their or when grafted on a tolerant root stock. Some of the clones of *Poncirus trifoliate* (L) and its hybrids, *S. buxifolia* and *S. glutinosa* are resistant to CTV.

In India, rough lemon (Jamberi) is the predominant root stock and yet citrus decline is rampant all over the country since 1950, perhaps due to synergistic effects of tristeza greening, psorosis and other viruses. Sweet orange on rough lemon are more prone to decline than mandarin on rough lemon (Phadnis, 1961). Since 1963, Rangpur lime is advocated as the best root stock for mandarin and sweet oranges. Experimental evidence at Indian Institute of Horticulture also, indicates that for acid lime, Rangpur lime is a very good root stock.

For the Punjab region, Jatti Khatti, Cleopatra mandarin and sweet orange, may be used as resistant root stocks. For Maharashtra, Andhra Pradesh and Karnataka, Rangpur lime is recommended as a root stock resistant to tristeza (Ahlawat and Raychaudhuri, 1982).

10.7.3   PROTECTION BY MILD VIRUS STRAINS (CROSS PROTECTION)

Preimmunization with mild virus strains (cross protection) has been tried in several countries for controlling citrus tristeza disease and showed satisfactory performance. Cross protection has been used widely to control the decline of sweet orange and limes in Brazil, grape fruits in Australia and South Africa, Hassaku Dwarf in Japan, etc. The performance of the protected tree was satisfactory and, with a new exception, no adverse effect has been reported (Bar-Joseph et al., 1989).

In India, Balaraman and Ramakrishnan (1977) isolated mild strains of CTV for use against severe strain to protect acid lime plants. Balaraman (1980) reported that the mild strains of tristeza virus completely protected against the severe strain when challenge inoculation was done 8 weeks after the protecting inoculation, confirming earlier report. Simultaneous inoculation of mild and severe strains caused a severe reaction, indicating the dominant nature of the severe strain. Observations recorded in cross-protection demonstration trial, in acid lime seedlings, budlings and also in preimmunized acid lime orchards in growers field indicated that growth parameters of the trees were statistically at par with each other in mild, mild + severe and uninoculated plants. In 10 years old acid lime plantations, there was no significant difference in the number of fruits in all the treatments. When the above plantations were 7 to 8 years old, fruit production was significantly less in severe strain inoculated plants. However, in subsequent years it did not differ from other treatments. Cross protection was a failure in Nilgiri district of Tamil Nadu due to intense severe strain infection pressure and cool environmental conditions.

It is well known that a single isolate of mild strain will not be able to protect against a range of severe strains of CTV present in the field of different species of citrus. Therefore, attempts are to be made to isolate mild strains from acid lime, sweet orange and mandarins and tested against a wide range of CTV strains isolated from the above hosts.

10.7.4   *PLANTING HEALTHY PLANTING MATERIAL*

By the use of certified healthy planting material many of the citrus virus/ viroid and mycoplasmal diseases can be avoided. Certified planting material requires legal regulation for the different steps of nursery operation and requirement for periodical indexing of the blocks of trees used at nurseries. Certification programs are always necessary to achieve an adequate control of disease.

For the multiplication of planting material, care should be taken to select outstanding field trees of the required citrus species and form them obtain virus/viroid /mycoplasma/greening pathogen free plants via heat and chemical therapy combined with micro STG. These plants are reindexed using any of the detection methods, that is, biological, ELISA, electron microscopy, nucleic acid probes to assure freedom from known pathogens in the new plants. Plants that are found healthy are to be propagated under protected cultivation to prevent reinfection of any of the pathogens. Form these healthy super plants, bud wood may be released for commercial production.

Citrus decline in India refers to a particular syndrome known as die-back. Citrus decline is characterized by defoliation of young shoots and die back of new twigs from tip to downwards, resulting in loss of plant vigor, general health, and decreased fruit production. The infected plants bears more number of fruits than healthy plants but size of the fruits are smaller than healthy one and not reached at proper maturity. The disorder is reported to have been present as early as the 18th century (Cheema and Bhat, 1929; Capoor, 1963) but is has assumed alarming proportions during the last century causing heavy losses and increasing concern to the growers. In the past, dieback has been attributed to several factors, including soil disorders, nutritional deficiencies, and parasitic agents such as fungi and greening pathogen (Raychaudhuri et al., 1969; Ahlawat and Raychaudhuri, 1998).

Citrus is infected with large number of viruses and virus-like pathogens worldwide and in India too (Ahlawat and Pant, 2003). Three decade studies on die-back of citrus revealed that a few virus and virus-like pathogens play the major role in citrus die-back or decline (Ahlawat, 1997) and these are briefly described in this Address.

10.8   FUTURE THRUSTS

Over the years, scientists discovered the pathogens infection citrus, some of them were new to Indian and a few of them were new to India and a few of them were new to the worlds. Diagnostic technology for most of the high risk pathogens have been developed which requires validation and availability to agencies involved in citriculture. Budwood certification programs in each citrus growing region should be mandatory. Biotechnological approaches to develop virus resistant material are the need of the day. Popularization and demonstration STG in citrus.

KEYWORDS

- *Citrus tristeza virus*
- symptomatology
- transmission
- management

REFERENCES

Ahlawat, Y. S. (1987). Virus and virus-like diseases of citrus, their detection and management. In: 5th Nat. Citurs Seminar, Jorhat. pp. 23.

Ahlawat, Y.S. (1997). Virus, greening bacterium and Viroids associated with citrus (Citrus species) decline in India. *Indian J. Agric. Sci.,* 67:51–57.

Ahlawat, Y. S. (2000). Induced resistance to viruses and pathogens in citrus. In: National Symposium on role of resistance in intensive agriculture, Feb. 15–17, 2000, DOWR, Karnal. pp. 21–22.

Ahlawat, Y. S. (2005). Virus and virus-like diseases affecting citrus in India – a step forward for Management of die-back complex. *Indian Phytopath.,* 58:257–268.

Ahlawat, Y. S., and Pant, R. P. (2003). Major virus and virus-like diseases of citrus in India, Their diagnosis and management. *Ann. Rev. Pl. Pathol.,* 2:447–474.

Ahlawat, Y. S., and Raychaudhuri, S. P. (1982). Virus and mycoplasma diseases of citrus in India. In: Problem citrus diseases in India, S.P. Raychaudhuri and Y.S. Ahlawat (Eds.), Surabhi Printers and Publishers, New Delhi. pp. 60–75.

Ahlawat, Y. S., and Raychaudhury, S. P. (1998), Status of citrus tristeza and dieback disease in India and their diagnosis in India and their diagnosis. In: Proc. 6th Int. Citrus Cong. pp. 871–879.

Ahlawat, Y. S., and Srivastava, K. P. (1997). Potential IPM tactics to grow virus-free citrus. In: Potential IPM tactics, D. Prasad and R. D. Gautam (Eds.), Westville. Publishing House, New Delhi, pp. 109–129

Bakshi, J. C., and Dhillon, J. S. (1964). A report on the decline of Sweet orange trees in arid irrigated region of Punjab. *Punjab Hortic. J.,* 4:15–22.

Balaraman, K. (1980). Interaction studies between a mild strain of tristeza on acid lime with mild and other virus or virus-like diseases of Citrus. In: Proc. of 8th International Organization of Citrus Virologists (IOCV). pp. 54–59.

Balaraman, K., and Ramakrishnan, K. (1977). Studies on strains and strain interaction in ctrus tristeza virus, *Technical Series Bulletin 19,* University of Agricultural Sciences, Bangalore. p. 62.

Balaraman, K., and Ramakrishnan, K. (1978). Cross protection of acid lime with mild strain of tristeza. *Indian J. Agric. Sci.,* 48:741–744.

Balaraman, K., and Ramakrishnan, K. (1979). Transmission studies with strain of citrus tristeza virus on acid lime. *Z. fur Pflkrankh. Pflschutz,* 86:653–661.

Bar-Joseph, M., Garnsey, S. M., and Gonsalves, D. (1979). The clostero viruses: a distinct group of elongated plant viruses. *Adv. Virus Res.,* 25:93–168.

Bar-Joseph, M., Ruth M., and Lee, R. F. (1989). The continuous challenge of citrus tristeza virus control. *Ann. Rev. Phytopath.,* 27:291–316.

Cheema, S. S., and Kang, S. S. (2000). Virus and phytoplasma diseases of Citrus. In: Diseases of fruit crops, V. K. Gupta and S. K. Sharma (Eds.), Kalyani Publisher, Ludhiana, pp. 169–181.

Capoor, S. P. (1961). Kagzi lime as indicator plant of Citrua decline virus in India. *Indian Phytopath.,* 14:109–112.

Capoor, S. P. (1963). Decline of citrus trees in India. In: T.S. Sadasivan (ed.) Symposium on plant and animal viruses. *Nat. Inst. Sci. India Bull.* pp. 48–64.

Capoor, S. P. (1965). Presence of seedling yellow complex in the citrus of South India. In: W. C. Price (ed.) Proc. of the Third Conference of International Organization of Citrus Virologist, pp. 30–35.

Capoor, S. P. (1966). Citrus viruses, their economic importance and mode of spread in India. Proc. of First Summer School in Plant Virology. 9th May–30th May, IARI, New Delhi, p. 80.

Capoor, S. P., and Rao, D. G. (1967). Trusteza virus infection of citrus in India. Proc. Symp. Sub-Trop and Trop. Hort., New Delhi, pp. 723–736.

Cheema, G. S., and Bhat, S. C. (1929). The dieback cause of citrus trees and its relation to the soil of western Indian. *Bull. Bombay Dept. Agric.,* 155:1–48.

Fraser, L. (1952). Seedling yellow, an unreported virus disease of citrusin Australia. *Agr. Gaz. N. S. Wales,* 63:125–131.

Garnsey, S. M. (1984). Separation of tristeza virus strain and strain variants by stem slash inoculation of citrus receptors (Abstract). *Phytopathology,* 74:848.

Garnsey, S. M., Barrett, H. C., and Hutchison, D. J. (1987). Identification of citrus tristeza virus resistance in citrus relatives and its potential application. *Phytophylactica,* 19:187–192.

Garnsey, S. M., and Muller, G. W. (1986). Efficiency of mechanical transmission of CTV. Proc. 10th Cong. Int. Organ. Citrus Virol. (IOCV) Riverside, California, pp. 46–54.

Guttwald, T. R. Garnsey, S. M., and Yokomi, R. K. (1994). Potential for spread of CTV and its vector, the brown citrus aphid. *Proc. Florida State Hortic. Soc.,* 106:85–94.

Hoa, N.V., Ahlwat, Y.S., and Pant, R.P. (2004). Production of virus free Kinnow mandarin and Mosambi sweeti orange nucleus planting material through shoot tip grafting *Indian Phytopath.,* 57:482–487.

Lal, T. B. (1953). Are your citrus orchard free from quick decline ? *Plant Proc. Bull.,* 5:91–94.

Manjunath, K.L. (1985). Studies on the distribution, transmission, strain and strain interaction of citrus tristeza virus. *Z. fur Pflkrankh und Pflschutz.,* 92:502–508.

Moreira, S. (1942). Observacoes sobre a "triatwza" dos citrus. *O. Biol.,* 12:285–287.

Muller, G. W., and Garnsey, S. M. (1984). Susceptibility of citrus varieties, species, citrus relatives and non-rutaceous plant to slash cut mechanical inoculation with citrus tristeza virus (CTV). Proc. 9th Conf. Organ. Citrus Virol., pp. 33–40.

Nagpal. R. L. (1959a). The cause of die-back disease of sweet orange (Musambi) in Bombay state and possible methods of control. *Sci. Cult.,* 24:478–480.

Nagpal. R. L. (1959b). Tristeza and other virus diseases of citrus found in Bombay state. *Citrus Ind.,* 40:14–15.

Naraiani, T. K., and Bhagabati, K. N. (1977). Purification and serology of citrus tristeza virus. Proc. Int. Sym Citriculture. Horticulture Society of India, Bangalore, pp. 47–48.

Naraiani, T. K., and Raychadhuri, S. P. (1970). Transmission of Citrus tristeza virus by dodder (*Cuscuta reflexa* Roxb.) *Ann. Phytopath. Soc. Japan,* 36:289–290.

Naraiani, T. K., Raychadhuri, S. P., and Bhalla, R. B. (1966). Citrus tristeza virus in northern and central India. *Indian Phytopathol.,* 19:397–399.

Nariani, T. K., Raychaudhury, S. P. and Sharma, B. C. (1968). Exocortis in Citrus in India. *Plant Disease Reporter*, 52:8341.

Naraiani, T. K., Raychadhuri, S. P., and Sharma, B. C. (1970). Citrus viruses in citrus Assam and Orissa. *Indian Phytopathol.,* 23:141–143.

Naraiani, T. K. and Raychadhuri, S. P. (1970). Transmission of *Citrus tristeza virus* by dodder (*Cuscuta reflexa* Roxb.) *Ann. Phytopath. Soc. Japan,* 36:289–290.

Naraiani, T. K. Sahambi, H. S., and Chona, B. L. (1965). Occurrence of tristeza virus in citrus in northen India. *Indian Phytopathol.,* 18:220.

Phadnis, N. A. (1961). Root stock trial with Nagpur mandarin orange (*C. reticulata* Blanco). Proc. 4th Horth. Res. Workers Conf. Poona.

Raychaudhuri, S. P., Nariani, T. K., and Lele, V. C. (1969). Citrus die-back problem in India. In: H.D. Chapman (Ed.). 1st Int. Symp. Vol. 3, Univ. Cal. Riverside, pp. 1433–1437.

Reddy, G. S., and Papa Rao (1960). Problems of sweet orange decline in Andhra Pradesh. *Andhra Agric. J.,* 7:175–178.

Reddy, G. S., and Rao, G. (1961). Is there tristeza in Andhra Pradesh. Proc. 2nd Conf. Int. Organ. Citrus Virologist. University of Florida Press, Gainesville, pp. 132–135.

Sharma, S. R. (1987a). Root pit in citrus- An additional symptoms of tristeza virus (Abstr.). Proc. 3rd Annual Convention, Indian Virological Society, Calcutta.

Sharma, S. R. (1987b). Additional vectors of tristeza virus (Abstr.). *Sci. Cult.,* 54:91–92.

Vijayakumari, N., and Singh, S. (1999). Invitro shoot tip grafting of citrus exotic germplasm. In: Proc. Internl. Symp. Citriculture. NRC for citrus, Nagpur, Nov. 23–27, 1999. Pp. 206–210.

Vasudeva, R. S. (1959). Plant virus research in India. *Indian Phytopathol.,* 12:1–7.

Vasudeva, R. S., and Capoor, S. P. (1958). Citrus decline in Bombay state. *Plant Pro. Bull. FAO,* 6:91.

Weathers, L. G., and Harjung, M. K. (1964). *Plant Dis. Reporter,* 48:102–103.

CHAPTER 11

DISEASES OF GRAPEVINE (*VITIS VINIFERA* L.) AND THEIR MANAGEMENT

S. T. INGLE[1,*], J. N. SRIVASTAVA,[2] and R. S. SHETE[1]

[1]*Department of Plant Pathology, Dr Panjabrao Deshmukh Krishi Vidyapeeth, Akola, Maharashtra, India*

[2]*Department of Plant Pathology, Bihar Agricultural University, Sabour, Bhagalpur, Bihar, India*

Corresponding author. E-mail: stingle_ngp@yahoo.co.in

ABSTRACT

Grapevine is attacked by many pathogens, which include nematodes, fungal, bacterial, and viral pathogens. Knowing the nature of these pathogens is essential to minimizing the losses. This chapter describes the major diseases of grapes and their management.

11.1 FUNGAL DISEASES

11.1.1 DOWNY MILDEW

Introduction and Economic Importance

Downy mildew is one of the serious diseases of grapevine and it is prevalent in many part of the world including India. The disease was originated from North America and introduced in Europe in 1875 and spread very rapidly through France and other European countries. In India, the disease is serious throughout South India, especially in coastal region and is a major factor in minimizing grape production. It can easily cause 50%–75% crop losses in one season (Pearson and Goheen, 1988a, 1988b, 1988c).

In the 1860s, an aphid-like pest was introduced to Europe from the United States causing the heavy infestation to the vineyards by the Phylloxera disease. Millardet and Jules Émile Planchon (1823–1888) suggest using American grape vines as grafting root stock which is resistant to Phylloxera, helps in controlling the infestation.

Prof. P. A Millardet (French botanist and mycologist), in October 1882, observed that mixture of copper sulfate, lime, and water sprinkled by farmer for protection of grapes from thieves also controlled the downy mildew. He made the improved concentration of mixture and suggested its application as a fungicide that was later known as "Bordeaux mixture." It was the first fungicide to be used in the world and is still an effective fungicide (Agrios, 2005a, 2005b, 2005c; Wilcox, 2006).

Symptoms:

Symptoms earlier appeared as small translucent, pale yellow spots with indefinite borders on the upper surface of leaves. A downy growth of the fungus also appears on the lower surface of leaves and it has directly under the spots. There are reddish lines in the tissue of the spot. Later, the infected portion gets killed and appears brown. Dirty grey growth appears on the lower surface of leaves. The irregular necrotic lesions in outline become large and coalesce, further forms larger necrotic areas and lastly defoliation of leaves. Diseased shoots become stunted. Infected parts of the plant are covered with whitish mycelial growth of the fungus. Flowers and berries are also affected with whitish mycelial growth of the fungus. Flowers may blight or rot after whitish mycelial growth of the fungus (Ellis et al., 2004).

During blossom or early fruiting stages, center clusters or part of them get attacked which quickly get covered with the whitish downy growth and eventually die. The fungus grows internally if infection takes place after the berries are half-grown. The berries develop a reddish marbling to brown coloration and become wrinkle and leathery. In severe attack, shedding of fruits occurs. The juice quality of fruit is hampered. Infection to the green young shoots, tendrils, stems, and fruit stalks cause stunting, distortion, and thickening of the tissues (Pearson and Goheen, 1988a, 1988b, 1988c; Wilcox, 2007).

Causal Organism: *Plasmopara viticola* (Berk. and Curt.) Berl. and de T.

Pathogen:

Mycelium of downy mildew fungus is hyaline, intercellular, coenocytic, and thin walled with spherical haustoria. Sporangiophores are formed from hyphae in the substomal spaces. Sporangiophores are 1 to 20 in number from each stoma. Sporangiophores may emerge directly through the cuticle. Sporangiophores can emerge through the lenticels in case of young berries of grapevine.

The sporangiophores are 300 to 500 µm long and 7 to 9 µm wide. Branching sporangiophores is almost at right angles to the main axis and at regular intervals. They are secondary branches arising from lower branches. From the apex of each branch, 2 to 3 sterigmata arise and bear sporangia singly. The sporangia of the fungus are thin walled, oval, or lemon-shaped. The zoospores of fungus are pear shaped and biflagellate. The oospores of fungus are thick walled (Ellis et al., 2004).

Disease Cycle:

The pathogen survives as oospores and dormant mycelium on the infected leaves and grapevines plant. The secondary infection of pathogen is through wind-borne sporangia and zoospores (Wilcox, 2007).

Epidemiology:

Warm and wet environmental condition is favorable for the development of disease. Disease development is favored during rainy season when there is heavy dew, relative humidity is above 80% to 100%, and temperature is between 23 °C and 27 °C but favorable temperature range for germination of sporangia is 10 °C to 23 °C (Agrios, 2005a, 2005b, 2005c).

Management:

- Sanitation in the field is important in terms of the management of the disease. Removing and burning of diseased plant parts which may contain hibernating oospores is useful in preventing the disease.
- Growing tolerant varieties such as Manjari Naveen, Arka Trishna, etc., and resistant varieties like Champa, Amber Queen, Champion, and Red Sultana.
- Pruning in April and May or September and October and burning of infected plant parts.
- After pruning, spray the vines with Bordeaux mixture 1% or Chloro-thalonil 0.2% and Difolatan 0.2%.

- When the flushes are in growth stage, spraying the Difolatan @0.2% or Chlorothalonil @0.2% or Metalaxyl @0.2% is effective with weekly intervals for the management of disease. When the nonsystemic fungicide is sprayed during humid and rainy period, spraying should be repeated for every two or three days.
- Three to five prophylactic sprays with B.M @1% or metalaxyl + mancozeb @0.3 to 0.4% or Fosetyl-Al (Aliette) @0.2% or Azoxystrobin or Dimethomorph can be effective for management of disease.
- Five chemical sprays in different stage, Grapevine crop with 1% Bordaux mixture.
  1. Instantly after the pruning of vines.
  2. When new flush formed in Grapevine after pruning (3–4 weeks after pruning).
  3. Before opening of buds.
  4. At the time of formation of bunches or berries.
  5. During growth of shoots.

11.1.2 POWDERY MILDEW

Introduction/Economic Importance:

This disease is devastating among the diseases of grapevine, distributed worldwide. The disease is widespread in different countries including India. The disease caused extensive damage in relatively dry areas. The disease was first reported from England and France. In India, the disease is major in central and South India, especially in Maharashtra, Gujarat, Karnataka, Tamil Nadu, Telangana, and is a major limiting factor in grape production. Downy mildew causes 50 to 75% crop losses in one season. About 40 to 60% yield losses in terms of fruits are caused due to this disease. Infected berries pose higher acid content comparison to healthy fruits which are unsuitable for wine making (Wilcox, 2003).

Symptoms:

The pathogen attacks on the grapevine in all the stages of the crop growth. The significant symptom of Powdery mildew disease is the appearance of white powdery mosses or patches on affected foliar parts of plant. Primarily, there are small, white patches on leaves as well as others foliar parts of plant which later become larger in size and powdery in appearance.

The stem becomes grey and later turns dark. Infected flowers get wither and dry up. Disease-affected plants remain dwarf and show wilted appearance. Whitish powdery growth is seen on the berries. If the pathogen-attacked fruits are near maturity, they become misshaped, crack, and only few among them ripen. Young fruits may not develop in early infection (Pearson and Goheen, 1988a, 1988b, 1988c; Wilcox, 2003).

FIGURE 11.1   Powdery mildew symptoms on leaves.

Causal Organism: *Uncinula necator* (Schw.) Burr. (syn. *Oidium tuckeri* Berk.).

Pathogen:

The mycelium adheres to the host surface by means of appressoria and superficial in nature. The hyphae are slender, branched, hyaline, and turn darker when conidia formation is completed. Conidiophores are simple, erect, and bear a chain of 3 to 4 conidia which are oval and ellipsoid to cylindrical. Cleistothecia are rarely formed. They are fixed in the mycelium and become are dark, globose with flattened top when fully mature (Ellis et al., 2004).

Disease Cycle:

Pathogen survives on the shoots and buds as dormant mycelium and as cleistothecia from season to season. The dispersion is done by the air-borne conidia (Rombough, 2002).

Epidemiology:

The disease is severe from October to November in North India and February to June in South India. Cloudy and warm weather is favorable for disease and decelerate by sunshine. Warm winter temperature from 20 °C to 33 °C is found to be suitable for epidemic. Disease development is adversely affected by rain (Pscheidt, 2007).

Management:

- Remove all diseased parts and burn it.
- Follow the sanitation practices for control of disease.
- Select resistant varieties like Skibba Red, Skibba White, Red sultana, Chholth Red, Chholth white, Saint George, and No. 1613, etc., for cultivation.
- Avoid the overcrowded growth of the vines.
- Dustings of vines with 300 mesh Sulphur (first when new shoots are 2 weeks old, second prior to blossoming, and third when the fruits are half ripe).
- Spraying of Wettable Sulfur @0.2% or Dinocap 0.25% or Carben-dazim @0.1% or Karathane or Calixin @0.1% at berry formation and development stage with a prophylactic treatment manner.
- Red sultana, Saint George, and No. 1613 are the highly resistant varieties.

11.1.3   ANTHRACNOSE/BIRD'S EYE SPOT

Introduction and Economic Importance:

Anthracnose of Grapevine disease also called "bird's-eye rot," is a mere important disease in regions with warm, humid, and rainy climates. The disease is first reported in India in 1903 and now prevalent in Haryana, Karnataka, Punjab, Rajasthan, Tamil Nadu, and Uttar Pradesh. Anthracnose reduces fruit quality and yield, as well as weakens the vine, hence, it is economically important (Pearson and Goheen, 1988a, 1988b, 1988c).

Symptoms:

The pathogen attacks on leaves, veins, stems, shoots, tendrils, and petioles of grapevine. Young shoots show numerous spots on them. These spots unite and girdle the stem, resulting into the death of the tips. These spots as small, irregular, dark brown color develop on petioles and leaves. The central tissue turns grey and falls off. Dark red spots appear on the berries. After sometime, these spots look circular, sunken, and ashy grey in color, and in late stages these spots get surrounded by a dark margin which gives the bird's eye appearance; hence, the disease is named bird's eye spot disease. The diameter of spots is about 7 mm but they may cover about half of the fruit (Rombough, 2002).

Causal Organism: *Elsinoe ampelina* Shear
(Syn. *Manginia ampelina* v. & p.).
Conidial stage is *Sphaceloma ampelinum* de Bary
[Syn. *Gloeosporium ampelophagum* (Pass.) Sacc.].

Pathogen:

Pathogen is formed of small and inconspicuous conidia and conidia are formed in pink Perithecia (pseudothecia). Ascospores are hyaline and asci are three-celled, globular, and 15–16 × 4–4.5 µm (Agrios, 2005a, 2005b, 2005c).

Disease Cycle:

Survival of pathogen as mycelium is in the cankers of the stem and on the infected twigs of the plant. Primary infection by mycelium and secondary infection occur through conidia which are carried by rain water and wind.

Epidemiology:

Disease severity is high during the months of July to August and November to December. New sprouts get infected more promptly in rainy season. Heavy rains after pruning are responsible for more incidences. In low lying and badly drained soils having warm wet weather, the disease incidence is comparatively high (Agrios, 2005a, 2005b, 2005c).

Management:

- Pruning and burning of diseased leaves and twigs.
- For effective control of this disease, take spray with Bordeaux mixture @1.0% or Carbendazim @0.1% or Mancozeb @0.25% or Difolatan @0.2% or copper oxychloride @0.3% with 10 to 15 days intervals.

- Grow-resistant varieties like Arka Trishna, Anab-e-shahi, Bangalore blue, Beauty seedless, Angur kalan, Bharat early, Delight, Golden Queen, Golden Muscat, Khalili, White Muscat, etc.

11.1.4   BLACK ROT

Introduction/Economic Importance:

The origin of this disease is in North America, and then it was introduced in France and different other European countries. Depending on the weather conditions, infective capacity of the inoculums and susceptibility of the host plants, disease can be cause loss to 50 to 80% of crops (Pearson and Goheen, 1988a, 1988b, 1988c; Wilcox, 2003).

Symptoms:

Black rot symptoms appeared on all parts of the vine but the most significant losses are seen by berry infection. If the control measures are not followed, susceptible varieties can experience complete yield loss in warm humid climates.

Black rot symptoms appeared on young leaves as small, circular spots which are tanned with a dark brown perimeter appearing in spring and early summer on young leaves. These spots appeared after about two weeks of the initial infection. After that formed small black fruiting bodies, pycnidia by the necrotic area in few days. Spots appeared on the petioles in spring and early summer. These spots may become large and girdle the entire petiole killing the leaf. Infected young shoots will be developing elongated black cankers throughout the growing season. If numerous cankers are formed, it will lead to blighting of the tips (RomBough, 2002).

Symptoms on berries firstly as the appearance of a small, whitish color dot which is immediately surrounded by a reddish, brown color ring. Within a day, this ring can enlarge 0.1 to 2 mm. In few days, the berry started drying, losing their spherical shape, and becoming flat on one side. These berries showed light or chocolate brown in color. At last, these infected berries got shriveled and became mummies that serve as a secondary inoculum for the pathogen (Pearson and Goheen, 1998a, 1988b, 1988c, 1988; Wilcox, 2003). Berries are likely to get infected before the bloom till four weeks after bloom.

Causal Organism: *Guignardia bidwelii*

Pathogen:

Black rot is caused by the ascomycete Guignardia bidwellii (Ellis) Viala and Ravaz (anamorph: Phyllosticta ampleicida (Englem.)). The principal characteristic of Phyllosticta species is the production of pycnidia containing aseptate, hyaline conidia that are usually covered by a mucoid layer and bear a single apical appendage.

Disease Cycle:

The pathogen survives on fallen mummies (Shriveled and diseased grapes) or in old shoots still on the vine. In the beginning of spring season, shortly after bud break (opening of buds), ascospore gets released. Ascospores needing water to germinate this process are accelerated by frequent rainfall. Ascospores are formed lesions on the leaves and inflorescences and young fruit. Fruit can become infected at any time from flowering period until change in color before harvesting. The fungus survives in the winter as pycnidia on the diseased plant residues (Ellis et al., 2004; Hartman and Hershman, 1988).

Epidemiology:

Suitable temperature for Pycnidia is between 5 °C and 35 °C and at 90 to 100% relative humidity, but more pycnidia are produced between 20 °C and 30 °C. At relative humidity <90%, no pycnidia is produced (Hartman and Hershman, 1988).

Management:

- Destroy the Mummies and bagging of fruit clusters.
- Spraying of Bordeaux mixture or any other suitable fungicides to manage the spread and development of the disease is effective.

11.1.5 BOTRYTIS BUNCH ROT/GRAY MOLD OF GRAPE

Introduction and Economic Importance:

Botrytis bunch rot is considered as miner significant disease problems of grapevine worldwide. *Botrytis* bunch rot is known as both, vulgar rot and noble rot when referring to its effect on wine. Both vulgar rot and noble rot are caused by *Botrytis cinerea*, and that type of rot is dictated by specific climatic changes. *Botrytis* bunch rot/gray mold of grape is responsible for yield loss, approximately 20% (Agrios, 2005a, 2005b, 2005c, Dharmad-hikari, 2007, Ellis et al., 2004, Genescope, 2002).

Symptoms:

Irregular necrotic spots appear on young as well as older leaves. In humid weather, these spots expand and coalesce. Infected flowers do not show any visible symptoms. Berries show characteristic symptoms as they become dark colored with typical grey mold symptom (Jackisch, 1985, Pscheidt, 2007).

Causal Organism: *Botrytis cinerea* Pers. fr.

Disease Cycle:

Sclerotia of the fungus are the major source of primary inoculum. They are found on dormant vines and on fallen berries (Agrios, 2005a, 2005b, 2005c, Rombough, 2002).

Management:

- Follow efficient cultural practices. It is most effective practice for control of *Botrytis* bunch rot.
- Sanitation and canopy management are major key approaches.
- Remove debris and shoots containing sclerotia.
- No highly effective fungicide has been recommended as fungus develops resistance.
- Fungus can be biologically controlled with the use of *Trichoderma harzianum*.

11.2   BACTERIAL DISEASES

11.2.1   BACTERIAL CANKER

Introduction and Economic Importance:

In India, the disease was first time reported in 1972 by Nayudu (1972). The disease causes about 60 to 80% loss in yield in severely infected vineyards.

Symptoms:

Symptoms are appeared as small water-soaked spots surrounded by yellowish color halo at lower surface of the leaves and also showed on vein infection. The spots enlarge in size and become dark brown in color and angular shape. Spots coalesce to form larger patches. After drying, infected leaves stay attached to the stem. Brown to black, elongated, and cankerous lesions are formed on petioles and canes. Stunting, cracking, and abnormal growth of canes are seen in advancement of disease. Berries show brown to black and cankerous lesions. Severely affected berries become small and shriveled (Araujo and Robbs, 2000; Chand and Kishun, 1990a; Chand and Kishun, 1990b).

Casual organism: *Xanthomonas campestris* pv. *viticola* (Nayudu.) Dye. Bacterium

Pathogen:

Bacterium is rod shaped with rounded ends, motile by single polar flagellum (monotrichous) and measures 0.4 to 1.2 × 2 to 3 μm and gram negative in nature.

Disease Cycle:

The alternate hosts are neem, mango, and *Phyllanthus maderaspatensis*. Survival of the bacterium is in the infected dry leaves up to 65 days. Secondary infection takes place through wind and splash of rain. Diseased cuttings may spread the disease to distant areas (Malavolta et al., 1999).

Epidemiology:

Favorable temperature range for disease development is 25 ° to 30 °C. Free water from dew, irrigation, or rain on leaves are responsible for pathogenesis (Chand and Kishun, 1990a).

Management:

- Removal and destruction of infected plant materials, regular inspection of vineyard, use of healthy and disease-free cuttings, and in-late October pruning are recommended for its management (Chand, 1996).
- Timely spray of copper containing chemicals and Bordeaux mixture. Effective rate of metallic copper is 4.5 kg/ha. The first spray on the opening of first female flower should be followed by subsequent sprays at 7 to 14 days interval according to weather conditions.
- Starting from two-leaf stage up to 70 days with an interval of 15 days spraying Streptocycline 300 ppm is also effective. Spray yellow

cuprous oxide (Yellow cuprocide) at the rate of 500 g in 380 L of water at pre-bloom stage (Chand, 1996).

11.2.2   CROWN GALL

Introduction/Economic Importance:

This disease is caused by the bacterium *Agrobacterium tumefaciens*. It is common where grapes are grown. The pathogen infects through injury to the lower trunk; hence, vines suffer more in cold winter temperatures due to winter injury which leads to gall development. Vines may carry the bacterium for many years without forming any galls if no trunk injury is present (Burr, 2004).

Symptoms:

Crown gall symptoms such as soft over growths or swelling appear on collar or root crown portion but occasionally on the trunk, branches, and lateral roots. Galls or tumors start increasing in size, due to the death of peripheral cells the surface becomes convoluted having dark brown or black color with roughened nature.

Casual Organism: *Agrobacterium tumefaciens*

Disease Cycle:

Perpetuation of the bacteria occurs in infested soil, where they can live as a saprophyte for several years. When susceptible host plants rise in such sites, the bacteria enter the crown region through the freshly wounds, either natural or caused by pruning, grafting, mechanical injury formed cultural practices, chewing insect, or the emergence of lateral roots.   Bacterium enters a wounded a small piece of its DNA is transferred in to the host trees DNA resulting in plant cell transformation (Burr, et al., 1998).

Epidemiology:

The dissemination of bacterium can also be done by splashing rain or irrigation water, tools, wind, insects, and part use for propagation.

Management:

- Use disease-free stock when growing a new vineyard.

- Infected vines should be removed from the site before replanting.
- Avoiding mechanical injury during cultural practices.
- Planting with cold hardy varieties because that are not susceptible to trunk splitting winter injury so that crown gall can be avoided.
- Crown gall can be controlled by surgery (in early stage of infection only) or by painting the galls with a chemical mixture consisting of 20% sodium dinitro cresylate (Elgetol) and 80% methyl alcohol (Methanol). Only 30% Elgetol is available, the mixture should consist of 1 part Elgetol to 6½ parts of Methanol (Martinson and Burr, 2012).

11.3 VIRAL DISEASES

11.3.1 FAN-LEAF DEGENERATION

Introduction/Economic Importance:

Fan-leaf degeneration/decline disease is one of the most severe viral disease complexes of grapevine worldwide. It is one of the oldest known viral diseases of grapevine firstly reported in Europe as early as 1841 (Hewitt et al., 1962). This disease is now known to affect grapevines in all temperate regions where grapevine and hybrid rootstocks are grown. About 10 to 80% crop losses are caused due to this disease (Martelli and Savino, 1990).

Symptoms:

The typical symptom of the disease is reduction in size of the leaves. The affected leaves turn light yellow transformed in cupping. The distance between internodes is reduced and internodes and nodes become thin and weak. Growth becomes more zig-zag in manner at the internodes but some lateral branches are produced. Infected plants do not produce flowers and fruits even at the age of three years. Plants become stunted in growth and produce very few and weak rootlets.

Affected young leaves showed variegated mottling manner. The malformed leaves have been open petiolar sinuses. Widening of the petiolar sinuses and reduction of areas between the veins gives the structure of a half-closed fan. Affected leaves stand upright along the axis of young shoots and become cup-like structure. The dark green areas in the mottled leaf bulge upward and leaf surface becomes rough (Demangeat et al., 2003; Hewitt et al., 1962).

Fruit clusters are reduced in size and number, and their ripening is irregular (Martelli and Savino, 1990).

Causal Organism: Grapevine fan leaf virus (GFLV)

Pathogen:

Grapevine fanleaf virus (GFLV) is nepovirus. Virus particles are isometric and 25 to 30 nm in diameter and thermal inactivation point is 60 °C to 65 °C with longevity in vitro is 15 to 30 days at 20 °C and the dilution end point is between 10^{-3} and 10^{-4}. The vectors are nematodes, *Xiphinema index,* and *X. italiae.* The virus is mechanically transmissible to *Cucumis sativus, Chenopodium amarantocolor, Gomphrena globosa, Nicotiana tabacumcv.* "White Burley," *Phaseolus vulgaris* cv. Prince (Dias, 1963).

Transmission:

The pathogen/disease is transmitted by grafting or budding of the infected scion onto the healthy rootstock. The pathogen/disease is not transmitted through pollen. The ectoparasitic nematode *Xiphinema index* is responsible for transmission of GFLV from grapevine to grapevine (Hewitt et al., 1958, Bovey et al., 1990).

Management:

- Soil application with nematicide is recommended for control of the nematode vectors and reduces the spread of the disease.

KEY WORDS

- **grapevine**
- **diseases**
- **symptomatology**
- **management**

REFERENCES

Agrios, G. N. (2005a). Plant Pathology, Fifth edition. Elsevier Academic Press, pp. 428–433.
Agrios, G. N. (2005b). Plant Pathology, Fifth Edition. Elsevier Academic Press, pp. 427–433.
Agrios, G. N. (2005c). Plant Pathology, Fifth Edition. Elsevier Academic Press, pp. 510–514.

Araujo, J. S. P. and Robbs, C. F. (2000). Symptomatology, pathogenicity and control of bacterial canker of grapevine (*Xanthomonas campestris* pv. viticola) in Brazil. Agronomia, 34(1/2): 83–86.

Bovey, R., Gärtel, W., Hewitt, W. B., Martelli, G. P. and Vuittenez, A. (1990). Soil-borne viruses transmitted by nematodes. In: Bovey R., Gärtel W., Hewitt W. B., Martelli G. P. and Vuittenez A. (eds.). Virus and virus-like diseases of grapevines, pp. 46–50. Editions Payot, Lausanne, Switzerland.

Burr, T. J. (2004). Grape Crown Gall Biology and Strategies for Control. Foundation Plant Services Grape Program Newsletter. October 2004. University of California-Davis, pp. 16–18.

Burr, T. J., Bazzi, C., Sule, S. and Otten, L. (1998). Crown gall of grape: Biology of Agrobacterium vitis and the development of disease control strategies. Plant Disease, 82: 1288–1297.

Chand, R. (1996). Grapevine bacterial canker and its management. In: Advances in Diseases of Fruit Crops in India, pp. 161–172.

Chand, R. and Kishun, R. (1990a). Effect of temperature on the growth of grapevine bacterial canker pathogen. Drakshavritta Souvenir, 6: 73–75.

Chand, R. and Kishun, R. (1990b). Outbreak of grapevine bacterial canker disease in India. Vitis, 29(3): 183–188.

Demangeat, G., Voisin, R., Minot, J. C., Bosselut, N., Fuchs, M. and Esmenjaud D. (2003). Survival of *Xiphinema index* and retention of Grapevine fanleaf virus in a nematode population from a naturally GFLV-infected vineyard. In: Proceedings of the 14th Meeting of the International Council for the Study of Virus and Virus-like Diseases of the Grapevine, Locorotondo, Italy, 2003 (http://www.agr.uniba.it/ICVG 2003).

Dharmadhikari, M. (2007). *Botrytis cinerea*in Winemaking, Iowa State University Extension,   http://www.extension.iastate.edu/NR/rdonlyres/173729E4-C734–486A-AD16–778678B3E1CF/56372/botrytiscinerea1.pdf

Dias H. F. (1963). Host range and properties of grapevine fanleaf and grapevine yellow mosaic viruses. Annals of Applied Biology, 51: 85–95.

Ellis, M., Doohan, D., Bordelon, B., Welty, C., Williams, R., Funt, R. and Brown, M. (2004a). Midwest Small Fruit Pest Management Handbook. The Ohio State University Extension. pp. 133–134. http://ohioline.osu.edu/b861/

Ellis, M., Doohan, D., Bordelon, B., Welty, C., Williams, R., Funt, R. and Brown, M. (2004b). Midwest Small Fruit Pest Management Handbook. The Ohio State University Extension. pp. 125–129. http://ohioline.osu.edu/b861/

Ellis, M., Doohan, D., Bordelon, B., Welty, C., Williams, R., Funt, R. and Brown, M. (2004c). Midwest Small Fruit Pest Management Handbook. The Ohio State University Extension. 123–125. http://ohioline.osu.edu/b861/

Genescope. (2002). *Botrytis cinerea* Estimated losses to vineyards in France, (Annual Report, UIPP, 2002).

Hartman, J. and Hershman, D. (1988). Black Rot of Grapes, College of Agriculture, University of Kentucky, http://www.ca.uky.edu/agc/pubs/ppa/ppa27/ppa27.htm

Hewitt, W. B., Goheen, A. C., Raski, D. J. and Gooding, G. V. (1962). Studies on virus diseases of the grapevine in California. Vitis, 3: 57–83.

Hewitt, W. B., Raski, D. J. and Goheen, A. C. (1958). Nematode vector of soil-borne fan leaf virus of grapevines. Phytopathology, 48: 586–595.

Jackisch, P. (1985). Modern Winemaking, Cornell University Press, pp. 173–174.

Malavolta, V. A. Jr., Almeida, I. M. G., Sugimosi, M. H. and Ribeiro, I. J. A. (1999). Occurrence of *Xanthomonas campestris* pv. viticola in grape in Brazil. Summa-Phytopathology, 25(3): 262–264.

Martelli, G. P. and Savino, V. (1990). Fanleaf degeneration. In: Pearson, R. C. and Goheen, A. (eds.). Compendium of Grape Diseases, pp. 48–49. APS Press, St. Paul, MN, USA.

Martinson, T. and Burr, T. J. (2012). Research Focus 2012–1: How close are we to Crown Gall-Free Nursery Stock? Appellation Cornell, Cornell University Viticulture and Enology Program, p. 6.

Nayudu, M. V. (1972). *Pseudomonas viticola*sp. nov., incitant of new bacterial disease of grapevine. Journal of Phytopathology, 73: 183–186.

Pearson, R. and Goheen, A. (1998a). Compendium of Grape Diseases, pp. 15–16.

Pearson, R. and Goheen, A. (1998b). Compendium of Grape Diseases, pp. 9–11.

Pearson, R. and Goheen, A. (1998c). Compendium of Grape Diseases, pp. 18–19.

Pearson, R. C. and Goheen, A. C. (1988). Compendium of Grape Diseases. St. Paul, MN, USA: APS Press, p. 93.

Pscheidt, J. (2007). Grape-Botrytis bunch rot, Oregon State University Extension, http://plant-disease.ippc.orst.edu/disease.cfm?RecordID=514.00000

Rombough, L. (2002). The Grape Grower, A Guide to Organic Viticulture, Chelsea Green Publishing, pp. 90–97.

Wilcox, W. (2003). Grape Disease Identification sheet, Black Rot, Cornell University Cooperative Extension, http://www.nysipm.cornell.edu/factsheets/grapes/diseases/grape_br.pdf

Wilcox, W. (2006). Grape Disease Control. Dept. of Plant Pathology, Cornell University, NY State Agric. Expt. Station, Geneva, NY, USA.

Wilcox, W. (2007). Grape Disease Control, Cornell University Cooperative Extension, pp. 4–9. http://blogs.cce.cornell.edu/grapes/files/2007/

CHAPTER 12

GUAVA (*PSIDIUM GUAJAVA*): KEY DISEASES AND THEIR MANAGEMENT

MANOJ KUMAR KALITA[1,*] and J. N. SRIVASTAVA[2]

[1]*Department of Plant Pathology, Biswanath College of Agriculture, Assam Agricultural University, Biswanath Chariali 784176, Assam, India*

[2]*Department of Plant Pathology, Bihar Agricultural University, Sabour 813210, Bhagalpur, Bihar, India*

Corresponding author. E-mail: manojpathoaau@gmail.com

ABSTRACT

Guava *(Psidium guajava* Linn.) an important fruit of subtropical countries is affected by about 177 pathogens of which 167 are fungal, 3 bacterial, 3 algal, 3 nematodes, and one epiphyte. Wilt of guava is the most important disease of guava. Besides this disease, fruit and postharvest diseases are also important which causes serious loss. The fruit diseases are of two types, that is, field diseases and postharvest diseases, which develop during transit and storage. In this chapter, important diseases are described with their symptoms, causal organisms, etiology, disease cycle, epidemiology, and disease management practices.

12.1 INTRODUCTION

Guava (*Psidium guajava*) thought to be originated in Mexico or Central America, is now cultivated in many countries of the world. In India also, a sizable area is under the cultivation of guava. The crop is attacked by many pathogens, besides it also suffered from physiological disorders and nutritional deficiencies. Misra (2004) reported that out of 177 pathogens attacking guava plants, the highest is fungi (167 numbers), three are bacteria, three are algae, three are nematodes, and one epiphyte. Among these, 91 attack fruits, 42 pathogens on foliage, 18 pathogens were recorded infecting twigs, 18 pathogens on roots, and a total of 17 fungi were isolated when the fruits surface wash was analyzed in vitro. Misra and Prakash (1990) recorded various diseases of guava like dry fruit rots, wet fruit rots, soft rots of fruits, sour fruit rots, anthracnose, ring rots, pink fruit, waxy fruit rots, canker, wilt, twig die back and drying, defoliation, various leaf spots, blightening of leaves, red rust due to algae, sooty mold, rust, seedling blight and damping-off, and so forth. Major diseases of guava are as mentioned below:

Sr. No.	Name of the Disease	Causal Organism
1.	Guava wilt	*Fusarium oxysporum* f. sp. *psidii*, *Fusarium solani*, *Macrophomina phaeseolina*, *Rhizoctonia bataticola*, *Cephlosporium* sp., *Gliocladium roseum*, and *Verticillium albo-atrum*
2.	Guava anthracnose	*Colletotrichum gloeosporioides*
3.	Fruit canker or scab	*Pestalotiopsis psidii* Pat.
4.	Algal leaf and fruit spot	*Cephaleuros virescens* Kuntze
5.	Guava rust	*Puccinia psidii* G. Winter
6.	Twig blight	*Phomopsis psidii* de Camara
7.	Stem canker and die back	*Hendersonula toruloidea* Nattrass
8.	Leaf blight	*Phoma jolyana* Priozy and Morg. *Alternaria alternata* (Fr.) Keissler
9.	Leaf spot	*Cercospora sawada* Yamamoto, *Pestalotia jodhpurensis* Bilgrami and Purohit *Discosia hiptage* Lib. *Curvularia siddiqui* Ahmed & Quraishi, *P. psidii* (Pat.) Mordue.
10.	Damping off	*Rhizoctonia solani* Kuhn.
11.	Root rot	*Clitocybe tabascens* (Scop.) Bres.
12	Storage rots	*Gloeosporium psidii, Pestalotia psidii, Phytophthora nicotianae* var. *parasitica* Dastur, *Botryodiplodia theobromae* Pat. *P. psidii* *F. solani* (Mart.)App. and Wollenw, *Fusarium moniliforme* var. *intermedium* *Phytophthora citricola* Sawada

12.2 WILT DISEASE

Introduction/Economic Importance:

Guava wilt is reported from almost all guava growing countries including India. Guava plants showing wilting symptoms were reported from Florida, USA for the first time (Webber, 1928). In 1935, it was reported from Babak-karpur (Allahabad) in India. Das Gupta and Rai in 1947 reported a very severe incidence of the disease in orchards of Lucknow. Likewise, Dey (1948) also reported its incidence from areas like Allahabad, Kanpur, and Lucknow. Prasad et al. (1952) made a detailed study on guava wilt and reported the spread of the disease in about 20,000 m^2 area in Uttar Pradesh. Singh and Lal (1953) from a detailed study in 12 districts of Uttar Pradesh reported that about 5%–15% loss was caused by guava wilt very year, amounting to almost 1 million rupees. In a similar way, Chattopadhyay and Sengupta (1955) reported that guava wilt reduced the yield of guava by 80% in West Bengal, that is, the yield of 113.5 q/ha in healthy plantations was reduced to about 18.16–22.7 q/ha in wilt affected orchards. Misra and Shukla (2002) reported that losses due to guava wilt varied from 5% to 60% around the Lucknow area in India. Since, chemical control of the disease proved to be unsatisfactory in field, the disease is a challenge for the coming millennium (Misra and Pandey, 1999b; Misra, 2006).

Symptoms:

At the initial infection, the plants become yellow, and leaves in the terminal branches curl slightly. Later on, they become reddish and subsequently leaves fall prematurely. No new leaves or flowers develop on the twigs and they dry up subsequently. The fruits in the affected plants do not grow further, mummify, and remain on the tree. In the dead wood bark, blisters having spores of white to salmon pink color develop. The infected plants become defoliated completely and die. In about 16 days the infected plants develop complete wilting, but it was also reported that some affected trees may live up to 252 days (Misra and Pandey, 2000b). There may be variations in the symptom development during a different time of the year (Misra and Pandey, 2000b). Wilt symptoms also as yellowing of the leaves with inter-venial chlorosis in the month of August followed by general drooping of the leaves in the month of September. In October, completely wilted plants with almost dried leaves were seen with small dried black fruits attached to the branch. They also observed partial wilting of the plants, a very common symptom of guava wilt. Wilting of variable degree is also observed in different months which may later escape or resist the wilting disease. These disease escaping plants start recovering from the month of December onward. Neeraj et al. (2005) and Gupta (2010) recorded that about 17% of the total infected guava plants with initial wilting symptoms, ultimately escaped or resisted wilting. The finer roots of the wilt infected plants show black streaks, which may be observed more prominently under the bark as reported by Das Gupta and Rai in 1947. Rotting of the basal region roots is also observed where the bark is easily detachable from the cortex. Vascular tissues become light brown. Splitting bark in the wilted plants occur at later stages. Although guava plant is susceptible to wilt in all the stages of plant growth, it is reported that aged plants are more susceptible to infection (Misra and Shukla, 2002).

Causal Organism: *F. oxysporum* f. sp. *psidii* and *Fusarium solani*

Pathogen:

Different species of the fungus *F. oxysporum* f. sp. *psidii* and *Fusarium solani* have been reported as pathogens of the disease (Das Gupta and Rai,1947; Dey, 1948; Prasad et al., 1952; Edward and Srivastava, 1957). At different times, various pathogens were also reported to be involved with the disease, namely, *M. phaeseolina*, *R. bataticola*, *Cephlosporium* sp., *G. roseum*, and *V. albo-atrum*, and so forth. *Macrophomina phaseoli* and *F. solani* were reported to cause the guava wilt either alone or by combined infection of both in West Bengal. In the case of *F. solani*, it enters the xylem vessels, then grows inside and blocks the vessels. *Macrophomina* also attacked the xylem vessels in a few cases as reported by Chattopadhyay and Sengupta (1955). *Gliocladium roseum* was reported as the pathogen by Misra and Pandey (1997, 2000a, b) as it reproduces symptoms of wilt when used as artificial inoculation. Pandit and Samajpati (2002) were of the opinion that *B. theobromae* was responsible for guava wilt in Midnapur (West Bengal). *Verticillium albo-atrum* was reported to cause guava wilt in Allahabad by Gupta et al. (2003). Out of various pathogens of guava wilt, those were reported at different times by different workers, *Fusarium oxysporum* f. sp. *psidii* and *F. solani* is the most important pathogen causing guava wilt (Misra and Gupta, 2007).

Disease Cycle:

Guava wilt pathogen is soil-borne in nature. Planting materials containing contaminated soil help it spread to new areas. During summer months, the fungus survives in association with root debris, while during the rainy and winter seasons it survives better on the roots (Dwivedi, 1990). Short distance spread occurs through water flowing from infected to healthy fields. The disease is more severe if the injury to the roots occurs during normal cultural operations and also by insects and nematodes.

Epidemiology:

Water stagnation in the guava field for a longer period of time facilitates the initiation of infection by the pathogen. Higher rainfall specially during July–September and high temperature, that is, 31.3 °C–33.5 °C (max. temperature) and 23 °C–25 °C (min. temperature) along with humidity of about 76% is favorable for severe disease.

Integrated Disease Management:

Cultural Management

Maintenance of proper crop sanitation in the orchard is the key to managing the disease. Digging of deep trenches around the infected tree trunk is helpful in reducing disease spread. The infected trees should be uprooted completely and destroyed by burning (Mathur, 1956). Plants should be manured timely and adequately, intercultural operations and irrigation should be done timely to maintain proper tree vigor, which will enable the plants in withstanding infection. Before transplanting the pits should be treated with formalin solution, keeping them covered for about 3 days to kill the pathogen. After pit treatment, transplanting can be done after 2 weeks (Edward, 1960). It is very much effective for controlling the wilt pathogens by soil solarization with transparent polyethylene sheet during May–June. Intercropping with turmeric or Marigold checks that the wilt pathogen is from causing the disease in guava by Misra et al. (2004). Guava varieties like Banarasi, Dholka, Sindh, Nasik white, and Luchnow49 were reported to possess resistance against the disease which may be utilized in managing the disease. Resistant root stock *Pisidium friedrichsthalianum* can be successfully utilized in managing the wilt disease of guava (Leu and Kao, 1979).

Management through Botanicals:

Botanicals can be an effective substitute for chemical pesticides for managing the guava wilt disease. Extracts of many botanicals such as Dathura, Isabgol and neem (Bhatnagar et al., 2004), *Achyranthes roses*, turmeric, marigold, and Cannabis (Gupta et al., 2007) proved to be very effective in managing *F. oxysporum*. Application of oil cakes such as kusum cake, mahua cake, neem cake, and so forth supplemented with 10 and 1 kg urea was found effective in checking the disease incidence.

Management through Chemicals:

From the results of different experiments conducted by many researchers in managing the wilt disease of guava, it is evident that different chemicals are effective in managing the disease. Severe pruning of guava plant, drenching with Bavistin or Benlate @0.2% solution four times in a year, and spraying with Metasystox and Zinc sulfate twice was reported to be effective in managing the disease, Suhag (1976). According to Misra and Pandey (1999b), although different fungicides are capable of checking various wilt pathogens effectively in the laboratory, but these pathogens increase their aggressiveness by the production of spore masses in large quantities in the soil, as soon as the effect of these fungicides starts diminishing in soil. When lime or gypsum was applied @1.82 kg/plant, a significant reduction of the disease was observed (Mathur et al., 1964). Likewise, the addition of 2 kg gypsum + 6 kg neem cake per tree was also reported to manage the disease (Misra and Pandey, 1994).

Management through Bio-Agents:

As the pathogen is soil borne, the chemical control of the disease often proved to be almost unpractical. Moreover, the application of chemicals is also hazardous for the soil, livestock, and also the environment. Due to the soil-borne nature of the pathogen, chemical application in managing the disease remains has also not been so effective for the last so many years (Misra and Pandey, 1999b). Hence, it is more desirable to utilize the bio-agents for the management of the disease. VAM symbiont @5 kg tree^{-1} was very effective in managing the guava wilt (Srivastava et al., 2001). Misra et al. (2000) reported that *Aspergillus niger* which is very fast growing was the most effective to manage wilt disease in the field. Besides, as growth enhancer, plants treated with *A. niger* grew faster having more height, higher thickness, and increased numbers of leaves. Misra and Gupta (2009) observed the effectiveness of *Trichoderma* and *Aspergillus* in the suppression of wilt incidence completely in standing crops in the orchard.

12.3 ANTHRACNOSE

Introduction/Economic Importance:

Anthracnose is one of the important diseases of guava which is most commonly observed affecting both pre- and postharvest stages of guava. Considerable postharvest losses are caused by the disease besides affecting the young developing flowers and fruits. It is very severe in the guava growing countries of the world having high rainfall and high humidity.

Symptoms:

Anthracnose is a major problem in guava-growing areas having high rainfall and humidity. In India, it is a serious disease in Uttaranchal, Uttar Pradesh, Punjab, and Karnataka. It is mainly a postharvest disease but also infects young developing flowers and fruits. Anthracnose symptoms are developed on all parts of the affected tree except the roots. Growing tips become dark brown in color, developing black necrotic areas which extend backward causing die back symptoms. As the first symptom of the disease, small, dark-colored sunken necrotic spots are developed on immature fruits in the field. The spots gradually enlarged up to 1–2 cm in diameter and due to the presence of black acervuli, the central portion of the spot becomes dark black in color. Numerous spots are developed which coalesce and lead to the eventual rotting of the affected fruit. In general, the infected portions of the fruits are harder rather than soft (Amusa et al., 2005). Under humid conditions, pinkish spore masses are produced by the fungus which covers the necrotic areas.

Causal Organism:

The fungus *C. gloeosporioides* is responsible for causing the anthracnose disease of guava. The fungus produces conidia which are unicellular, hyaline, cylindrical with obscure ends or ellipsoidal bearing a rounded apex having a narrow, truncate base. Hyaline to faintly brown and irregularly shaped measuring approximately 500 μm in diameter are produced in the acervuli. The setae are septate, having one to four septa, brown in color with slightly swollen base and tapered apex.

Disease Cycle:

The anthracnose fungus infects several crops (Grover and Bansal, 1970; Waller, 1992). The wide host range of the pathogen was reported by Wahid

(2001) infecting many fruit trees. According to Morton (1987), in the rainy season, guava fruits are infected by *Colletotrichum* to cause the disease. *Colletotrichum* penetrates the seed cavity which is evident by the presence of the pathogen in the guava seeds (Amusa et al., 2005). High humidity favors disease development and also guava orchards with closer planting and without canopy management experience higher disease incidence. Prevailing dew or rains highly encourage the production of spore and its dispersal around the tree canopy. The disease can appear in temperatures ranging from 10 °C to 35 °C, but the best temperature favoring the disease is from 24 °C to 28 °C.

Integrated Disease Management:

- Rotten and mummified fruits falling on the ground or those attached at the tree top must be collected and destroyed or buried deep into the soil to reduce the source of inoculums.
- Adequate provision for draining out of excess rain or irrigation water from the orchard should not be undertaken. Accumulation of water around the tree trunk should be discouraged.
- Dipping fruits in Aretan (50 ppm) solution for 2 min after harvest helps in protecting the fruits from the disease.
- Infected parts of shoots/branches should be removed by pruning and sprayed with 2:2:250 Bordeaux mixture or copper oxychloride @ 3 g/L of water. Spray should be repeated after 20 days of the fruit set and continued till maturity of the fruit.
- Gamma radiation can be applied to prevent postharvest decay due to *Colletotrichum*.

12.4   FRUIT CANKER OR SCAB DISEASE OF GUAVA

Introduction/Economic Importance:

The fruit canker disease is also named as fruit necrosis (Montiel, 1997) or fruit scabby canker disease. It is one of the most common diseases of guava prevalent in all the guava-growing areas across the globe. It occurs in several countries like India, Australia, Ecuador, Burma, Mozambique, Malaysia, and many other countries which attacks fruit at all stages of development (Mordue, 1969). The disease is responsible for causing losses in the field and also during postharvest storage.

It affects all the developmental stages of guava fruit. Fruit canker has the potential of drastically reducing fruit yield during the preharvest stage and also fruit losses during postharvest storage. In India, *P. psidii* was found to cause postharvest damage of ripe guava fruits.

Symptoms:

Symptoms are visible on leaves during nonfruiting seasons and also in the exocarp of young fruits of a pinhead size. The size of the canker increases as the fruits get larger. Typical symptom on fruit begins as tiny, water-soaked spots, slowly the spots darken in color and become necrotic. Gradually the tiny spots expand to discrete, circular, dark brown to black spots. As the fruit develops, the small, corky lesions often tear open, giving rise to raised, corky scabs. Fruits infected with the disease remain smaller due to underdevelopment, which are hard in texture, get malformed, mummified, and finally drop down. Small dark brown spots are developed on the leaves which expand to become gray or light brown circles and are surrounded by a dark brown border. Keith and Zee (2010) also reported the appearance of dark brown to black necrotic spots that develop into lesions having a corky appearance.

Causal Organism: *P. psidii*

Pathogen:

P. psidii is the fungus causing scab or fruit canker in guava. The fungus produces spindle-shaped or clavate spores. The spores are five-celled with three central colored cells and two hyaline cells. The upper hyaline cell bears two to three cellular appendages having a short hyaline pedicel below the lower hyaline cell. Spore germination is favored by a temperature around 30 °C with relative humidity above 96% at which maximum germination of spore occurs. Spores generally do not germinate below 15 °C or above 40 °C.

Integrated Disease Management:

- Rouging of diseased plant parts.
- Growing resistant varieties like Allahabad Sofeda or Apple color is effective.
- Storage of fruit at low and dry temperature helps in reducing the disease incidence.
- Spraying of extracts of neem or tulsi is effective in managing the disease (Radha and Methew, 2007)

- Spraying of 1% Bordeaux mixture or Zineb 75% WP @ 2.5 g/L water or Captafol 80% DS @ 2.5 g/L of water at 30 days interval during June–October.

12.5 GUAVA RUST

Introduction/Economic Importance:

This is one of the economically important diseases of guava affecting guava and other plants, belongs the Myrtaceae family (Anonymous, 1985). The disease was reported for the first time from Brazil in 1884 (Anonymous, 1985). Recently, Rayachhetry et al. (1997) has reported the disease on melaleuca in Florida. There is an unconfirmed report of occurring of this disease on *Eucalyptus* spp. in India (Anonymous, 1985).

Symptoms:

The infection of guava rust is evident by the development of conspicuous orange to reddish pustules on the foliage, young shoots, flowers, and fruits. The disease causes a reduction in the growth of the plant, distortion, and severe defoliation which ultimately kill the infected plants under severe disease condition (Burnett and Schubert, 1985).

Causal Organism: *Puccinia psidii* G. Winter

Pathogen:

Puccinia psidii produces *p*ale yellow amphigenous uredia of 0.1–0.5 mm in diameter which arise in groups on the spots (Laundon and Waterson, 1965). The uredospores are ellipsoid to obovoid, 21–26 × 16–19 µm, with finely echinulate cell walls that are 1.5–2.5 µm thick. Ellipsoid to cylindrical teliospores measuring 30–48 × 19–22 µm in size are produced which are rounded above and slightly constricted at the septum. T h e c ell walls are smooth, buff colored, 1–1.5 µm in thickness toward the side and 2–4 µm thick above.

Disease Cycle and Epidemiology:

Germination of urediospores is favored by temperature ranging from 18 °C to 22 °C and 8 h of dark period. Young shoots are most susceptible during the onset and development stages. As the leaves get matured, their resistance capacity increases and leaves with an age of 40 days or more have been reported as more resistant (Holliday, 1980). Splashing rain helps in the dissemination of the urediospores.

Integrated Disease Management:

- It is advisable to scout the orchard and observe the guava plants for any initiation of the disease or at times during the year when favorable environmental conditions prevail for proper and timely fungicide applications against the disease.
- Proper cultural operations viz. recommended dose of fertilizers, irrigation, proper pruning, and orchard sanitation should be followed to facilitate healthy, vigorously growing trees which will be less vulnerable to disease development.
- Spraying of Triadimenol or Tebuconazole is effective in disease management (Martin et al., 2011).
- Other than the above diseases, there are many minor guava diseases which are location specific and in general cause economically insignificant losses to the growers.

12.6 ALGAL LEAF SPOT OR RED RUST

Introduction/Economic Importance:

Distribution of this disease is worldwide in tropical and subtropical countries. This is a foliar disease most commonly seen inside the greenhouses and also under warm humid climates.

Symptoms:

Infection of the disease is exhibited on both epidermal layers (abaxial and adaxial leaf surfaces) as orange, rust-colored spots (Figure 12.1). When these spots are scraped, a thin, grayish white to dark-colored, necrotic crust is seen that remains on the leaf surface. These spots usually coalesce together forming large patches which are irregular in shape. The mature spots look dull, grayish green in color. Infection of the twigs and branches resulted in the cracking of barks due to the growth and expansion of the pathogens.

FIGURE 12.1   Algal spot on leaves.

Causal Organism: *C. virescens* Kunze (algae)

Pathogen:

The pathogen is a green algae. It is both aerophilic and terrestrial. The algae require a film of water that is, humid conditions to live and complete its life cycle. The genus *Cephaleuros* belong to Trentepohliales and the order Chlorophyta. The algae contain branched filaments comprising a thallus composed of irregular discs. The algae multiply below the cuticle or epidermis of the infected plants. This thallus is pigmented that is, orange to red-brown in color consisting of a prostrate portion that branches irregularly with irregular cells and an erect portion of unbranched hairs containing cylindrical cells that are sterile or fertile and protrude through the cuticle. They produce haustoria for the absorption of nutrients. Gametangia are produced on the prostrate cell filaments terminally or as intercalary. The asexual stage of algae is reported to be of great importance in causing infection and disease.

Disease Cycle:

Cephaleuros prefers humid conditions and proliferates within the trees canopy if such conditions prevail. Sporangia and biflagellate zoospores which are produced by the alga are carried through the agencies like water splash and wind. The pathogen causes primary infection with the help of biflagellate zoospores. Once the sporangia or thallus fragments of the alga carrying the sporangia fall on the host, infection process gets started.

Zoospoores produced inside the sporangia are released under moist conditions. They penetrate the host cuticle and start the infection process. The infection resulted in the reduction of the photosynthetic area resulting in less production of food, defoliation, twig dieback, tissue necrosis, and reduction of the market value of the fruits. The pathogen survives and also reproduces on infected leaves, stems, and also in fallen plant debris.

Epidemiology:

The growth and development of the pathogen are favored by frequent rains and warm weather. Guava plants growing under poor plant nutrition, inadequate soil drainage, more humidity within crop canopy, and stagnant water predispose the plants for infection by the algae.

Integrated Disease Management:

- Maintenance of proper tree vigor, practicing necessary cultural operations such as balanced fertilization and irrigation, need-based pruning for enhancing air circulation and sunlight penetration within the tree canopy, controlling weeds periodically, and providing wider spacing within guava plants helps in managing the disease incidence.
- Adequate cultural, mechanical, physical, biological, and need-based chemical management practices should be adopted to manage other maladies to maintain proper tree vigor and resistance of the plants.
- Copper-based fungicide like COC should be applied periodically to control the alga.

KEYWORDS

- **Guava**
- **diseases**
- **symptomatology**
- **management**

REFERENCES

Anonymous (1985). Guava rust. *Australia Commonwealth Department Primary Industry Plant Quarantine Leaflet No. 45.*

Ansar, M., Saleem, A., and Iqbal, A. (1994). Cause and control of guava decline in Punjab (Pakistan). *Pak. J. Phytopath.* 6: 41–44.

Bhatnagar, K., Sharma, B. S., and Cheema, H. S. (2004). Efficacy of plant extract against *Fusarium oxysporum* f. sp. cumini wilt in cumin. *J. Mycol. Plant Pathol.* 34: 360–361.

Burnett, H. C. and Schubert, T. S. (1985). *Puccinia psidii* on allspice and related plants. Division of Plant Industry, Florida Department of Agriculture and Consumer Services, Tallahassee, Plant Pathology Circular No. 271.

Chattopadhyay, S. B. and Bhattacharjya, S. K. (1968). Investigations on the wilt disease of guava (*Psidium guajava* L.) in West Bengal I. *Indian J. Agric.* 38: 65–72.

Chattopadhyay, S. B. and Sengupta, S. K. (1955). Studies on wilt of *Psidium guajava* L. in West Bengal. *Indian J. Hortic.* 12: 76–79.

Das Gupta, M. K. and Ghoshal, B. K. (1977). Is it possible to control guava wilt through oil cake amendments. *Sci. Cult.* 43: 131–133.

Das Gupta, S. N. and Rai, J. N. (1947). Wilt disease of guava (*P. guajava*). *Curr. Sci.* 16: 256–258.

Dey, P. K. (1948). Plant pathology. *Adm. Rep. Agric. Dept. UP* 1945–1946, 43–46.

Dwivedi, S. K. (1990). Efficacy of some antibiotics on *Fusarium oxysporum* f. sp. *psidii* causing wilt disease of guava. *Hindustan Antibiotic Bull.* 32: 88–90.

Edward, J. C. (1960). Variations in the guava wilt pathogen, *Fusarium oxysporum*i f. *psidii*. *Indian Phytopathol.* 13: 30–36.

Edward, J. C. and Srivastava, R. N. (1957). RE 502—Studies on guava wilt. *Allahabad Farmer*, 31: 144–146.

Gupta, P., Dubey, V., David, A., and John, D. (2003). *Verticillium albo-atrum*—a new pathogen associated with guava wilt. *Indian Phytopathol.* 56: 503–504.

Gupta, V. K., Misra, A. K., Gaur, R. K., Jain, P. K., Gaur, D., and Sharma, S. (2010). Current status of *Fusarium* wilt disease of guava (*Psidium guajava* L.) in India. *Biotechnology,* 9: 176–195.

Gupta, V. K., Misra, A. K., Pandey, B. K., and Chauhan, U. K. (2007). *In vitro* evaluation of leaf extracts against *Fusarium* wilt pathogens of guava (*Psidium guajava* L.). *J. Eco-Friendly Agric.* 2: 167–169.

Hamiduzzaman, M. M., Meat, M. B., and Ahmad, M. U. (1997). Effect of *Fusarium oxysporum* and nematode interaction on guava wilt. *Bangladesh J. Plant Pathol.* 13: 9–11.

Holliday, P. (1980). Fungus Diseases of Tropical Crops. Cambridge University Press, Cambridge. p. 401.

Hsieh, S. P. Y., Liang, W. J., Kao, C. W., and Lau, L. S. (1976). Morphological and physiological characters of *Myxosporium psidii*, the causal organism of guava wilt. *Plant Prot. Bull.* 18: 309–317.

Junqueira, N. T. V., de Andrade, L. R. M., Pereira, M., Lima, M. M., and Chaves, R. C. (2001). Diseases of guava (*Psidium guajava* L.) cultivated in *Brazilian* Cerrdos. *Circ. Tech. Embrapa. Cerrados.* 15: 31–31.

Keith, L. M., Velasquez, M. E., and Zee, F. T. (2006). Identification and characterization of *Pestalotiopsis* spp. causing scab disease of guava, *Psidium guajava*, in Hawaii. *Plant Dis.* 90: 16–23.

Leu, L.S. and Kao, C.W. (1979). Artificial inoculation of guava with *Myxosporium psidii*. *Plant Dis. Rep.* 63: 1077–1079.

Lim, T.K. and Manicom, B.Q. (2003). Diseases of Guava. In: Diseases of Tropical Fruit Crops, Ploetz, R.C. (Eds.). CABI Publication, Wallingford, UK. pp. 275–289.

Logani, R., Pandey, R. C., Dwivedi, B. K., and Singh, S. P. (2002). An ecofriendly management of wilt disease of guava by wiltnema (a botanical fungicide). *Bioved.* 13: 47–50.

Martins, M. V. V., Silveira, S. F., Maffia, L. A., Rocabado, J. M. A., and Mussi-Dias, V. (2011). Chemical control of guava rust (*Puccinia psidii)* in the Northern Region of Rio de Janeiro State, Brazil. *Australasian Plant Pathol.* 40(1): 48–54.

Mathur, R. S. (1956). Guava diseases in India. *Indian J. Hort.* 13: 26–29.

Merida, M. and Palmateer, A.J. (2013). Florida Plant Disease Management Guide: Guava (*Psidium guajava*). University of Florida IFAS Extension.

Misra, A. K. (2004). Guava Diseases—Their Symptoms, Causes and Management, pp 81–119 In: Diseases of Fruits and Vegetables: Diagnosis and Management. Volume II. Naqvi, S. A. M. H. (eds). Klower Academic Publishers.

Misra, A. K. (2006). Wilt of guava—a disease of national importance. *Indian Phytopath.* 59 (3): 269–280.

Misra, A. K. and Gupta, V. K. (2007). Variability in *Fusarium solani*—a causal organism of wilt of Guava. *CISH Newslett.*, 8: 2–2.

Misra, A. K. and Gupta,V. K. (2009). *Trichoderma*: biology, biodiversity and biotechnology. *J. Eco-Friendly Agric.* 4: 99–117.

Misra, A. K. and Pandey, B. K. (1997). Pathogenicity and symptom production of wilt disease of guava by a new potent pathogen *Gliocladium roseum*. Proceedings of the International Conference on Integrated Plant Disease Management for Sustainable Agriculture, November 10–15, IARI, New Delhi, India. pp. 319–319.

Misra, A. K. and Pandey, B. K. (1999). Natural wilting of guava plants during different months. *Indian Phytopathol.* 52: 312–312.

Misra, A. K. and Pandey, B. K. (1999). Pathogenicity and evaluation of fungicides against guava wilt pathogens. *J. Mycol. Plant Pathol.* 29: 274–275.

Misra, A. K. and Pandey, B. K. (2000). Progressive natural wilting of guava plants during different months. *Indian Phytopathol.* 53: 423–427.

Misra, A. K. and Prakash, O. (1990). Guava Diseases (An annotated bibliography 1907–1990). Bishen Singh Mahendra Pal Singh, Dehradun, India, p. 132.

Misra, A. K., Prakash, O., and Sen, B. (2000). Biological control of guava wilt by *Aspergillus niger* strain AN17 (Pusa Mrida). Proceedings of the National Seminar on Hi-tech Horticulture, June 26–28, Bangalore. pp. 149–149.

Misra, A. K., Prasad, D., Prasad, B., and Shukla, S. K. (2004). Effective management of wilt disease of guava. Proceedings of the National. Symposium on Crop Surveillance: Disease Forecasting and Management, Feb. 19–21, Division of Plant Pathology, IARI New Delhi. pp. 92–93.

Misra, A. K. and Shukla, S. K. (2002). Assessment of loss due to guava wilt around Lucknow. Proceedings of the National Seminar on Production and Post-Harvest Technology of Guava, January 9–10, Department of Horticulture CSAUA&T, Kanpur. pp. 34–35.

Montiel, C. A. (1997) *Pestalotiopsis psidii* (Pat.) Mordue causante de necrosis de fructos de guayabo (*Psidium guajava* L.) en plantaciones de loss municiplios Baralt y Mara del estado Zulia. *Revista de la Facultad de Agronomia, Universidad del Zulia*.14, pp. 341–347 (in Spanish).

Mordue, J. E. M. (1969) *Pestalotiopsis psidii. CMI Descriptions of Pathogenic Fungi and Bacteria No. 515*. Commonwealth Mycological Institute, Kew, UK.

Neeraj, M., Joon, S., and Bhatia, S. K. (2005). Guava wilt. An investigative review of its occurrence causes and cures. *Haryana J. Hort. Sci.* 34: 201–207.

Pandit, P. K. and Samajpati, N. (2002). Wilt disease of guava (*Psidium guajava* L.) by *Botryodiplodia theobromae* Pat. *J. Mycological Res.* 40: 71–73.

Prakash, O. and Pandey, B. K. (2007). Current scenario of guava diseases in India and their integrated management. International Guava Symposium. ISHS *Acta Horticult.* 735: 1.

Prasad, N., Mehta, P. R, and Lal, S. B. (1952). *Fusarium* wilt of guava (*Psidium guajava* L.) in Uttar Pradesh, India. *Nature*, 169: 753–753.

Radha, T. and Methew, L. (2007). Fruit crops In: Horticulture Science Series. Peter, K. V. (eds). New India Publishing Agency, Pitam Pura, New Delhi.

Rayachhetry, M. B., Elliot, M. L., and Van, T. K. (1997) Natural epiphytotic of the rust *Puccinia psidii* on *Melaleuca quinquenervia* in Florida. *Plant Dis.* 81: 831.

Rodriguez, F. M. E. and Landa, J. B. (1977). Chemical soil disinfection against parasitic nematode in guava nurseries. *Centrol Agricola Facultod Cieneias Agricolos.* 4: 57–77.

Rodriguez, N. J., Robbs, C. F., and Yamashino, T. (1987). A bacterial disease of guava (*Psidium guajava*) caused by *Erwinia psidii* sp. nov. *Fitopathologia Brasileira*, 12: 345–350.

Singh, B. and Lal, S. B. (1953). Wilt of guava. *Agric. Anim. Husb.* 3: 78–79.

Suhag, L. S. (1976). Observations in guava decline in Haryana and its control. *Pesticides.* 10: 42–44.

Tokeshi, H., Valdebenito, R. M., and Dias, A. S. (1980). Occurrence of a bacterial disease of guava in Sao Paulo State. Summa *Phytopathologica.* 6: 85–87.

Vos, J. E., Schoeman, M. H., Berjak, P., Watt, M. P., and Toerien, A. J. (1998). *In vitro* selection and commercial release of guava wilt resistant rootstocks. *Acta Hortic.* 513: 69–79.

Webber, G. F. (1928). Plant pathology. Annual Report of Florida Agricultural University Exp. Stat.

CHAPTER 13

MAJOR DISEASES OF HAZELNUT (*CORYLUS AVELLANA* L.) AND THEIR MANAGEMENT

SONIKA JAMWAL[1,*], UPMA DUTTA[2], and J. N. SRIVASTAVA[3]

[1]*Advanced Centre for Rainfed Agriculture-Dhiansar, Sher-e-Kashmir University of Agricultural Sciences and Technology-SKUAST, Jammu, Jammu and Kashmir, India*

[2]*Division of Microbiology, Sher-e-Kashmir University of Agricultural Sciences and Technology-SKUAST, Jammu, Jammu and Kashmir, India*

[3]*Department of Plant Pathology, Bihar Agricultural University, Sabour, Bhagalpur, Bihar, India*

Corresponding author. E-mail: annajamwal@gmail.com

ABSTRACT

Hazelnuts fruits are grown in Azerbaijan, Georgia, Italy, and United States. Turkey is the largest producer of hazelnuts in the world. Near about 70% of total production is given by Turkey. Hazelnut (*Corylus avellana* L.) is affected by many diseases caused by fungi, bacteria, viruses, nematodes, and parasitic plants. These diseases as well as pathogen either alone or in combination cause substantial damage to crops resulting in heavy economic losses every year. This chapter deals with important diseases of hazelnut and their management strategies.

13.1   INTRODUCTION

Hazelnut (*Corylus avellana* L.) is a deciduous, monoecious, multistemmed bush tree grown as a single trunk tree for commercial purposes. According to species of Hazelnut (*C. avellana* L.), it is known as filbert nut or cobnut. The shape of the filbert nut is more elongated, just double of its diameter. After pollination, 7–8 months fruits are ripe, when it is ripe, nut falls out of the husk. The shape of the cob nut is spherical to oval with 15–25 mm in length and 10–15 mm in diameter. Cob nut is covered with a fibrous husk and surrounded by a smooth shell. The nut is edible, rich in protein, mono-unsaturated fat, vitamin B, vitamin E and manganese, and rarely amount of other essential nutrients. The kernels/seeds have a dark brown thin skin (Josiah, 2000).

The kernel seed of hazelnut is edible, kernel seed is used as raw or as roasted or as paste after grinding. Hazelnuts are also used in bakery products with the combination of chocolate. Hazelnut oil, after it is pressed from hazelnuts, is strongly flavored and used as edible oil (Olcott-Reid and Reid, 2007).

Hazelnuts fruits are grown in Azerbaijan, Georgia, Italy, and United States. Turkey is the largest producer of hazelnuts in the world. Near about 70% of total production is given by Turkey (Stebbins and Olsen, 2007).

Hazelnut (*C. avellana* L.) is affected by many diseases caused by fungi, bacteria, viruses, nematodes, and parasitic plants. These diseases as well as pathogen either alone or in combination cause substantial damage to crops resulting in heavy economic losses every year (Pschiedt and Ocamb, 2008). This communication deals with important diseases of hazelnut and their management strategies.

The crop is affected with various diseases

Fungal Diseases:

1. Anthracnose: Casual organism: *Gloesporium corvli, Monostichella corvli*
2. Armillaria root disease: Casual organism: *Arnillaria* spp.
3. Powdery mildew: Casual organism: *Microsphaera hommae*
4. Rust: Casual organism: *Pucciniastrum coryli*

Bacterial Diseases:

1. Bacterial blight: Casual organism: *Xanthomonas arboricola* pv. *corylina*
2. Bacterial Canker: Casual organism: *Pseudomonas syringae* pv. *avellance*
3. Crown gall: Casual organism: *Agrobacterium tumefaciens*

13.2 FUNGAL DISEASES

13.2.1 ANTHRACNOSE

Casual Organism: *Gloesporium corvli*

Anthracnose disease may be caused by various fungi including *Apiognomonia errabunda, Apiognomonia veneta, Colletotrichum gloeosporioides, Discula fraxinea, Stegophora ulmea, Glomerella* spp., *Gnomonia* spp., *Marssonina* spp., and *S. ulmea.*

Symptoms:

- Anthracnose disease is very common in Hazelnut (*C. avellana* L.) tree. Anthracnose disease attacked the hazelnut (*C. avellana* L.) tree by two means: (1) In first means that caused spot on leaves and blossoms, (2) In the second means that caused canker, it is more harmful because it disrupts and can destroy the vascular system of trees (McCain, 1983).
- The disease attacked the hazelnut trees any time in the year and showed symptoms as small, necrotic spots with varying degree of size and color on infected branches, twigs, and leaves. In this stage, symptoms appeared as blight on leaves too. In advance stage, fungus of disease contaminated entire vascular system of trees and bark showed symptoms. In this stage symptoms appeared as dark-colored lesions or indents can begin to show on tree branches, tree twigs, and tree trunks. Lastly, cankers can develop and trees branches show dieback indications (Nameth and Chatfield, 1996).

Disease Cycle:

The pathogen survives mainly *in* infected leaves, defoliated branches, and mummified flowers. The disease is spread through rain splash or air current rainwater. In favorable circumstances, conidia are dispersed and invade young twigs causing twig dieback in twigs (Pataky, 1997).

Integrated Disease Management:

Cultural Control:

- Use resistant planting material for the establishment of a new orchard.

- Maintain proper distance plant to plant and line to line for air ventilation and sunlight.
- Pruning of infected tree twigs and tree branches during fall or winter and after that pruned material destroy them.
- Avoid excess irrigation system because those provide the wet condition of leaves.

Chemical Control:

- Spraying of Carbendazim @0.1%, Chlorothalonil @0.2% with 12 days interval manages the disease (Dimova and Arnaudov, 2008).
- Application of thiophanate methyl (systemic fungicides) to be effective against the disease (Berry and Frederick, 1997).
- Spraying of copper oxychloride @0.3% for minimization of foliar disease infection.
- Two sprays of Strobilurin fungicides with 15 days interval is beneficial in reducing anthracnose infection.

13.2.2   ARMILLARIA ROOT ROT

Casual Organism: *Armillaria spp.*
This Armillaria root rot disease is caused by various species of the genus *Armillaria.*

Symptoms:

- Symptoms appeared as reduced size leaves, discolored leaves which defoliate very early. Further, disease infection caused death of branches and plants, formation of clusters of honey-colored mushrooms which should be grown at the base of trees (*Williams et al., 1989*).
- Contaminated trees exhibit symptoms similar to a white rotting fungus. The infected host also showed light or bleached wood after the degradation of essential cell wall compounds such as lignin and hemicellulose (*Worall, 2010*).
- In severe conditions, disease showed chlorosis symptoms in leaves as well as showed dieback symptoms in twigs and branches.
- In severe condition of disease, the trunks of the tree will also exude more amount of resin. That process is called resinopsis.
- After removal of the bark of hazelnut trees, white mats of mycelium are visible along with rhizomorphs, rhizomorph is a distinctive

reproductive structure. Mushroom-like structure will also grow at the base of the infected tree and melanized fungal cells may be exhibited within the contaminated wood (Shaw and Kile, 1991).

Disease Cycle:

Pathogen survives as mycelium or rhizomorphs in diseased trees or in decaying roots. Pathogen produces basidiocarp by sexual reproduction under the mating procedure of hyphae at the base of the infected host trees. After that, the basidiocarp produces basidiospores which are capable infecting new hosts. Hence, *Armillaria* root rot disease rarely spreads by basiodiospores. In the second way, more chances the disease will spread either by rhizomorphs or direct contact to host trees by mycelia (Hartig, 1874). Rhizomorphs are string-like masses of hyphae capable of infecting a new host and spread through the soil with regard to the uninfected roots (*Williams et al., 1989; Coetzee et al., 2001;* Lachance, 1996).

The conditions like shady condition, normal temperature, poor air ventilation around trees favored the disease development (*Davari and Askari, 2005;* Guillaumin et al., 1993).

Integrated Disease Management:

Cultural Control:

- The important way of controlling the disease is limiting the spread of the pathogen (Harrington et al., 1992).
- Transplanting resistant cultivars to newly established gardens is very helpful to manage the disease.
- Follow removal of dead trees for the management of the disease and infected stumps, and it does destroy them.
- Sanitation of orchards by removing all old roots and stems of tree debris and solarization of soil in a hot climate is important to manage the disease.

Chemical Control:

- *Armillaria* root rot can be managed through soil fumigation. However, the management of disease depends upon the complete eradication of the fungus.
- Injections of systemic fungicides have been shown to reduce the mortality of Armillaria infected trees. For example, postplant passive

injections of propiconazole into 7- to 8-year old infected hazelnut trees.

13.2.3   RUST OF HAZELNUT

Casual Organism: *Pucciniastrum coryli*

Symptoms:

- The symptoms appeared mostly on leaves, but these disease symptoms can also be found occasionally on leaf stalks (petioles), twigs of trees, and sometimes on flowers and fruits (Anonymous, 1938).
- The symptoms appeared as spots on the leaf surface eventually into spore-producing structure called pustules. The spore pustules of the rust pathogen are determined by color, namely, orange, yellow, brown, black, or white but mostly in rusty color hence given the name of the disease (Perlbeger, 1943).
- The spore pustules of rusts are mostly established on the lower leaf surface and produce large numbers of rust spores.
- Affected leaves of the infected trees often turn yellow and fall prematurely in severe condition of disease.
- Heavy infection often reduces the vigor of the plant (Meparishvili et al., 2019).

Disease Cycle:

The inoculum survives in the form of uredospores and teliospores in the host trees during the off season, which provides an excellent source of primary inoculum and disseminated by wind (Perlbeger, 1943).

Integrated Disease Management:

Cultural Control:

- Transplanting resistant cultivars for newly established orchards.
- The important effective way of controlling the disease is limiting the spread of the pathogen.
- Planting resistant cultivars in newly established gardens is very helpful to manage the disease.
- Following removal of dead material over the year and destroying them.

- Facilitate circumstances that encourage growth and development of trees, but avoid an excess dose of nitrogen fertilizers because it provides soft, lush growth that is favorable for rust pathogen in infection.

Chemical Control:

- Two or three sprays of Zineb @0.25% or Mancozeb @0.25% or Plantavax @0.1% at 15 days intervals can manage the disease.

13.2.4 POWDERY MILDEW OF HAZELNUT

Casual of Organism: *Phyllactinia guttata* (Wallr.:Fr.) Lév

Symptoms:

- Symptoms showed on the lower surface of leaves particularly, but the powdery mildew disease can also show on any above-ground part of the plant.
- Symptoms showed as white powdery patches on leaves and fruit which can enlarge and cover the entire leaf surface and fruit surface. Signs of powdery mildew disease are white mycelial growth and accompanying chasmothecia on abaxial leaf surfaces (Teviotdale, 2002).
- Sometimes symptoms appeared as chlorotic areas visible on adaxial leaf surfaces. The chasmothecia are globose and convex on the ventral side, and nearly plane on the dorsal side; each is with a gelatinous mass of penicillate cells and acicular appendages.

Disease Cycle:

- Powdery mildew causing pathogens are most prevalent where cool to warm, humid weather persists with little rainfall. Primary infection of disease by ascospores released from overwintered cleistothecia on infected fallen leaves. Conidia can cause reiterated infections during the growing season (Hartney et al., 2005).
- Disease favored high moisture and moderate temperatures.

Integrated Disease Management:

Cultural Control:

- Disposal of contaminated leaves, stems, twigs, flower, and fruits and alternate/collateral hosts from orchards.

- Summer ploughing practice can help in reducing the inoculum.

Chemical Control:

- Spraying with Karathane @0.1%, Calixin @0.1% proved the best control.
- Fungicides, namely, Bayleton 25 WP @0.05%, Karathane 48 EC @0.01%, and Sulfex 80 WP @0.25% are applied during flowering to after fruiting reduces the disease severity.

13.3   BACTERIAL DISEASES

13.3.1   BACTERIAL BLIGHT OF HAZELNUT

Casual Organism: *Xanthomonas arboricola*

Symptoms:

- Pathogen attacks the buds, leaves, branches, trunk, and occasionally the nuts.
- Symptoms appeared previously as necrotic spots on leaves which are very ordinary symptoms in nurseries as well as rarely in fields (Pulawska et al., 2010; Lamichhane et al., 2012a). Leaf buds and flower-bearing buds in leaf axils on shoots are often affected. However, the staminate buds are rarely attacked (Lamichhane et al., 2012b).
- Symptoms exhibit on leaves of trees, symptoms showed as small angular or irregularly circular, pale yellow-green, water-soaked lesions. The lesions turn reddish-brown over time. Usually, individual leaf lesions are small, 2–3 mm in diameter, but often they merge near the leaf tip (Miller et al., 1940).
- Symptoms appeared on stems, the presence of infection by a pathogen can be easily noticed as dark green, water-soaked areas in the bark. The infection area further turns reddish brown over time.

Disease Cycle:

The disease cycle of bacterium starts on the contaminated leaves of the trees where the bacteria live as an epiphytic stage (absorbed nutrients and water from the air) until mid-to-late spring season when sufficient rain splash.

Further, bacteria are transmitted to new buds and fruits where they become virulent and also cause disease. Bacterium with less rainfall in a late season that is infection caused in fruit and nuts also. After the infection of fruit and nuts in the autumn season (Miller et al., 1940).

Integrated Disease Management:

Cultural Control:

- Removal and disposal of infected plant part from orchards reduces the disease.

Chemical Control:

- Spraying with copper oxychloride 50% @0.3% + *Streptocycline* @0.06% with a spreader sticker is very effective for control of the disease. First application should be in August/September before rainfall, the second application should be when three-fourth of leaves have fallen, and the third application should be in early spring before bud set.

13.3.2 BACTERIAL CANKER OF HAZELNUT

Casual Organism: *Pseudomonas syringae* pv. *avellanae*

Symptoms:

- It appears that the pathogen can disperse throughout the late spring, summer, and fall within a block of trees. Movement of cuttings and nursery stock would allow long-distance dispersal.
- Symptoms appeared primarily visible in spring as twig die-back and eventually cankering.
- Sometimes, trees will leaf out despite the case of bacterial canker, but these incipient leaves expeditiously wilt and die (Jence et al., 1995).

Disease Cycle:

- The disease cycle of canker bacterium begins, is initiated in spring by the colonization and development of large populations on blossoms. The primary inoculum originates from bacteria located in dormant buds or in cankers. Further, bacteria are transmitted to new buds where it also causes disease. Bacteria are primarily wind-borne but can be spread by rain splash and also insects (Marco Scortichini, 2002; Martin, 1998).

Integrated Disease Management

Cultural Control:

- Pathogen is copper sensitive however, the speed and lethality of this pathogen require strict quarantine and rouging.

Chemical Control:

- Spraying with copper oxychloride 50% @0.3% + *Streptocycline* @0.06% with a spreader sticker is very effective for control of the disease. First application should be in August/September before rainfall, the second application should be when three-fourth of leaves have fallen, and the third application should be in early spring before bud set.

13.2.3   CROWN GALL DISEASE OF HAZELNUT

Casual Organism: *Agrobacterium tumefaciens*

The disease is caused by soil-inhabiting bacterium namely *A. tumefaciens* and was first studied by the American plant pathologist Erwin Frink Smith in 1891–1894. It seriously affects susceptible plants, causing significant financial loss in commercial nurseries where up to 80% of the stock has been reported diseased (McGuire et al., 1991).

Symptoms:

- Galls are occurring in 3 inches diameter or more and usually occur below the soil line. Galls sometimes are found on stems and on the pruning wounds. Gall is determined in size, often growing many times larger than the diameter of the roots or stem on which they are outgrowths (Ophel and Kerr, 1990).
- Newly formed galls are soft on the surface, light or tan-colored, and frosty appearance. Further galls become older, they become darker, turning almost black, and are commonly hardy and woody (Burr and Otten, 1999).

- The bacterium enters the tree only through wounds. Many wounds caused on roots from digging convey and also pruning practices, all give an entry point for the disease. Mechanical injuries of crown roots zone by cultural equipment and insects are also consequential ingression points (Kado, 1976).
- Infected plants usually exhibit a weak appearance and unproductive because of their damaged root system and ultimately may die (Kado, 2002).

Disease Cycle:

A. tumefaciens persists in infested soil for long period even without host plant presence because *A. tumefaciens* is saprophytic in nature. Bacteria must enter into host tissue via wounds or natural openings of roots or stem near the ground. Further, that pathogen causing excess growth of plant tissue leads to gall formation on roots or stem near the ground. Breakdown of the soft tissue/young gall leads to the release of *A. tumefaciens* into the soil, they are capable to restart the disease process with a new host plant (Kado, 2002; Burr and Otten, 1999).

Integrated Disease Management:

- Preventative treatment of seeds or transplanting materials with the nonpathogenic biocontrol organism, K-84 strain of *A. tumefaciens* (formerly *Agrobacterium radiobacter*), is a relatively inexpensive and effective means to control the development of crown gall in commercial operations.
- Use of this antagonist by soaking seeds or dipping transplanting materials can prevent infection by most strains of *A. tumefaciens* due to the production of the antibiotic agrocin 84 by strain K-84 of *A. radiobacter* (Kado, 2002; Burr and Otten, 1999).

KEY WORDS

- **Hazelnut**
- **diseases**
- **symptomatology**
- **management**

REFERENCES

Anonymous (1938). The rust of Stone fruits. *Phytopathology.* 4: 411–427.

Berry, L. I. and Frederick, H. (1997). Control of walnut anthracnose with fungicides in a black walnut plantation. *Plant Dis. Rep.* 61: 378–379.

Burr, T. J. and Otten, L. (1999). Crown gall of grape vine: biology and disease management. *Annu. Rev. Phytopathol.* 37: 9004.

Coetzee, Martin, P. A., Wingfield, Brenda D., Harrington, Thomas C., Steimel, J., Coutinho, T. A., and Wingfield, M. J. (2001). The root rot fungus Armillaria mellea introduce in to South Africa by early Dutch settlers. Mol. Ecol. 10 (2): 387–96.

Davari, M., and Askari, B. (2005). Armillaria mellea as a cause of oak decline in Hatam-baigh forest of Iran. Agricult. Appl. Biol. Sci. 70 (3): 295–304.

Dimova, M. and Arnaudov, V. (2008). Control of the over-wintering stage of Anthracnose (*Gonomonia leptostyla* (Fr.) Ces. and de Not) in walnut. *Rastenievdin-nouki.* 45: 32–35.

Guillaumin, J.-J., Mohammed, C., Anselmi, N., Courtecuisse, R., Gregory, S. C., Holdenrieder, O., Intini, et al. (1993). Geographical distribution and ecology of the *Armillaria* species in Western Europe. *Eur. J. Forest. Pathol.* 23: 321–341.

Harrington, T. C., Worrall, J. J., and Baker, F. A. (1992). *Armillaria.* In L. L. Singleton, J. D. Mihail and C. M. Rush, eds., Methods for Research in Soilborne Phytopathogenic Fungi. American Phytopathological Society Press. St. Paul, MN, USA. pp. 81–85.

Hartig, R. (1874). Important Diseases of Forest Trees. English translation by W. Merrill, D.H. Lambert and W. Liese, (1975). Phytopathological Classics No. 12. American Phytopathological Society. St. Paul, MN, USA.

Hartney, S., Glawe, D. A., Dugan, F., and Ammirati, J. (2005). First report of powdery mildew on *Corylus avellana* caused by *Phyllactinia guttata* in Washington State. *Plant Health Progress*, doi:10.1094/PHP-2005–1121-01-BR.

Pulawska1, J., Kaluzna1, M., Kolodziejska A., and Sobiczewski1 P. (2010). Identification and characterization of *Xanthomonas arboricola* pv. *corylina* causing bacterial blight of hazelnut: a new disease in Poland, *J. Plant Pathol.* 92 (3): 803–806.

Janse, J. D., Rossi, P., Angelucci, L., Scortichini, M., Derks, J. H. J., Akkermans, A. D. L., De Vrijer, R., and Psallidas, P. G. (1995). Re-classification of *Pseudomonas syringae* pv. *Avellanae* as *Pseudomanas avellanae* (spec. nov.), the Bacterium Causing Canker of Hazelnut (*Corylus avellana L.*). *Syst. Appl. Micribiol.* 19: 589–595.

Josiah, S. (2000). Commercial hazelnuts in Minnesota. *Fact sheet* (FO-07280-GO). University of Minnesota Extension. Minneapolis, MN, USA.

Kado, C. I. (1976). The tumor-inducing substance of *Agrobacterium tumefaciens. Annu. Rev. Phytopathol.* 14: 265–308.

Kado, C. I. (2002). Crown gall tumors. In: Encyclopedia of Genetics. S. Brenner and J. H. Miller, eds. Academic Press, San Diego, CA, USA. pp. 1–3.

Lachance, D. (1996). Armillaria Root Rot. Information Leaflet LFC 14E, Natural Resources Canada, Canadian Forest Service, Quebec.

Lamichhane, J. R., Fabi, A., and Varvaro, L. (2012a). Severe outbreak of bacterial blight caused by *Xanthomonas arboricola* pv. corylina on hazelnut, cv. Tonda di Giffoni, in central Italy. *Plant Dis.*, 96: 1577.

Lamichhane, J. R., Fabi, A., and Varvaro, L. (2012b). Bacterial species associated to brown spots of hazelnut in central Italy: Survey, isolation and characterization. *Acta Hort.*, 2014(1): 243–251.

Scortichini, M. (2002). Bacterial Canker and Decline of European Hazelnut Council for Agricultural Research and Agricultural economy. *Plant Dis.* DOI: 10.1094/PDIS.2002. 86.7.704

Martins, J. M. S. and Scortichini, M. (1998). Spatio-Temporal Spread of a Bacterial Disease in a Hazelnut Orchard. Actas Da 2 Reunio Bienal Da Sociedade Portuguesa De Fitopatologia. pp. 82–88.

McCain, A. H. (1983). *Anthracnose.* Oakland: Univ. Calif. Agric. Nat. Res. Publ. 2618.

McGuire, R. G., Rodriguez-Palenzuela, P., Collmer, A., and Burr, T. J. (1991). Polygacturonase production by *Agrobacterium tumefaciens* biovar 3. *Appl. Environ. Microbiol.* 57: 660–664.

Meparishvili, G., Gur, L., Frenkel, O., Gorgiladze, L., Meparishvili, S., Muradashvili, M., Koiava, L., Dumbadze, R., Reuveni, M., and Jabnidze, R. (2019). First report of powdery mildew caused by *Erysiphe corylacearum* on hazelnuts in Georgia, *Plant Dis.* 10.1094 / PDIS-05–19-1053-PDN, (PDIS-05–19-1053).

Miller, P. W., Bollen, W. B., and Simmons, J. E. (1940). Filbert bacteriosis and its control. Oregon Agricultural Station Technical Bulletin 16.

Nameth, S. and J. Chatfield. (1996). *Anthracnose Leaf Blight of Shade Trees.* Columbus: Ohio State Univ. Ext. Factsheet HYG-3048–96. Accessed December 9, 2009.

Olcott-Reid, B. and W. Reid (2007). Fruit and nut production. Stipes Publishing. Champaign, IL. Snare, L. 2008. Hazelnut production. *Fact Sheet* (Primefact 765). New South Wales Department of Primary Industries. Sydney, NSW, Australia.

Ophel, K., and Kerr, A. (1990). *.Agrobacterium vitis,* new species for strains of *Agrobacterium* biovar 3 from grapevines. *Int. J. Syst. Bacteriol.* 40: 236–241.

Pataky, N. R. (1997). Anthracnose Disease of Shade Trees. Urbana-Champaign: Univ. Illinois. Ext. Rep. on Plant Dis. N. 621. Accessed December 9, 2009.

Perlbeger, J. (1943). The rust disease of stone fruit trees in Plestine. *Rahobath Agric. Exp. Stn. Bull.* 34.

Pschiedt, J., and C. Ocamb. (2008). Pacific Northwest disease management handbook. Oregon State University Press. Corvallis, OR, USA.

Shaw, C. G. and Kile G. A. (1991). Armillaria Root Disease. USDA Forest Service, Agricultural Handbook No. 691.

Stebbins, R. and Olsen J. (2007). Growing tree fruits and nuts in the orchard. Bulletin (EC819). Oregon State University Press. Corvallis, OR, USA.

Teviotdale, B. L., Michailides, T. J., and Pscheidt, J. W. editors. (2002). Compendium of Nut Crop Diseases in Temperate Zones. American Phytopathological Society, St. Paul, MN, USA.

Williams, R. E., Shaw, C. G., Wargo, P.M., Sites, W. H. (1989). Armillaria Root Disease. Forest Insect and Disease Leaflet, 78. Washington, D.C.: USDA Forest Service. Archived *from the original on 26 October 2010.* Retrieved 30 October 2010.

Worall, James J. (2010). Armillaria Root Disease. Forest and Shade Tree Pathology. Archived from the original on 21 November 2010. Retrieved 30 October 2010.

CHAPTER 14

IMPORTANT DISEASES OF JACKFRUIT (*ARTOCARPUS HETEROPHYLLUS* LANK.) AND MANAGEMENT STRATEGIES

K. JAYALAKSHMI[1,*], J. RAJU[2], S. RAGHU,[3] and PRITI S. SONAVANE[4]

[1]*National Bureau of Agriculturally Important Microorganisms (NBAIM), Kushmaur, Mau Nath Bhanjan, Uttar Pradesh 275103, India*

[2]*Plant Quarantine Station, Ministry of Agriculture and Farmers Welfare, Government of India, Mangalore, Karnataka 575011, India*

[3]*Crop Protection Division, National Rice Research Institute, Cuttack, Odisha 753006, India*

[4]*Indian Institute of Horticultural Research, Hesaraghatta, Bengaluru, Karnataka, India*

*Corresponding author. E-mail: jayalakshmipat@gmail.com

ABSTRACT

Jackfruit is grown in various subtropical as well as warmer countries. In India, it is grown especially in Assam, Bihar, Chhattisgarh, Kerala, Madhya Pradesh, Uttar Pradesh, and West Bengal. In Kerala, it is grown on commercial scale. The jackfruits are used as green vegetable or when ripe is eaten as fresh fruit. Many preparations (jam, chips) are made from jackfruit and the seeds are also eaten as tasty vegetables. The jackfruit tree is regarded as hardy tree with a reputed durable and valuable wood for making furniture. Various diseases have been reported in different countries also in India mainly from fungal origin affecting the leaves, twigs, branches, stem and roots except bacterial wilt reported from Indonesia. This chapter deals with important diseases of jackfruit and their management strategies.

Jackfruit is a tropical fruit crop especially in Uttar Pradesh, West Bengal, Bihar, and Kerala, in Kerala it is grown on commercial scale. The following diseases are the major ones causing huge loss of jackfruit, namely, *Corticium salmonicolor* (Pink disease); *Rhizopus artocaipi* (rotting of stem, fruit, and inflorescence) and *Phomopsis artocarpina, Pestalotia elasticola, Colletrotrichum lagenarium, Septoria artocarpi, Gloeosproium* sp, and so forth (leaf spot/blight), *Botrytis cinerea* (gray mold), *Pythium splendens, Phytophthora* sp., *Fusarium* sp., and *Rhizoctonia* sp (root rot.), *Ustilana zonata* (charcoal rot), *Rosellinia arcuata* (collar rot), and *Uredo artocarpi* (rust; Morton et al.,1987).

14.1   PINK DISEASE OF JACKFRUIT

The disease commonly seen in the tropical and subtropical regions causes the death of individual branches of the whole tree.

Symptoms:

- On stem initially pinkish powdery growth is seen. Actually, this pink color coating or growth is basically profuse production conidia of the causal fungus.
- In affected young woody branches lose their leaves and produce die back symptoms and pink color powdery growth is seen on the underside.
- Later it causes girdling cankers which disrupt the physiological processes of the tree and leading to defoliation, finally death of the distal parts of the tree.

Causal Organism:

Erythricium salmonicor (Corticum salmonicolor) or Boryobasium salmonicolor is a basiomycetous fungus

Disease Cycle:

Pink disease is presumably transmitted by basidiospores and conidia. Under warm, moist climatic conditions, the pink fungus mat produces great numbers of conidial spores which spread through rain-splash and/or wind.

Epidemiology:

Pathogen growth favors 26 °C–27 °C temperature and more moist humid condition.

Management:

- Avoid dense or closely spaced planting.
- Pruning of affected and dried branches and pasting of the cut ends with chaubatia or copper oxychloride or Bordeaux paste.
- Disinfection after pruning of dried and affected branches.
- During May and August, spray 1% Bordeaux mixture or 0.75% Thiram or 0.1% propiconazole.

14.2 RHIZOPUS ROT OF JACKFRUIT

The disease was first reported from Pusa and Banaras (Buttler and Bisby, 1931). In 1929, it was reported in Andaman island by Mitra. Gupta and Pandey (1985) reported that the disease is prevalent and severe in western Uttar Pradesh. In 1979, Pandey et al. showed *Rhizopus nigricans* causes stem and fruit rot.

Rhizopus rot has most prevailed in high rainfall regions. During flowering and fruiting, coincide of warm, humid wet climate causes more damage leading total fruit loss in trees.

Symptoms:

- The fungus affects only male inflorescence and unripe fruits.
- Fungus initially produces gray mycelial growth and later turns black by producing black-headed fruiting structures.
- Later fungus advances the entire inflorescence and fruit becomes rotten, and drooping occurs.

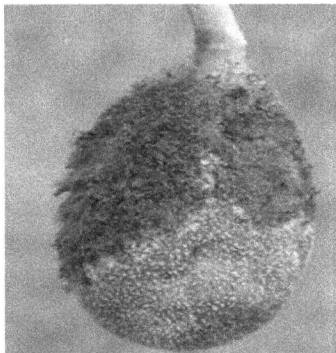

FIGURE 14.1 Rhizopus rot.

Causal Organism: *Rhizopus artocarpus, R. stolanifer, R. oryzae*

Disease Cycle:

Rhizopus survives in soil, compost, and other decaying plant material. It produces tiny fungal spores spread through wind, rains, and insects.

Epidemiology:

25 °C temperature, moist, warm, and humid conditions favor the Rhizopus infection and development.

Management:

- Removal and destruction of affected unripe fruits and male inflorescence.
- Rouging and cleaning of tree basin region.
- Avoid water stagnation in the basin region.
- Spraying trees with 1% Bordeaux mixture or 0.3% copper oxychloride.
- During harvesting, avoid fruit injuries or damage after harvesting wash fruits, and sun-drying before packing.

14.3　GRAY MOLD OF JACKFRUIT

Gray mold disease caused by *Botrytis cinerea L*. also blossom blight or fruit blight or botrytis rot. In India, it is one of the most serious disease causing severe yield loss and it is more common during flowering and fruiting season.

　　This disease was first time observed causing the premature fall of young fruits in FRI campus, Dehradun, Uttarakhand India and the loss was estimated up to 50%–55%. (Pandey et al., 1981).

Symptoms:

- Young fruits, tender buds, and stalk of fruits are mainly affected. Affected fruits become yellowish to brown.
- Necrotic spots appear on young buds and stalk later that spot covers with grayish green powdery mass.
- Affected buds get shriveled and premature drooping is observed. When the fungus attacks the peduncle and the portion below peduncles, the young shoots along with the leaves and fruits droop down and finally get desiccated.

Causal Organism: *Botrytis cinerea*

Disease Cycle:

The *Botrytis* survive on affected fallen fruits, dead tissues, and other plant hosts. Whenever fungus gets congenial conditions, it can produce *conidia* from hard resting structures like *sclerotia*.

Epidemiology:

Fungus can require temperatures between 15 °C and 20 °C and >90% high relative humidity.

Management:

Spraying Benlate (0.1%) or 1% Bordeaux mixture or 0.3% copper oxychloride to young fruits, young buds, and stalk at 3 weeks interval during January to March months.

14.4   LEAF SPOT/DIE BACK OF JACKFRUIT

Rao and Deshmukh (1986) reported a new leaf spot of jackfruit from Pune, India. This disease may be noticed during October–November.

Symptoms:

- The onset of the disease shows discoloration and darkening from the tip to downwards.
- Necrosis of infected twigs or branches and drooping.
- Affected branches ooze gum which is attacked by shoot borers leading to internal discoloration.
- In severe cases, on infected twigs, erumpent pycnidia can be seen.
- Infected leaves produce light to gray brown spots with dark brown margins and upward curling.
- Male inflorescences and young fruit are covered with black spore mass, and white mycelium causes poor fruit set and premature drop and it can also affect overripe fruit.

Causal Organism: *Lasidiplodia theobromae (Botryodiplodia theobromae)*

Disease Cycle:

- Primary spread: Dormant mycelium survives in affected debris.
- Secondary spread: Soil and air-borne conidia.

Epidemiology:

Pathogen requires an optimum temperature of 25 °C–30 °C with low relative humidity (80%–85%).

Management:

- Pruning of affected plant parts.
- After pruning, paste copper oxychloride or spray 0.3% copper oxychloride or 0.1% thiophanate methyl or 0.2% chlorothalonil.

14.5   PHYTOPHTHORA ROT OF JACKFRUIT

Under humid wet conditions Phytophthora palmivora causes significant losses.

Symptoms:

- Infection occurs through the wounded region.
- Initially produces water-soaked lesions and later lesions become enlarged to form light brown spots on fruits and finally become soft and lead to rotting.

Causal Organism: *Phytopthora palmivora*

Epidemiology:

Phytopthora palmivora thrives best at 25 °C–28 °C temperature and high soil moisture.

Management:

- Orchard sanitation should be a very necessary practice.
- Spraying 0.1% Benomyl and 1% Bordeaux mixture.
- Drenching of tree basins with 0.2% copper oxychloride
- Tree basin application of 1kg Trichoderma harzianum with farmyard manure.

14.6 LEAF SPOT OF JACKFRUIT

The disease was reported by Tandon and Bilgrami (1957).

Symptoms:

- Initially produces white spots with a brown center. Later turns to dark red on both the surface of the leaf in center of the mature spots seen black raised headed fruiting structure acervulus.
- Pathogens remain viable on diseased leaves for 7–8 months and serve as a source of inoculum.

Causal Organism: Petalotioposis artocarpina, P. clostica

Disease Cycle:

The primary source of inoculum conidia in plant debris persists saprophytically in plant debris. Secondary source of inoculum are conidia carried by wind splashed rain water.

Epidemiology:

Lower temperature (25 °C), long periods of high relative humidity, and rainfall favors the disease.

Management:

Spraying 1% Bordeaux mixture or 0.1% carbendazim or 0.2% thiophanate methyl.

14.7 RUST DISEASE

Symptoms:

- Small yellowish pustule appears on the upper surface of the leaves. Later these spots enlarge and turn a reddish-brown color. Corresponding to that on the lower surface of the leaf, pustules are reddish-brown in color and later produce a slightly raised, blister-like appearance.
- In severe infection, leaves may turn yellow or brown and premature drooping or defoliation occurs.

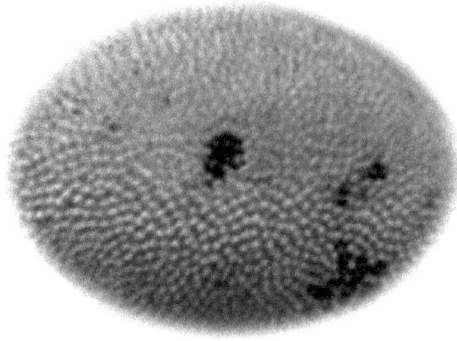

FIGURE 14.2   Rust symptoms.

Causal Organism: *Uredo artocarpii*

Disease Cycle:

Primary spread: Infected soil and plant debris.
Secondary spread: Air-borne uredospores.

Epidemiology:

Wet weather coupled with 25–30 °C temperature and >90% relative humidity.

Management:

- Removal and destruction of affected plants or plant parts.
- Spraying 0.2% oxycarboxin (Plantvax) or mancozeb or 0.05% triad-imefon (Bayleton).

14.8   ANTHRACNOSE

Symptoms:

- Initially small brown spots on leaves, twigs, and fruit cause blossom blight.
- Affected young leaves and tender twigs may wither and dry and may fall off from the tree.

- Brown spots appear on fruits near the stem end and the spots coalesce and become black. In severe case, pathogen enters the flesh and causes decay which is confined to the skin of the fruit.

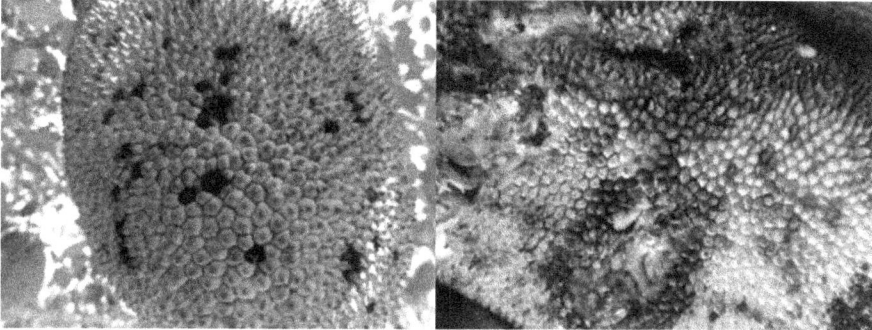

FIGURE 14.3   Anthracnose symptoms on fruits.

Causal Organism: *Colletotrichum gloeosporioides*

Disease Cycle:

Primary spread: Affected plant debris
Secondary spread: Air-borne conidia

Epidemiology:

20–25 °C and >95% RH continuous wet weather during flowering causes serious blossom blight.

Management:

- Diseased leaves, twigs, and fruits lying in the orchard should be collected and all the infected twigs should be pruned and burnt.
- Spraying of 0.2% zineb or 1% Bordeaux mixture twice at flowering in 15 days interval.
- 0.2% copper oxychloride + Zineb followed by 0.2% carbendazim at pea stage and 0.2% Zineb before stone hardening reduces the infection.

14.9   OTHER FUNGAL DISEASES

- *Alternaria* sp., *Botryodiplodia theobromae, Cercospora sp. Phomopsis sp. causes leaf spot.*
- *Pythium spendens*, *Phytophthora* spp., *Fusarium sp., Rhizoctonia* sp. infect roots and stem base of seedlings causes root rot.
- *Corynespora cassiicola* causes spots on leaves, stem, roots and flowers.
- *Diplodia theobromae* causes *Diplodia* fruit and collar rot, lesions on fruit, twigs, and branches.
- Brown rot and crown rot: *Phellinus noxius* (roots and base of the tree become encrusted by mycelium. Causes cracking of bark and gummosis. Infected wood becomes discolored, dry, friable, and honeycombed. Basidiocarps fruiting structure develops on base of the tree causing girdling and yellowing of the foliage and finally tree death.)
- Pinglap disease: Unknown pathogen (severe wilt or dieback syndrome resulting in the death of tree).

14.10   NEMATODE DISEASES

Symptoms:

- Stunting of the tree.
- Necrotic brown lesions on the root surface.
- Galling of roots.
- Necrosis and root-knot association occurs when secondary pathogen like fungi enter the root.

Causal Organism: *Meloidogyne* sp., *Pratylenchus* sp., *Aphelenchoides* sp., *Helicotylenchus dihystera, Helicotylenchus multicinctus, Hemicriconemoides cocophilus*, and *Ximphenema brevicoll.*

Management:

- Soil application of castor or neem cakes at 1 kg/tree.
- Application of carbofuran 3G 1 kg a.i./ha, or combination of neem cake at 1 kg/tree.

KEYWORDS

- jackfruit
- diseases
- symptomatology
- epidemiology
- management

REFERENCES

Alvarez, A. M. and Nishijima, W. T. (1987). Postharvest diseases of papaya. *Plant Dis. 71:* 681–686.

Alvarez, A. M., Hylin, J. W., and Ogata, J. N. (1977). Postharvest diseases of papaya reduced by biweekly orchard sprays. *Plant Dis. Rep.* 61: 731–735.

Basak, A. B. (1995). Fruit rot disease of jackfruit caused by *Colletotrichum gloeosporioides* Penz in Chittagong. *Bangladesh J. Botany.* 24: 197–199.

Butani, D. K. (1978). Pests and diseases of jackfruit in India and their control. *Fruits.* 33: 351–357.

Buttler, E. J. and Bisby, G. R. (1931). Fungi of India. Monograph Imp. *Council Agric. Res.* 1

Crane, Jonathan H., Balerdi, Carlos F., and Maguire, I. (2002). Jackfruit Growing in the Florida Home Landscape. This document is Fact Sheet HS-882, one in a series of the Horticultural Sciences Department, UF/IFAS Extension.

Gupta, O. and Nema, K. G. (1979). Effect of different temperature and relative humidity on the development of fruit rots of papaya caused by *Botryodiplodia theobromae* and *Colletotrichum papayae*. *Indian Phytopathol.* 32: 106–107.

Guta, J. H. and Pandey, I. C. (1985). Chemical control fruit rot disease of jackfruit. *Prog. Hort.* 17: 361–362.

Haq, N. (2006). Jackfruit: *Artocarpus heterophyllus*. Southampton Centre for Underutilised Crops, Southampton.

Holliday, P. (1980). Fungus Diseases of Tropical Crops. Cambridge University Press, Cambridge.

Hunter, J. E. and Buddenhagen, I. W. (1972). Incidence, epidemiology and control of fruit diseases of papaya in Hawaii. *Trop. Agric.* (Trinidad). 49: 61–71.

Hunter, J. E. and Kunimoto, R. K. (1974). Dispersal of *Phytophthora palmivora* sporangia by wind-blown rain. *Phytopathology.* 64: 202–206.

McGrath, M. T. (1994). Fungicides provided insufficient suppression of Phytophthora fruit rot of cucurbits when disease pressure was high. (Abstr.) *Phytopathology.* 84: 1373.

McMillan, J. R. (1974). Rhizopus artocarpi rot of jackfruit (Artocarpus heterophyllus).

Mitra, M. (1929). Some diseases of crops in Andaman island. Agri. Res. Inst. Pusa Bull. 195.

Morton, J. (1987) Breadfruit. In: Fruits of Warm Climates. Morton Collectanea. University of Miami, Coral Gables, FL, USA.

Morton, J. (1987). Jackfruit. In: Fruits of Warm Climates. Julia F. Morton, Miami, FL, USA, pp. 59–64.

Nelson, S. (2005). Rhizopus Rot of Jackfruit. Extension Service/CTAHR, University of Hawai at Mänoa, Honolulu, Hawai.

Nishijima, W. T., Fernandez, J. A., and Ebersole, S. (1990). Factors influencing development of postharvest incidence of Rhizopus soft rot of papayas. pp. 495–502. In R.E. Paull, Ed., Symposium on Tropical Fruit in International Trade, Honolulu, Hawaii, *Acta Horticulturae,* 269.

Pandey R. S., Bhargava, S. N., Shukla, D. N., Khati, D. V. S. (1979). Control of *Rhizopus* rot of Jackfruit. *Indian Phytopathl.* 32: 479–480.

Putter, C. A. J. (1998). *Phellinus noxius*. Global Plant Protection Information System, FAO, Rome.

Saldana, M. I., Marquez, M., and Ruiz, P. (1985). Identificacion de enfermedadesfungosasdelcultivo de la papaya (*Carica papaya* L.) en el estado de Tabasco. *Revista Mexicana de Fitopatologia*, 3: 14–17.

Sangchote, S., Wright, J., and Johnson, G. (2003). Diseases of breadfruit, jackfruit and related crops. In: R.C. Ploetz (Ed.), Diseases of Tropical Crops. CABI Publishing, Wallingford, Oxon, UK, pp. 135–145.

Sangchote, S., Wright, J. G., and Johnson, G. I. (2003). Diseases of breadfruit, jackfruit and related fruit crops. In: Ploetz, R. C. (Ed.) Diseases of Tropical Fruit Crops, CAB International Oxon, UK.

Ploetz, R. C. (ed) Diseases of Tropical Fruit Crops, CAB International Oxon, UK.

Shaw, D. E. (1984). Microorganisms in Papua New Guinea. Research Bulletin No. 33. Department of Primary Industry, Port Moresby.

Sirayoi, A. (1993). Diseases of Fruit Crops, Spices and Control. Takansak Foundation, Kasetsart University.

Srivasta, M. P. and Mehra, R. (2004). Diseases of minor tropical and sub-tropical fruits and their management. In: Naqvi SAMH (Ed.) Diseases of Fruits and Vegetables, vol. II. Springer, Netherlands, pp. 559–632.

Tandon. R. N. and Bilrami, K. S. (1957). *Proc. Natl. Acad. Sci. India.* 278: 204–209.

Trujillo, E. E. (1971). A list of diseases of economic plants in the Trust Territory of the Pacific Islands. Trust Territory of the Pacific Islands Department of Resources and Development, Division of Agriculture.

Vevai, E. J. (1971). Know your crop its pest problems and control—minor tropical fruits. Pesticides. 5: 33–54.

Zhang, J., Tong X., and Qixin, G. (2003). Notes on Pestalotiopsis from southern China. Mycotaxon, 85: 91–99.

CHAPTER 15

IMPORTANT DISEASES OF KARONDA (*CARISSA CARANDAS* LINN.) AND THEIR MANAGEMENT

PRITI S. SONAVANE* and V. VENKATARAVANAPPA

Central Horticultural Experiment Station, ICAR-IIHR, Chettalli 571248, Kodagu, Karnataka, India

Corresponding author. E-mail: Priti9883@gmail.com

ABSTRACT

Karonda is an evergreen shrub or short stature tree which is grown in various subtropical countries. It is a well suited to arid climate and grown well at higher temperature. It is found wildly in India, Malaysia, and South Africa. In India, it grows in Bihar, Gujarat, Karnataka, Maharashtra, Rajasthan, Uttar Pradesh, West Bengal, and other states. It is commonly used for making hedge for orchards. Karonda fruit is a rich source of iron and contains a fair amount of vitamin C. Mature fruit contains high amount of pectin. The fruits are used for pickle making, jam. jelly, squash, syrup, chutney, etc. Karonda plant as well as fruits attracted to various plant diseases which loss of yield. This chapter deals with important diseases of Karonda and their management strategies.

15.1 FUNGAL DISEASES

15.1.1 ANTHRACNOSE

Introduction/Economic Importance:

The disease is more common in tropical Asia and different parts of the world. In India, it is reported from UP, MP, AP, Maharashtra, Karnataka, Orissa,

Haryana, and West Bengal. It is widespread in UP areas wherever Karonda is cultivated.

FIGURE 15.1 *Carissa carandas* (Apocynaceae) plant with fresh fruits.

Symptoms:

The Karonda plants infected with anthracnose showed typical symptoms like water-soaked lesions, which are round to oval in shape have brownish red to black spots on the leaves (Figure 15.1). The disease also affects fruits and branches. The orchards sanitation like the burning of fallen leaves and fruits helps to reduce the inoculum.

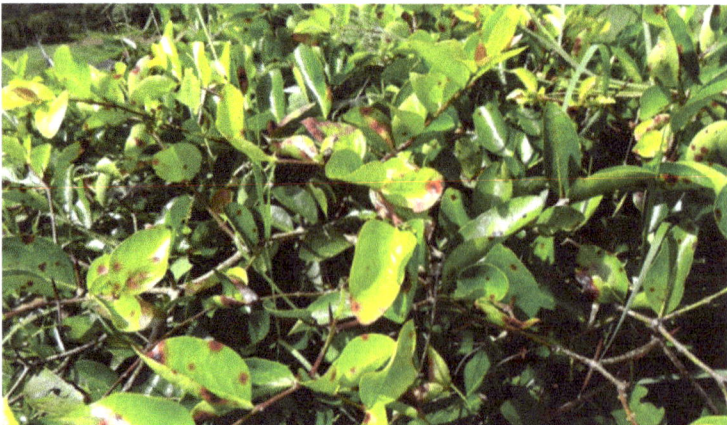

FIGURE 15.2 Symptoms: of anthracnose on leaf.

Casual Organism: *Collectotrichum inamdrii*

Etiology:

Collectotrichum inamdrii belongs to the Phyllachoraceae family of the division Ascomycota. The fungus is a facultative parasite. The colonies are pale orange with greenish gray or black, and pink, or reddish-purple underside of the leaf. The fungus produces hyaline, one-celled, slightly curved, or dumbbell shaped conidia is about 15–20 μm in length and 5–7 μm in width. The fungus produces nonseptate conidia having two or more oil globules at the center, which measure about 11–26 μm × 3.0–5.6 μm in length. The setae average 30.0 μm. Appressoria is light to medium brown, clavate to ovate with smooth margins. Acervuli are superficial and develop in lesions on leaves, branches, and fruits and remain viable for long periods (Gautam, 2014). The fungus is heterothallic in nature and the teleomorph can be readily induced in culture, it is observed rarely in the field.

Epidemiology:

The pathogen will survive under adverse conditions in the infected fallen twigs and injured plant tissues. Conidia are disseminating relatively short distances by overhead irrigation or rain splash. Ascospores are airborne and move to long distances. The fungus primarily invades into injured tissues of plants, produces specialized structures during infection. These specialized structures that will be produced during the interaction between host and pathogens are conidia, acervulli, setae, and appressoria. Moderate temperature of 30 °C and moisture are needed for production spores, germination, and infection to host. The conidia will be germinating immediately after contact with leaves, twigs, and fruit in turn produces appressoria, which infect the host (Lal and Singh, 1953). The pathogen is subsequently colonized in the tissue, then acervuli are formed, thus completing the pathogen's life cycle. The primary sources of inocula for infection are dead wood and infected plant debris. Quiescent infection of the pathogen on fruits will remain asymptomatic before harvesting.

Management:

The critical phases for disease control are flowering, fruit setting stage, and after harvest. It is most severe during wet weather when new growth flushes are particularly susceptible. However, prevention against the leaf spot symptom on new growth flushes requires regular spraying.

Cut and remove the diseased twigs; the wound is protected with 10% Bordopaste is quite effective in containing the disease.

- Spraying of Hexaconazole 5% EC (0.5 mL/L of water) or Carbendazim 50% WP alone or in combination with other fungicides such as Carbendazim (0.16%) are effective when floral buds begin to swell and throughout the fruit development period (intervals of 14–28 days; Peterson and Inch, 1980; Darvas and Kotze, 1987).

15.1.2   LEAF BLOTCH

Introduction/Economic Importance:

The pathogen was first described as *Cercospora angolensis* by De Carvalho and Mendes (1952), causing a leaf spot disease on *Citrus sinensis* in Angola. Later, based on the morphology of conidiophore and conidia production, the pathogen has been renamed as *Pseudocercospora*. The disease was first observed by Singh and Mukherjee (1979) under Faizabad conditions in India.

Symptoms:

A circular spot will appear on the infected leaves, later turning into light brown or grayish in color. The sporulation will take place during the rainy season. The spots are surrounded by a dark brown margin with a prominent yellow halo; occasionally the center of the spot will fall out, which are looking like a shot-hole symptom. The younger leaves are more susceptible than the older leaves. The leaf spots on younger leaves coalesce together and produce chlorosis, followed by premature abscission and defoliation of the infected leaves. Young leaves and fruit are more susceptible than older leaves. This may depend on the host species, variety, and location (Sief and Hillocks, 1993). On fruit, the lesions are circular to irregular resulting in hyperplasia, producing raised tumor-like growths. The spots on mature fruit are flat but sometimes have a slightly sunken brown center and ripen prematurely drop or dry up (Kuate, 1998).

Casual Organism: *Pseudocercospora angolensis*

Etiology:

The Leaf blotch disease is caused by *Pseudocercospora* sp. belongs to ascomycete fungi, which is an anamorphic stage of genus *Mycosphaerella*.

Conidiophores may be single, fasciculate, or forming loose synnemata 12–45 μm septate, smooth, brown, 120–240 μm × 4.5–7 μm in size, usually arising from a dark stroma 30–60 μm diam. Conidiogenous cells are terminal, geniculate, scars with slightly pigmented. Conidia are single or in simple or branched chains of 2–4, hyaline to brown, 3–6 septate, 24–79 μm × 4–5 (6.5) μm in size (Chalkley, 2016).

Epidemiology:

During the rainy season, the pathogen may be present on older leaves. Under favorable conditions, sporulation will start from the infected leaves after a period of 3–5 weeks and new symptoms appear on young leaves 2–3 weeks after the infection (Emechebe, 1981). The infected leaves present in the soil act as a primary source of inoculum for the next season. The fungal spore moved very long-distance through the wind (De Carvalho and Mendes, 1952); local dissemination of conidia by rain-splash and windborne. Undoubtedly human interference also helps in the dissemination of the fungal conidia through infected plant material or infected fruits from one area to another, where the pathogen was not present earlier (Seif and Hillocks, 1993). The fungus will survive in infected material as dormant mycelia until the onset of conditions are conducive to sporulation.

Management:

- Preharvest application of carbendazium 50% WP (0.1%).
- Preharvest sprays of *B. subtillis* (isolate B246, at a concentration of 10^7 cells mL) alone or in combination with different fungicides (Korsten et al., 1992).

15.2 BACTERIAL DISEASES

15.2.1 BACTERIAL CANKER

Introduction/Economic Importance:

The canker disease of Karonda is caused by X. *campestris* pv. *carissae*, was first described from Maharashtra by Moniz et al. (1964). Later, this disease was reported from Rajasthan (Kumar and Patel, 1968) and Uttar Pradesh (Garg and Gupta, 1980).

Symptoms:

Initial symptoms were observed on the under surface of the leaves as brown water-soaked, translucent lesions bordered by a clear halo, which becomes blister-like. Gradually, these spots expand and turn dark brown. Finally, they become raised, measuring 4–5 mm in diameters and surrounded by chlorotic halos. The central tissue of the lesion collapses and forms a shot-hole. More often the spots are produced on wounds and scars created by thorns during the rainy season (Kumar and Patel, 1968).

Casual Organism: *Xanthomonas campestris* pv. *carissae*

Etiology:

The canker is caused by *Xanthomonas campestris* pv. *carissae* (Moniz et al.) dye. Bacteria are positive to Gram-negative, rod-shaped, 0.4–1.0 μm × 1.2–3.0 μm in size with a single polar flagellum. The colony characteristics are mucoid, convex, and produce yellow pigment on yeast dextrose chalk agar and nutrient dextrose agar, colonies are large, smooth-domed, mucoid-fluidal, and yellow with entire edges; nonfluorescent on King's medium and sensitive to triphenyl tetrazolium chloride and oxidative test (Prasad et al., 1931).

Epidemiology:

The pathogen entry into the plant is through natural openings of stomata and hydathodes. The pathogen also enters through the wounds created by windblown soil and insects. Water soaking of the leaves, as caused by high-pressure sprays, greatly enhances bacterial spot infection. The splashes of rains are most favorable for bacterial dissemination under field conditions. The pathogen can also survive infected plant debris present in the soil for at least 1 year (Pohronezny et al., 1990; Bashan, 1986).

Management:

- Remove and destroy diseased plant parts in the orchards.
- Pruning of infected plant parts before monsoon and the pruned branches protected with 10% Bordeaux paste to prevent the spread of the disease.
- The disease will spread more on immature young leaves, immediately after noticing, spray 2-bromo-2-nitro-1,3-propanediol (Bronopol; 0.1%, 0.5 to 1 g/L) in combination with copper oxychloride 50% WP @ 2.5 g/L of water to effectively contain the disease.

15.3 ALGAL DISEASE

15.3.1 ALGAL LEAF SPOT

Introduction/Economic Importance:

The disease is also called algal leaf spot, algal fruit spot, green scurf, and red rust. This pathogen is aerophilic, terrestrial, and filamentous algae. They require moisture on the leaves to complete their life cycles (Nelson, 2008). The disease was first reported from India in 1906. The available information indicates that *Cephaleuros* sp. is limited in occurrence due to variation in temperature and humidity, which is not suitable for their growth and reproduction. Its major distribution in India is Tarai and other humid regions specially Bihar, Karnataka, West Bengal, Maharashtra, Gujarat, UP, Punjab, Haryana, and other states (Chowdhury, 1975; Prakash and Singh, 1979).

Symptoms:

Round, orange to red, raised velvety spots are produced on leaves, branches, or stems. The spots turn to white, gray, dark brown to nearly black as they age. A yellow halo may surround the spots (Pegg, 1991). Spore crusts from spots on older leaves are spread by wind and rain to new leaves. Spores are produced continuously during high rainfall and prolonged humid conditions that encourage the development of the disease (Figure 15.2). Reducing canopy humidity by judicious pruning, and applying copper fungicides in the spring controls the disease (Pegg, 1991; Vock, 2001).

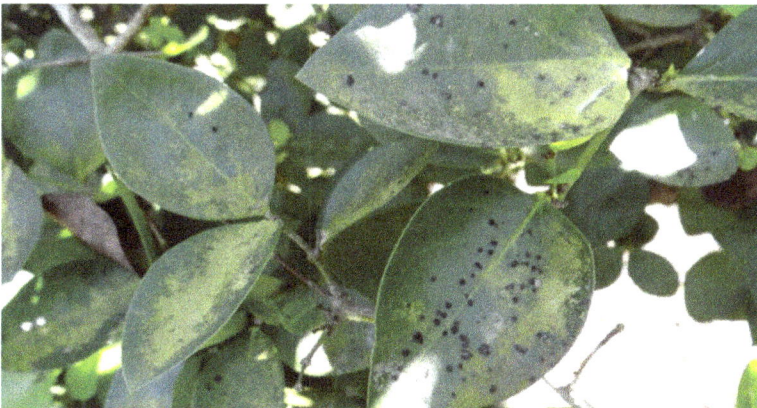

FIGURE 15.3 Symptoms of algal leaf spot.

Casual Organism: *Cephaleuros virescens*

Etiology:

The causal agent of the disease is *Cephaleuros virescens* that belongs to the order Trentepohliales and phylum Chlorophyta, which contains green algae is the most common of pathogens on tropical fruits. The species consist of branched filaments thalli to form irregular discs (Brooks, 2004; Chapman and Henk, 1985). The fungal thalli will grow below the epidermis or cuticle of the host plant. The thalus is orange to red-brown in color containing prostrate portion, which is branched irregularly with irregular cells and unbranched hairs produced from cylindrical cells. The haustorial is present inside the host's tissue. Sporangiophores have more than one head cells originated from sporangiate-laterals. The sexual reproduction through gametangia contact is terminal or intercalary (Chapman et al., 2007).

Epidemiology:

The mycelia fragments or sporangia that will deposit on host tissues may act as a primary source of inoculum for the spread of the algae. Under favorable condition, the motile zoospores are released from algae sporangia, which penetrate into the host cuticle and generate disc-like algal thalli from the algal filaments. The algal leaf spot reduced photosynthetic activity of the leaves leading to defoliation. The pathogens will survive on leaves or stems and in fallen plant host debris (Nelson, 2008).

Management:

- Remove infected leaves by hand.
- Prune branches, which are infected with the disease and discard all fallen leaves.
- Reduce stress to plant by applying the fertilized and improve soil drainage.
- Remove the unwanted plants around affected plants.
- The main stress for controlling alga is laid on correcting cultural malpractices and alleviating nutritional deficiencies and other limitations imposed by soil management. The direct link between host vigor and damage caused by the alga has been noticed. Such stress will bring about suboptimal plant growth and indirectly increase host susceptibility.

- Spraying of 1% Bordeaux mixture has been found effective in checking the algal infection.

15.4 POSTHARVEST DISEASES OF KARONDA FRUITS

15.4.1 FRUIT ROT OF KARONDA

Introduction/Economic Importance:

Fruit rot of Karonda was observed as a common and serious disease in local orchards in the Allahabad region of UP. These diseases rendered fruits unfit for human consumption (Kumar and Tandon, 1978). The pathogen infects both mature and immature fruits. The fruit rot of Karonda was caused by *Myrothecium roridum,* which was first reported on tomato in America (Stevenson and McColloch, 1947). Subsequently, the fruit rot was reported on coffee and many commercial crops (Schieber and Zentmyer, 1968; Tulloch Li, 1972; Zhao, 2009).

Symptoms:

The pathogen produced small circular, tan colored spots with broad violet to brown margin. Fructifications of the fungus appeared as dark green sporo-dochia. Depressed pin head size translucent spots were formed on tender and older fruits and turned to necrotic, which lead to rotting and produced shriveled/misshapen the quality fruit become very poor.

Casual Organism: *Myrothecium roridum*

Etiology:

The fruit rot disease of Karonda is caused by *Myrothecium roridum* Tode belongs to the Stachybotryaceae family (Kumar and Tandon, 1978). The sessile or stalked, viscid sporodochia having horny when it was dry and later it will turn to green to the black mass of conidia, usually bordered by a white zone with flocculent hyphae. The presence of stroma and the absence of hyphopodia. Setae are unbranched, colorless, or pale in color. Conidio-phores are macro and mononematous, closely together to form sporodochia. Monophialidic conidiogenous cells are cylindrical or discrete and clavate. The fungal spores are aggregated in black slimy masses, semi-endo, or acrogenous in nature (Ellis, 1971; Chen et al., 2016).

Epidemiology:

The fungus survived better in infected plant tissues to the next season of crop either debris buried or kept on the surface of the soil. The spores are probably disseminated by wind, water, and insects.

Management:

- Preharvest chemical treatments with different fungicides such Chlorothalonil 75% WP (0.2%) or Carbendazim 50% WP (0.1%) followed by a combination of Mancozeb 63% + Carbendazim 12% WP (0.2%) are effective when floral buds begin to swell and throughout the fruit development period (intervals of 28 days).

KEYWORDS

- **Karonda**
- **diseases**
- **symptomatology**
- **epidemiology**
- **management**

REFERENCES

Bashan, Y. (1986). Field dispersal of *Pseudomonas syringae* pv. *tomato*, *Xanthomonas campestris* pv. *vesicatoria* and *Alternaria macrospora* by animals, people, birds, insects, mites, agricultural tools, aircraft, soil particles and water sources. *Can. J Bot.*, **64**:276–281.

Brooks, F. (2004). Plant parasitic algae (Trentepohliales, Chlorophyta) in American Samoa. *Pacific Sci.*, **58**:419–428.

Chalkley, D. (2016). Angular leaf spot of citrus - *Pseudocercospora angolensis*, *Syst. Mycol. Microbiol. Lab.*, ARS, USDA. Invasive Fungi.

Chapman, R. L. and Henk, M. C. (1985). Observations on the habit, morphology, and ultrastructure of *Cephaleuros parasiticus* (Chlorophyta) and a comparison with *C. virescens*. *J. Phycol.*, **21**:513–522.

Chapman, R. L., Henk, M. C. Lopez-Bautista, J., Rindi, F., and Casamatta, D. (2007). The systematics of subaerial algae. In: Algae and Cyanobacteria in Extreme Environments. Springer, Netherlands, pp. 599–617.

Chowdhury (1975). Occurrence of red rust in Madhya Pradesh. *Indian Phytopathol.*, **28**:261–262.

Chen, Y., Ran, S. F., Dai, D. Q., Wang, Y., Hyde, K. D., Wu, Y. M., and Jiang, Y. L. (2016). Mycosphere Essays 2. *Myrothecium*. *Mycosphere*. **7(1)**:64–80.

Darvas, J. M. and Kotze, J. M. (1987). Avocado fruit diseases and their control in South Africa. South African Avocado Growers Association Yearbook. **10**:113–116.

De Carvalho, T. and Mendes, O. (1952). Uma cercospo- riose em Citrinos. *Motambique*, **72**:1–8.

Ellis, M. B. (1971). Dematiaceous hyphomycetes. Commonwealth Mycological Institute, Kew, 552–556.

Emechebe, A.M. (1981). Brown spot of citrus caused by *Phaeoisariopsis* sp. *Ann. Appl. Biol.*, **97**:257–262.

Garg, P. K. and Gupta M. N. (1980). *Indian Phytopath.*, **33**:481.

Gautam, A. K. (2014). *Colletotrichum gloeosporioides*: Biology, pathogenicity and management in India. *J. Plant Physiol. Pathol.* 2:2. http://dx.doi.org/10.4172/2329–955X.1000125.

Korsten, L., Lonsdale, J. H., De villiers, E. E., and De Jager, E. S. (1992). Effect of *Bacillus subtilis* and fungicide sprays for control of pre-harvest diseases of avocado. South African Avocado Growers' Association Yearbook. **10**:117–119.

Kuate, J. (1998). Citrus fruit and leaf spot disease caused by *Phaeoramularia angolensis*. *Cahiers Agric.*, **7**:121–129.

Kumar, S. and Tandan, M. P. (1978).A new fruit rot of Carissa. *Curr. Sci.*, **51**:119.

Kumar, V. and Patel, P. N. (1968). Symptoms and host range in bacterial disease on *Carissa carondas* L. *Indian Phytopathology.*, **21**:315–317.

Lal, A and Singh, R. P. (1953). Anthracnose disease of *Carissa carandas* Linn, caused by *Colletotrichum inamdarii*. *J. Indian Bot. Soc.*, **32**:54–63.

Li, B. J. and Zhao, Y. J. (2009). The identification of paint spot pathogen on eggplant (in Chinese). *China Veg.*, **15**:20–21.

Moniz, L., Sabeya J. E., and More, W. P. (1964). *Indian Phytopathol.*, **17**:256.

Nelson, S. C. (2008). *Cephaleuros* species, the plant-parasitic green algae. College of Tropical Agriculture and Human Resources (CTAHR), http://www.ctahr.hawaii.edu/freepubs.

Pegg, K. G. (1991). Causes of disease, In: Broadly, R.H. (ed.) Avocado Pests and Disorders. Queensland Deparment of Primary Industries. Brisbane, pp. 1–7.

Peterson, R. A. and Inch, A. J. (1980). Control of anthracnose on avocados in Queensland. *Queensl. J. Agric. Anim. Sci.*, **37**:79–83.

Pohronezny, K., Moss, M. A., Dankers, W., and Schenk, J. (1990). Dispersal and management of *Xanthomonas campestris* pv. *vesicatoria* during thinning of direct-seeded tomato. *Plant Dis.*, **74**:800–805.

Prakash, O. M. and Singh, U. N. (1979). Fungicidal control of red rust of mango. *Indian J. Myco. Pl. Pathol.*, **9**:175–176.

Prasad, P., Pathak, V. N. Goyal, J. P., and Szarma, H. C. (1931). Control of bacterial canker of karonda. Secont Int. symposium on Plant Pathology orgaised by IPS, 1971.

Schieber, E. and Zentmyer, G. A. (1968) *Myrothecium* stem necrosis and leaf spot: important coffee disease in Guatemala. *Plant Dis. Rep.*, **52**:115–117.

Seif, A. A. and Hillocks, R. J. (1993). Phaeoramularia fruit and leaf spot of citrus with special reference to Kenya. *Intl. J. Pest Manage.*, **39**:44–50.

Seif, A.A. and Hillocks, R.J. (1997). Chemical control of Phaeoramularia fruit and leaf spot of citrus in Kenya. *Crop Prot.*, **16**:141–145.

Singh, N. and Mukerji, K. G. (1979). Stuides on Indian Coprophilous fungi. In. some rare records. *J. Indian Bot. Soc.*, **58:**163–167.

Stevenson, J. A. and McColloch, L. P. (1947). Myrothecium as a tomato fruit rot organism. *Plant Dis. Rep.*, **31:**147–149.

Tulloch, M. (1972).The genus *Myrothecium Tode* ex Fr. *Mycol. Pap.*, **130**: 1–42.

Vock, N. (2001). Avocado information Kit. Agrilink series Qal. Dpi, Queensland.

CHAPTER 16

DISEASES AND PHYSIOLOGICAL DISORDER SPECTRUM IN LITCHI (*LITCHI CHINENSIS* SONN.)/LONGAN (*DIMOCARPUS LONGAN* LOUR.) AND THEIR MANAGEMENT

J. N. SRIVASTAVA[1,*], A. K. SINGH[2], RAKESH KUMAR SHARMA[3], and V. B. SINGH[2]

[1]*Department of Plant Pathology, Bihar Agricultural University, Sabour, Bhagalpur, Bihar, India*

[2]*Division of Plant Pathology, Sher-e-Kashmir University of Agriculture Sciences and Technology Jammu, Jammu and Kashmir, India*

[3]*Division of Fruit Science, Sher-e-Kashmir University of Agriculture Sciences and Technology Jammu, Jammu and Kashmir, India*

[*]*Corresponding author. E-mail: j.n.srivastva1971@gmail.com*

ABSTRACT

Litchi *(Litchi chinensis* Som.) is a subtropical evergreen tree which is grown in various subtropical countries and belonging to the family Sapindaceae. Litchi is grown in Assam, Bihar, Punjab, Tamil Nadu, Tripura, Uttar Pradesh, and West Bengal. In India, the major litchi growing area is Bihar, particularly in its northern part especially Muzaffarpur and Darbhanga districts. Other important areas of litchi are Dehradun (Uttarakhand), Saharanpur and Muzaffarnagar in Uttar Pradesh. Litchi is soft and juicy fruits with delicious flavor. Several diseases have been reported to occur on litchi but none of them causes any major disease of economic importance. Some postharvest rots commonly occur after the injury on fruits causing major economic loss in marketing and export of litchi fruits. This chapter deals with important diseases of Litchi and their management strategies.

16.1 INTRODUCTION

Litchi or lychee (*Litchi chinensis* Sonn.) is a juicy fruit belonging to the family Sapindaceae. "Lychee or litchi (*Litchi chinensis* ssp. *chinensis*) is an evergreen subtropical fruit tree native to southern China that is now grown in a number of subtropical and tropical countries for its desirable high-value fruit" (Nakasone and Paull, 1998).

Litchi or Lychee (*Litchi chinensis* Sonn.) is one of the most important evergreen fruit trees mostly grown in Assam, Bihar, Jharkhand, Uttar Pradesh, Punjab, West Bengal, and some irrigated pockets of Jammu and Kashmir subtropics. These Indian states produce 78% of the total production. Litchi originated in China, where it was cultivated for 300 years and was introduced by the end of the 17th century to Burma and India. It is a highly priced, common, and important table fruit in India known as "fruit queen" which is grown in around 84,000 hectares with a production of around 5.85 lakh metric tons and productivity of around 7.0 metric tons/ha (NHB, 2015).

Diseases are one of the main threats in litchi production, with many diseases affecting the cultivation of litchi sometimes. Diseases caused indirectly and adversely affect the quality and quantity of fruit yield. An effective disease management strategy depends mainly on selecting good planting materials, sanitation in the field, balanced dose of manures and fertilizers, scientific cultivation and management of crops, effective monitoring of diseases through regular seasonal scouting, pruning and destruction of affected plant parts, and ultimately the use of pesticides as required and, finally, a harvesting and shipping procedure that maximizes the shelf life and fruit quality.

Litchi is infected with many microorganisms that cause diseases in various parts of plants, such as trunks, roots, twigs, leaves, flowers, and fruits. There are records that are not available of bacterial or viral infections of litchi plants, but there are also few records of algal and nematode diseases. Litchi has great export potential, but many diseases affect its fruit quality and therefore impede its export. Postharvest diseases are also threats to its export. It is therefore necessary to know about the treatment and control of these diseases.

16.2 MAJOR DISEASES OF LITCHI

16.2.1 LEAF SPOT OR TWIG BLIGHT

"Twig blight" is very important among emerging disease issues. Disease was first time reported by Prasad (1962, 1967), from Muzaffarpur, Bihar in India. Twig blight is one of the emerging problems with disease in litchi orchards today (Kumar et al., 2014a,b, 2011).

Symptoms:

Visual symptoms emerged on leaves that grew on new shoots. Influenced leaves showed as blighted and further dieback. Just as occurred before the time process of the disease improvement tip dieback. The leaf blight appears on the leaves like tanning spots. The troubled leaves look like they are scorched from the sun. The severity of the disease increased several times when the plants were infested with the foliage feeding pest complex, in particular the new weevil virus, leaf miner, leaf cutting weevil, and shoot borer not only seriously impeded the growth of young plants, but also decreased the potential for fruit-bearing flushes in the orchards cultivated (Awasthi et al., 2005).

Casual Organism: *Colletotrichum gloeosporioides* Penz. (*Glomerella cingulata* in the sexual state)

Pathogen:

Colletotrichum gloeosporioides Penz., a facultative parasite has a position with the Melanconiales order. The fungus produces hyaline, single-celled, ovoid to oblong, slightly bent, or dumbbell-shaped conidia, 10–15 µm long and 5–7 µm wide. Masses of conidia tend to be pink, or hued with salmon. The waxy acervuli formed in the infected tissue are subepidermal, usually with setae, and conidiophores are smooth, flat, and erect.

Disease Cycle:

C. gloeosporioides get several acervules and conidia on dead twigs and damaged tissues and structures of plants. Conidia shall spread with a down-pour sprinkle or overhead water system over typically short separations. Ascospores are airborne and important at large distances of dispersion. Conidia, which interact with leaves, twigs, and natural products, grow to create appressoria and calm pollution that causes tissue rot. As a result, this tissue is colonized, acervuli is produced along these lines, completing the pathogen's life cycle. Dead wood and plant flotsam and jetsam are important inoculum springs. Natural products with peaceful contaminations stay asymptomatic before reaping. Wounds and tissues debilitated by different variables bring about additional advancement of tranquil diseases to frame sores at postcollecting.

Epidemiology:

Hot temperatures (28 °C) and high humidity promote the growth and production of pathogens. Spores germination is very weak below 97% relative humidity. In order to release the spores from acervuli and ensure their germination, a high moisture content is required. Rain splashing is a common method of spreading.

Integrated Disease Management::

- Pruning and destruction of affected leaves and twigs can effectively manage the disease.
- Copper oxychloride (0.25%) or carbendazim (0.1%) spray may be used when the severity of the disease increases.
- Spray carbendazim @ 0.1% or chlorothalonl @ 0.25% or difenconazole @ 0.05% thiophanate-methyl @ 0.1% or mancozeb @ 0.2% at fortnightly intervals.

16.2.2 LITCHI BROWN BLIGHT/LITCHI DOWNY BLIGHT/LITCHI BLOSSOM BLIGHT/LITCHI DOWNY MILDEW

Litchi brown blight/litchi downy blight/litchi blossom blight/litchi downy mildew is a major fruit disease of litchi. It is popular in Tawan, China, Thailand, and Vietnam (Coates et al., 2003). Chen first isolated the causal pathogen from lithium fruit in Taiwan in 1934, and described it in 1961 in a new genus (Chen, 1961; Kao and Leu, 1980). The disease has been documented by several workers from China (Chi et al., 1984), Taiwan (Kao and Leu, 1980), Thailand (Sittigul, 2000—Staff Communication), and Vietnam (Vien et al., 2001). The disease is caused by *Peronophythora litchii* and attacks on young and mature fruits, pedicels, and leaves, but flower panicles are especially susceptible to the pathogen. The infection lowers fruit production and shelf-life. Litchi brown blight or downy blight is the most important disease which causes severe economic losses. This disease is caused by *Peronophythora litchi* (Ou, 2001).

Symptoms:

The disease affects young leafy foods with natural products such as seeds, flowers, and pedicels of litchi trees (Chi et al., 1984). The disease is described as earthy colored wounds and the production of white wool on leaves and young organic products or an untimely natural decrease in the substance. Juvenile organic products darkened by *Peronophythora litchi* have a brownish appearance, while age-old natural products have a white appearance due to the white mold that grows on the skin. Chen (1961) accounted for small, light, dark earthy wounds caused by natural product skin (restricted to the point of contact with nearby organic products). Such lesions are about 1–1.5 cm wide and covered with light yellowish aeronautical mycelia. The exocarp's inner surface is paler than the outer surface with the stem's tissue—the same deep dark earthy hue. Rain during the late contamination stage transforms the flowers and panicles with earthy colors, which are protected with white masses with sporangiophores and sporangiophores. Finally, the panicles evaporate. Unpredictable earthy-colored wounds with a muddled fringe are also seen on young and ready organic goods, pedicels, and leaves (Ann and Ko, 1984). The stained natural product could fall rashly.

Casual Organism: *Peronophythora litchii* (Pythiales: Pythiaceae)

In the Peronosporaceae family, higher temperatures during the day are suitable for sporulation, germination, and pathogen disease, and lower temperatures and high humidity around Chen (1934). Ho et al. (1978) reinvestigated the microorganism's morphological properties and created the autonomous Peronophythoraceae family, in the middle of the path between Peronosporaceae and Pythiaceae, to compel the genus (CMI, 1979; Kao and Leu, 1980). Oh, Huang et al guy (1983) revised the family portrait in order to understand the multideterminate existence of the sporangiophores.

Pathogen:

Chromistan *Peronophythora litchii* is a facultative necrotroph that produces dull, 4–6 μm-wide mycelium aseptate, unpredictably extended at the right or extreme point (Hall, 1989). Peronophythora litchii is an oomycete that is a diploid eukaryotic parasite. The causal living being creates different sporangiophores with dichotomous branchlets. At the point where the growth of sporangiophores had ended, sporangia had simultaneously spread and increased at each tip of the branchlet. Determined sporangiophore is natural for molds of wool. Peronophythora litchii takes after wool molds as far as sporangiophore morphology, while mycelium and sexual generation qualities, just as the way the species can be effectively refined, and makes it like the species *Phytophthora* and *Pythium*.

The chromistan's nutritional requirements are close to those of Phytophthora and mycelium has detected amylase activity (Huang et al., 1983).

Disease Cycle:

The pathogens, as zoospores or inactive mycelium, live in soil or in plant waste. Higher daytime temperatures are ideal for sporulation, germination, and pathogen disease, while lower evening temperatures and high humidity promote zoosporic discharge and spread. Ceaseless downpour and recontamination are the most critical components of the broad dissemination of this disease. The pathogen lives in the skins of the soil or contaminated organic product overwintering mostly in the spores. Spores form and spread like sporangia. Sporangia are not released by moving air but are quickly dispersed in sprinkle beads, indicating a downpour sprinkling device (CMI, 1989), and can be further spread by creepy creeps.

Epidemiology:

Kao and Leu (1980) found that the optimal temperatures for mycelial production, sporulation, and germination of *P. litchii* sporangia are consistent with the organic litchi development process, where temperatures range from 20 °C to 28 °C. Higher daytime temperatures are ideal for sporulation, germination, and pathogenic disease, and lower nighttime temperatures and high humidity facilitate the discharge and transportation of zoospores. *Peronophythora litchii* has an optimal temperature for growth of 30 °C, an average temperature of 36 °C, and a base temperature of 12 °C (Chen et al., 1998a). The pathogen creates the most sporangia at 25 °C, yet no sporangia are delivered at temperatures higher than 35 °C or lower than 15 °C with blustery spring days during contamination causing genuine misfortunes.

The pathogen lives mainly in spores overwintering in the soil or on infected fruit skins.

Integrated Disease Management:

- Sanitation of orchards is an important process for the control of the disease.

- Removing infected and dead branches of the crops after harvest.
- In spring, higher temperatures and higher soil dampness combine two splashes of copper sulfate @ 0.3% with a surfactant on the ground accompanying the application of lime, which also tends to minimize inoculum levels (CIQ, 2000; Zhang, 1997). Additionally, copper oxychloride shower in winter and copper sulfate in the spring help to reduce inoculum levels.
- Winter orchard washing and winter spraying of trees with a 50% solution of copper oxychloride (1:600; Li, 1997).
- Chemical control options suggested in the growth stage, in the fruiting stage, and in the pready stage of the natural product are 90% aliette arrangements (1:400–500) and 58% Ridomil MZ or 64% Sadofan (1:600). Where pollution is serious, two applications are needed in mid to late March at the sprouting of blooms (Li, 1997).
- During flowering and fruit development, spray Ridomil MZ @ 0.25% reduces the onset of disease.
- Applications of mancozeb can help control diseases (Ou, 1999).
- Zentmyer and Mitchell (1985/86) suggested that a substantial number of control measures for *Phytophthora* diseases of tropical natural product trees would also apply to control *Peronophythora* organic product decline due to similarities between the two genera.

16.2.3 ALTERNARIA LEAF BLIGHT OR ALTERNARIA INFLORESCENCE BLIGHT OR ALTERNARIA FRUIT BLIGHT/FRUIT ROT DISEASE

In India, *Alternaria* leaf blight/*Alternaria* inflorescence blight and *Alternaria* fruit blight was first reported from Bihar, India (Kumar et al., 2018). Since the last 20 years, *Alternaria* spp. has been introduced and reported on several new host fruit trees. In consequence, the disease *Alternaria* leaf blight/ *Alternaria p*anicle blight caused by *Alternaria alternata* was first recorded on apple trees by Johnson et al. in the year 2000. Similarly, *Alternaria* leaf blight/*Alternaria p*anicle blight caused by *Alternaria alternata* was first recorded on citrus trees by Peever et al. in the year 2002 and *Alternaria* leaf blight/*Alternaria p*anicle blight caused by *Alternaria alternata* was first recorded on Pomegranate trees by Ezra et al. in the year 2010. Fruit blight in litchi trees and fruit rot in postharvest stage are also caused by *Alternaria alternate*. Postharvest litchi fruits decay was primarily reported in Australia

(Johnson et al., 2000), then in India (Kumar et al., 2016a) and in Pakistan (Alam et al., 2017). Postharvest losses of Litchi fruits up to about 35%–44% have been reported by Kumar et al. (2016a).

Symptoms:

Disease symptoms basically appeared on nursery plants but further symptoms also appeared on inflorescence and developing fruits of heading Litchi plants. The disease has been producing same symptoms in nursery plants and in orchard trees. A primary symptom looks like potassium deficiency and further, the symptom starts from the tip of the leaf as light brown colored to dark brown colored necrotic spots. In severe stage of disease, both the margins of the leaf leading to complete necrosis and affected leaves showed as dry up. At the last stage of the disease, the same pathogen also causes litchi fruit blight and postharvest litchi fruit decay/rots.

Casual Organism: *Alternaria alternata*

Pathogen:

The fungus *Alternaria helianthi* produces pale gray-yellow colored cylindrical conidiophores which are straight or curved as well as geniculate, simple or branched, septate, and bear single conidium (straight or slightly curved, cylindrical to long ellipsoid, multiseptate, pale gray-yellow to pale brown).

Disease Cycle:

The pathogen survives by conidia in infected leaves/fallen leaves and also in litchi canopy. Infected leaves act as primary inoculum and wind-borne conidia act as a secondary spread.

Epidemiology:

Hot weather and frequent rain during the flowering and fruit setting stage favor the infection.

Management:

- Destruction of infected leaves is an essential practice.
- Follow sanitation practices.
- Spraying of copper oxychloride 3 g/L water, mancozeb 2.5 g/L water, carbendazim 2.5 g/L water reduces the disease.

16.2.4   ANTHRACNOSE DISEASE

Anthracnose is an important disease of litchi It is also considered as a postharvest disease, especially in regions with high rainfall. Anthracnose is an essential litchi fruit disease caused by *Colletotrichum gloeosporoides* in India. Symptoms usually occurred on the fruit, often affecting the leaves, twigs, and flowers of the litchi. Owing to anthracnose fungus, the quality of fruit has decreased and become unmarketable.

Symptoms:

Anthracnose can grow on immature litchi fruits and cause premature abscission (Nakasone and Paull, 1988), but it affects both harvested fruits and ripened fruit in the field more frequently (McMillan, 1994b). The disease usually appears at the top of the semimature fruit first as a brown pinhead lesion. Dark-brown to black, sunken lesions on mature fruits are readily visible in the early stages of circulatory disease. Across the top and side of the fruit, there are more lines, which slowly cover the fruit surface (Kumar, 2016).

The symptoms on the leaves begin at the tip of the lamina, extending toward the base. These spots in the diagram are sporadic, and a buckthorn of earthy colored shade. Anthracnose of litchi organic products was defined as inner and outer earthy round wounds with much-characterized edges on pericarp, but not on aril or seed (Prasad and Bilgrami, 1974). The highest latent disease incidence of natural product anthracnose induces increasingly genuine postharvest rot and caramelization of fruits. The pace of inactive contamination of organic products could be obviously constrained by coordinated administration of the illness in the developing season and the postharvest rot and searing of the natural products are viably diminished. Normally, shallow skin imperfections do not influence creation and fruit quality, yet attractiveness is influenced (Awasthi et al., 2005).

Casual Organism: *Colletotrichum gloeosporioides* Penz. (*Glomerella cingulata* in the sexual state)

Pathogen:

Colletotrichum gloeosporioides Penz., a facultative parasite belongs to the order *Melanconiales.* The fungus produces conidia produced by single-celled, ovoid to oval, slightly bent or free weight, 10–15 μm long and 5–7 μm wide. Masses of conidia tend to be pink, or hued with salmon. In the tinted tissue, the waxy acervules are subepidermal, usually with setae, and simple, thin, erect conidiophores.

Disease Cycle:

C. gloeosporioides causes a plenitude of acervuli and conidia due to dead twigs and damaged plant tissues and structures. Conidia will spread with a downpour sprinkler or overhead water device over generally short separations. Ascospores are airborne and important at large distances of dispersion. Conidia, which interacts with leaves, twigs, and organic materials, develops to create appressoria and calming pollution that causes tissue rot. As a result, this tissue is colonized, acervuli are produced along these lines to complete the pathogen's life cycle. Dead wood and plant flotsam and jetsam are essential wellsprings of inocula. Organic products with peaceful contaminations stay asymptomatic before collecting. Wounds and tissues debilitated by different variables bring about additional improvement of tranquil diseases to frame sores at postcollecting.

Epidemiology:

High temperatures are the natural conditions favored by the pathogen, 28 °C being an optimal and high mugginess. Spores must have free sprouting water; germination below 97% relative humidity is negligible. Spores can be discharged from acervuli when humidity is abundant. Sprinkling with downpour is a common form of distribution. Severe infection is associated with the environment and the parasite is usually idle in the dry climate.

High temperature and high relative humidity favored the disease. Outbreaks of the disease are normal following warm wet weather.

Integrated Disease Management:

- Diseased leaves, twigs, nerve midge contaminated leaves, and natural products ought to be gathered and consumed.
- Prevent obstruction of planted trees and branches.
- Pruning and processing of affected plants have been suggested to reduce the risk of new contaminations.
- Sprinkling of copper oxychloride (0.25%) or carbendazim (0.1%) or chlorothalonil (0.15%) or difenconazole (0.05%) or azoxystrobin (0.3%), respectively. Sprinkling of fungicides helps to extend post-gathering life but, at the same time, care must be taken to keep the substance toxic.

16.2.5   WILT DISEASE

Young litchi trees, often below 5 years of age, will wilt in less than a week (Kumar et al., 2011).

Symptoms:

Young litchi trees, typically less than 5 years old, decrease in less than 7 days. The main symptoms occur as the foliage yellowing, the hanging leaves followed by gradual withering and drying, resulting in the plant being completely killed within 4–5 days. Some earthy-colored spots appear on the crown root and the phloem of parallel roots, which later spread to the xylem. Essentially the shrink's causal living being is Fusarium solani. In soil, chlamydospore is the source of endurance. Through the development of weakened soil, the pathogen spreads from the infected plant to the sound plant rhizosphere.

Casual Organism: *Fusarium solani*

Three types of abiogenetic spores were produced by *Fusarium solani*: microconidia, macroconidia, and chlamydospore. Microconidia are the most prevalently distributed spores. They are oval, circular, or formed on aeronautical mycelia, or kidney. Macroconidia, which has three to five cells and gradually has pointed or bent edges, is found on the outside of an unsanitary plant on sporodochia. Chlamydospores are generally represented separately or two by two, but may be found in groups or in short chains once in a while. They are round, thick-walled spores delivered on more developed mycelium or macroconidia, either inside or terminally. Chlamydospores not at all like different spores can make due in the dirt for an extensive stretch of time (Agrios, 2005).

Disease Cycle:

Growth is a typical soil pathogen and saprophyte which benefits from a natural problem which is dead and rotting. This creates soil garbage as mycelium and other forms of spores, most often recovered as chlamydospores from the soil. This pathogen spreads in two fundamental ways: spreading short separations by water sprinkles and planting equipment and spreading long distances through infected transplants and seeds.

Microconidia will go into the stream of the sap and be carried upward where sap's motion slows the microconidia's growth. In the long run, the vascular vessels block spores and mycelia which prevent the plant from taking up and translocating supplements. Finally, the plant goes beyond what it can bring, closing the stomata, cleaning the leaves, and removing the bucket from the field. The organism destroys all tissues, sporulates, and keeps contaminating adjacent plants after the plant bites.

Epidemiology:

Warm soil temperature and low soil moisture favor Fusarium wilt infection and disease development (Lewis, 2003). 27–31 °C temperature and 28%–75% soil moisture with a pH ranging between 6 and 8 are optimum for disease development. In comparison to the heavier mud soil, the disease is increasingly severe in sandy soil. The pathogen reaches the plant via the tips of the root and can live in the soil for as long as 30 years. Mycelium grows in xylem tubes, where the water is flexibly cut off and begins to shrink.

Integrated Disease Management:

- Use castor cake as manure or neem cake.
- Apply *Trichoderma harizanum*, *Trichoderma viride*, and *Pseudomonas fluorescens*, and so forth as biocontrol agents that also serve as promoters of plant growth.
- Drench rhizosphere soil with hexaconazole or carbendazim (0.1%) in the absence of a biocontrol agent.
- Litchi farmers are also advised not to plant litchi trees on waterlogged soils or in low-lying areas which often receive floodwater.

16.2.6   *PHYTOPHTHORA LEAF BLIGHT/FRUIT ROT/ROOT ROT*

In Thailand, longan infections that cause natural product spoilage and lychee often cause root decay are considered to be relatively important (DOA, 2003a,b). It is conceivable that low-hanging longans or lychees or dropping organic products may be polluted and damaged. *Phytophthora* leaf scourge and natural product decay do not cause longan disease where organic products are stimulated to grow during the slow time of year (Visitpanich et al., 2000).

Symptoms:

Indications are produced during the cool climate, particularly after 2–3 days of precipitation (Bhavakul et al., 1998). Side effects in litchi and longan incorporate dull corruption of young shoots, an earthy colored scourge on panicles, bloom drop, sporadic earthy-colored sores on soil products develop natural product drop and in some cases cause root spoilage. During wet weather, the fruit will crack and lesions will be covered by white sporangia and in litchi and longan, infection occurs by zoospores, released when there is free moisture. The disease then develops during extended periods of wet weather (DOA, 2003a, b).

Casual Organism: *Phytophthora palmivora*

Disease Cycle:

Phytophthora palmivora has four types of spores that can cause disease, either directly or by implication. Sporangia are formed on polluted natural products, leaves, stems, or roots. They can legitimately sprout on the surface of the plant or in the soil. They can also grow to deliver tiny, swimming zoospores. Zoospores float in soil water or on a wet plant surface before they reach the plant in the long run. Sporangia and zoospores are dispersed by downpour sprinkles, wind-blown downpour, soil, and surface water.

P. palmivora may undergo dry periods such as torpid chlamydospores, oospores, or lethargic mycelium, and may produce sporangia and zoospores when returning from the downpour. Chlamydospores are also present in organic tissue and have the most important structure for durability (CABI, 2002). Chlamydospores and oospores are significant endurance structures in this way. They can live in soil or dead plant material during periods when host plants are missing.

Epidemiology:

Rain and wind are the two focal points of the *P. palmivora* disease transmission research. For the release of *P. palmivora* sporangia from the outside of the infected organic product into the air and for the projection of inoculum dirt into the air, downpour sprinkles are needed. Wind is necessary to disperse the inoculum once it enters the air. Wind-blown downpour is therefore important for the onset of a major disease and for the advancement of pests in papaya plantations. The pathogen responds at a temperature of 18–20 °C amid high stickiness. Wind dispersal of inoculum and wind blow downpour grants distribution and enhancement of scourges (CABI, 2002).

Integrated Disease Management:

- Removal of infected fruits and plant material.
- Sanitation of orchards is an important process for the control of the disease.
- Prophylactic spray of mancozeb @ 0.2% at canopy can reduce the disease.
- Application of cupper oxychloride @ 0.3% can reduce the disease.
- As an effective chemical regulation of this disease, the soil at the base of the tree trunk can be drilled with metalaxyl (DOA, 2003a).

16.2.7   LEAF SPOT DISEASES

Infection is typically combined with various pathogens. Older leaves indicate higher infections. From Muzaffarpur (Bihar) India, namely *Pestalotia pauciseta*, *Botryodiplodia theobromae*, and *Colletotrichum gloeosporioides*, three pathogenic fungi causing litchi leaf spots have been identified according to Prasad (1962). In nature, infection incidence may be either all three or, more commonly, *Botryodiplodia theobromae*, or *Colletotrichum gleosporiodie*. The details of the leaf spots are given below.

16.2.7.1   BOTRYODIPLODIA LEAF SPOTS

Prasad (1962, 1967) documented infection with Muzaffarpur, Bihar. The disease is caused by the *Botryodiplodia theobromae*. Disease manifestations start with the tip or edge of the lamina. Those spots in the shade are dark chocolate. The spot's small edges are impredictable. Pycnidia however appear on both surfaces of the leaf at the top of the leaf surface. The time-frame for hatching varies from 3 to 7 days.

Coates et al. (2003) indicate that by cutting out dead leaves and twigs in the overhang, the inoculum from the collection of growths that cause the stem end of longan, lychee, and rambutan decay can be decreased, thereby raising the airflow and providing a less good condition.

16.2.7.2   COLLETOTRICHUM LEAF SPOTS/LEAF NECROSIS

The disease is caused by *Colletotrichum gloeosporioides*. The symptoms start at the tip of the lamina and spread out to the base. The spots are tiny and brown in color, surrounded by prominent brown marsh. For *Colletotrichum gloeosporioides* the incubation period is 4–8 days.

16.2.7.3   GLOEOSPORIUM LEAF SPOTS/BLIGHT

Gloeosporium leaf spots/blight disease is due to the *Gloeosporium* sp. fungus. And it generally happens from late May till August. The spot of the leaf is light brown, sometimes overlapping, with the leaves being scorched. The impact can be confused with the burning of leaves associated with hot weather. There may be some defoliation during the rainy season.

16.2.7.4   MICRODIPLODIA LEAF SPOTS

Microdiplodia litchii is responsible for the illness. According to Pathak and Desai (1971), Microdiplodia leaf spot was first recorded from Udaipur (Rajasthan) in India. The diseased leaves show yellow brown to brick red areas mainly to the margins. The colored areas slowly turn light brown and show a black spot close to pycnidia. Pycnidia are ostiolate and have a maximum diameter of 108 μm. Conidia are 8.2–10.9 μm olive, uniseptate, and 2.4–4.6 μm.

16.2.7.5   PESTALOTIA LEAF SPOTS

Prasad (1962, 1967), in Muzaffarpur, Bihar, reported the illness. *Pestalotia pauciseta's* spots are finely spotted and occur on both sides of the leaf. The spots vary in scale from 0.5–2 cm to 0.2–2 cm. The spots then mix together to form larger wounds and the shade of the spots shifts from earthy to reddish-brown or earthy. Brooding time: 5–10 days (Mishra and Pandey, 2001).

16.2.7.6   INTEGRATED MANAGEMENT OF THE LEAF SPOT DISEASES

- Infected leaves are destroyed: survives on the leaves throughout the year and develops vigorously after the onset of the monsoon, so that infected leaves can be destroyed as much as possible.

- Use mancozeb (0.25%) or copper oxychloride (0.25%) or thio-phanate methyl (0.15%) foliar spray to increase the severity of the disease.

16.2.8   FRUIT ROTS

Litchi organic product decay has been a major issue. Litchi is host to a number of postreap pathogens, with a periodic cessation of various disease methods. In India, Prasad and Bilgrami (1973) accounted for a few natural products of decay. A wide range of fungi *viz.*, *Alternaria, Colletotrichum, Botryodiplodia, Aspergillus, Fusarium*, and *Penicillium* sp. have been reported from India by Prasad and Bilgrami (1973) and Kumar et al. (2016) that cause postharvest fruit rots if fruits are not handled properly. A mean pathological at 23.2% in 2012 and 17.9% in 2013 was recorded in the supply chain of litchi in India (Kumar et al., 2016), the highest at the retail level. At first, the malady side effects are distinguishable on the harmed part of the organic products. With the development of the infection, the rotted territories get discouraged. The spoil continuously infiltrates profound into the mash. At last, skin of tainted organic products breaks off uncovering the mash which in this manner is secured with thick cottony mycelium. Such influenced natural products emanate a scent of aging.

16.2.8.1   BOTRYODIPLODIA FRUIT ROT

Botryodiplodia theobromae as well as *Colletrotrichum gleosporiodie* are the most well-known causes of stem decay on longan, lychee, and rambutan (Coates et al., 2003). Stem-end spoils begin as the earthy colored discoloration of the skin at the end of the stem and spread quickly and are outwardly vague without segregation. Manifestations typically occur at the end of the stem but can occur anywhere on the organic product at first as water-drained sores. Decay develops deep into the material. White mycelium with different dark pycnidia can be visible at the cutting edge stages (DPI, 1993).

Coates et al. (2003) suggest that inoculum from the collection of parasites that make stem ends of longan, lychee, and rambutan spoil can be reduced by cutting out dead leaves and twigs in the shelter, thereby improving airflow and giving a less good condition.

16.2.8.2 COLLETOTRICHUM GLOEOSPORIOIDES ROT

Organic products are defenseless against blooming time contamination until they are about half grown. The vast majority of the rot created by organic products is caused by idle contamination when the natural product is small. Such contaminations, which could be nail-pointed to the size of evolving natural products, further generate as small, large, slightly deterred, and doused water. In this way, they combine to frame larger earthy-colored spots and become intermittent fiddles in the production of organic products. After 5–6 days, acervuli appear as a dim shaded dab in the disease trial. The acervuli are filled with large multicell setae.

During transportation or refrigerated storage, the infected fruits will develop a whitish mycelia mat over the skin.

16.2.8.3 ASPERGILLUS NIGER ROT

At the end of the fruit, the disease first occurs as a light brown lesion surrounding the leaves, which then turns dark brown. The conidial black heads are distinct 4–5 days after infection. In infected regions, some whitish deposits are frequently observed.

16.2.8.4 ASPERGILLUS FLAVUS ROT

The diseased fruits show brownish discoloration at the border of the infected regions at the early stages of the infection. The diseased region slowly becomes velvety and deprimed. Green conidial heads appear after 6–8 days, masking the main portions of the fruit. In advanced stages, dirty yellowish-green powdery occurs on the infected area (Awasthi, et al., 2005).

16.2.8.5 PESTALOTIOPSIS FRUIT ROT

Pestatiopsis fruit rot is one of the most serious litchi postharvest illnesses. Symptoms: on the fruit rind showed up as a dark brown lesion. Under humid conditions, the surface of the lesion may develop a whitish mycelial growth of causal fungus. The disease may be handled because the fruit should be kept at 10 °C.

16.2.8.6   INTEGRATED MANAGEMENT OF FRUIT ROT DISEASES

Colletotrichum sp. causes the occurrence of fruit rot. Reduces the dose between 75 and 300 Gy litchi fruit irradiation with no adverse quality impact when stored for 3 weeks at 5 °C (McLauchlan et al., 1992). According to Brown et al. (1984), litchi rot is also significantly reduced by 5 minutes at room temperature (22 °C) dipping the fruit in a solution of 0.125–0.25 g/L. Low-temperature storage is the most effective way to decelerate rot production. The fruit stored at 22 °C for example rotted three times faster than the fruit stored at 5 °C. The recommended management strategies for fruit rots are as under:

- 15–20 days before the meeting, one splash of carbendazim. Keep off physical wounds while collecting.
- Prompt precooling and maintenance of the ideal temperature and relative mugginess of organic products during capacity and transport.
- Fumigation of the sulfur whenever allowed by bringing in nations. For this, natural products are placed in a shut-down chamber where 50–100 g of sulfur is sung for 20–30 minutes for every m^3 of air.
- Using 2 kg corrugated fiber boxes correctly unitized for stacking for transport.

16.2.8.7   ARMILLARIA ROOT ROT

The disease also known as the oak root fungus, the mushroom's root rot is a recurring problem in many woody plants, which ultimately results in death. *Armillaria* root litchi rot was first recorded in 1955 by Cohen of Florida, America (Cohen, 1955). The disease can cause tree death or slow tree decline or extreme tree loss in a limited number of orchards (McMillan, 1994b; Menzel et al., 1988).

Symptoms:

Trees with *Armillaria* root spoil have marginal shades of yellow leaves and twigs, shoot, or dieback appendage. Crown and foliar signs of *Armillaria root* disease are the most common root ailments of the mill; poor shoot growth, congested leaves, untimely yellowing, defoliation, and branch dieback. It is necessary to confirm the proximity of the causal organism by exhuming the root crown and the relevant sidelong roots.

The foliage can turn yellow in shade and fall rashly (McMillan, 1994b). Influenced lithium trees are nonenthusiasts and do not contain new growth (Darvas, 1992). Branches have demonstrated mild decline and trees are occurring in the long term. *Armillaria* root spoil disease recognized by various types of dieback due to the presence of a shaded white or light mycelium layer between the bark and the wood of the roots of the crown (McMillan, 1994b). Basic wood is stained, and nectar mushroom clusters are so often produced from the base of contaminated trees, especially in the cool and stormy climate.

Casual Organism: *Armillaria mellea* and *Armillaria socialis* (Basidiomycetes)

Pathogen:

The pathogen has basidiomycetes and a number of structures on the surface of infected plants, including basidioms (mushrooms), basidiospores, mycelia, white tangles or plaques between the bark and wood, pseudosclerotic tissues and rhizomorphs (earthly colored to black, circular, and hollow infectious fibers that look to some degree like roots). Typically we are clustered and have tanning tops in color and 1–15 cm in diameter to the cove. The tops also have a Knob or a Knob in sight and are lined with small tufted light yellow scales. The stem is buff to brown, 3–30 cm long, 30–40 cm wide, and has a shaggy, ring-like skirt that covers it just below the tip. The stem frequently has focus scales beneath the rings. The basidiospores are borne on shaded gills of light yellow.

Armillaria mellea is sometimes referred to as nectar fungus, nectar agricus, oak root organism, or shingle growth (Shaw and Kile, 1991). The big host races, and gets through as a saprophyte as well. Contamination occurs when the soil includes roots that contract infected plants or parasite rhizomorphs. Tainted tree stumps and large underlying foundations of the dead have natural inoculum springs, and the parasite can live long or longer saprophytically (Cohen, 1955; Darvas, 1992). The pathogen grows up the root along the moving ground after the roots are breached, and the root crown burns and causes death at the end of the blaze.

Armillaria socialis and *Armillaria mellea* are distinguished by the lace of a ring on the stem and the rare production of mushrooms and rhizomorphs.

Disease Cycle:

Basidioms are often developed in cool and rainy weather around the base of infected trees, but no known role in the disease cycle (Shaw and Kile, 1991).

Epidemiology:

Armillaria root rot disease preferred wet, rainy weather conditions. The effects of infection occur when roots contract infected plants or fungal rhizomorphs in the soil. Infected tree stumps and large roots of dead hosts are common inoculum sources, and the fungus can saprophytically persist for 10 years or longer.

Integrated Disease Management:

- Pathogen:s have a wide host range and also survive saprophytically for 10 years or more in the soil so, it is difficult to control.
- Deep ploughing may help remove as many infected roots after all the tree stumps are collected and removed from the site.
- Soil fumigation with bisulphide or methyl bromide is the most effective and good practice for control of inoculums.

16.2.9   ALGAL LEAF SPOT/RED RUST/RED LEAF SPOT

It is a parasitic alga that sometimes attacks trees which cause loss of vigor. In Punjab, Bihar (Pusa), and Uttar Pradesh (Saharanpur), Sharma et al. (1972), Mishra et al. (1973), and Gupta (1992), respectively, identified algal leaf spot in litchi. The infection decreases the plant's productivity by impeding photosynthetic activity. The plant's growth is retarded due to defoliation that has an indirect impact on the yield. The disease's causative organism is *Cephaleuros virenscens*. This disease's emergence is often in the rainy season and continues until early winter.

Symptoms:

The disease is characterized by dark, circular to semicircular fog spots (3.0–6.5 mm) that often occur on the leaves, and occasionally on the litchi's tender stems. This alga, however, has never been seen on twigs or branches of twigs. On litchi leaves, orange-yellow to pink velvety coatings, cushiony patches are formed with sporangia of the algae, but in the absence of sporangia, the leaf spots remain greenish-brown. In older leaves, the lesions turn light brown to brick red.

Casual Organism: *Cephaleuros virescens* **(an alga)**

Pathogen:

Cephaleuros virescens has long unbranched rough earthy colored, thick, erect, fibers measured at 52.05–235.96 × 10.41–15.61 μm, growing through the plate of thalloids, 5–12 followed. Sporangia are purple earthy colored, subspherical to oval, smooth measuring 19.08–27.76 × 15.61–20.82 μm (Mishra et al., 1973). *C. virescens* is a parasite on litchi, mango, citrus, and ber, and so on, whatever it may be, *C. parasiticus* is referred to as a parasite on tea and *vanilla planifolia*. In addition, *C. mycoide* is epiphytic or saprophytic in nature. Away from the death of mesophyll cells to the lower epidermis due to the dissatisfaction of the plug arrangement, it is known to be a solid indication of the pathogenic idea of the algae, correspondingly distinct from the saprophytic condition. To stay away from this disarray and pending any genuine ordered submission, it may be fair and commercial to refer to all species likely to be *Cephaleuro* spp. unveiled by Jouebert and Rijkenberg (1971).

A minor examination of the infected portion of the leaf indicates the presence of a thick and reduced tangle of algae underneath the surface of the fingernail in the underlying disease development cycle. The thallus is a shell consisting of cells that are similarly mastered and that emerge dichotomously from the inside out. Within, there are a few layers of cells in the plate and out there is only one. The parasite normally remains between the skin of the fingernail and the epidermis, and spreads back and forth to the layer of palisade. Rarely, as in magnolia, the thallus is extraordinarily young and may be shallow. Still, the intracellular entry is not well known. A huge air space between the alga-circle and the host's necrotic epidermal cells lies under the subcuticular disease in the leaf (Agnihothrudu, 1964).

The types of rhizoids or cells enter through airspace into the hidden host cells. This provides a critical safe haven and provides a basis on which to draw water and supplements. Ultra-basic tests show that the auxiliary divider content considerably thickens the young algae cell dividers, in direct contact with the host fingernail surface. The agamic architecture of *C. Parasiticus* is a sporangiophore species with a few sporangiophores, septate, and orange shaded, blended with setae that emerge as the skin of the fingernail or epidermis is separated toward the edge of the thallus circle.

The further development of the vegetative cells below reveals sporangiophores as if they emerge from the disc either subterminally or even centrally. The 105–3991 mm × 9.4–45 mm sporangiophores, mostly 238.2 mm ×

29.91 mm, end up in a swollen apical cell called a vesicle. Typically four but eight or more protrusions or sterigmata, which are light orange, unicellular, and measure 16–45.5 mm × 5–15 mm, come out of the vesicle. Everyone of them grows into a 19–30.4 mm × 19–22.8 mm white, oval to globose zoosporangium and is connected to the apex by an oddly folded stalk through the ostiole in a suitable temperature and moisture regime. Per sporangium the number of zoospores released varies from eight to 30 (Chapman 1984; Ponmurugan et al., 2010).

The filaments of *C. parasiticus* are very thin, erect, and green in color which is usually 50.31–23 mm and 240.82 mm. This alga's exterior thallus is stronger than those in the inner algae. Outside thallus bears separate erect, sterile fibers while at the tip the ripe fibers bear groups of five to six sporangia. Fiber cells are typically earthy ruddy in color since they contain β-carotene. There are seven to eight septate sporangiophores, which are spread at equal separation. Sporangia create an algal production which is about 15 days old. These are oval and curved, 22.12 mm × 18.18 mm wide. Sporangia develop apart from the fibers and are spread with water and wind (Chapman, 1984; Ponmurugan et al., 2010).

Sexual proliferation, as seen on tea, occurs by the combination of isogametes, delivered within large, sessile, cup-formed cells, in a homothallic way. The zygote creates a tiny human sporophyte following a mixture consisting of an appended tail containing at least one cell. Each cell has a tiny amount of microsporangi. Completed the meiosis, four quadric flagellate microzoospores are released at each microsporangium. The flask-shaped cells open by fissure in the presence of free water and release the swarm spores (zoospores), stated by Chapman (1984) and Ponmurugan et al. (2010).

Disease Cycle:

Sporangia developed on fine hair, germinated in the wet summer weather, develop zoospores that enter the host tissue through the algal cell stomata and mycelium chain. The disease spreads most quickly from June to October in periods of strong and abundant rainfall. The disease is common in plantings nearby.

Epidemiology:

High humid conditions allow fruiting bodies to grow.

Integrated Disease Management:

- Training and pruning to the tree for avoiding overcrowding and proper aeration in the plant canopy.
- Copper oxychloride foliar spray (0.3%) may occur in July and October when the severity increases.
- Bordeaux mixture foliar spraying 5:5:50 during autumn (September–October) and spring (February–March) at intervals of 15 days depending on the degree of infestation.
- Mancozeb spraying 0.25% also reduces the severity of the disease.
- Fytolan/Blitox spraying 0.5% or lime sulfur 0.1% also decreases the severity of the disease.

16.2.10   NEMATODE DISEASES

Nematode disease in litchi is not a major problem but some workers have reported plant-parasitic nematodes associated with litchi from India. These are *Rotylenchulus reniformis*, *Helicotylenchus indicus*, *Hemicriconemoides litchi*, *Tylenchorrhynchus leviterminalis*, *Meloidogyne incognita*, and *Xiphinema* sp. Species of the *Trichodorus* and *Hemicriconemoides* were the most frequently encountered nematodes in Doon Valley, India (Srivastava et al., 2000). In South Africa, nematodes were associated with severe dieback and decreased litchi disorder (Milne et al., 1971). There are many uncovered twigs and buds over the field occurrences, leaf chlorosis and tip use, low blooming, overabundance of natural product drops, and whimsical leaf flushes. Hampered and distorted by affected sources. The feeder root mass is reduced leading to water and nutrient uptake decrease and ultimately dieback of the plant's above-ground portion. Preplant soil fumigation and postplanting application of nematicides have shown great promise in controlling these nematodes (Milne et al., 1971; Cohn and Duncan, 1990).

16.3   PHYSIOLOGICAL DISORDER OF LITCHI

16.3.1   FRUIT CRACKING

The development of fruit in litchi is a serious problem. Split fruit is unmarketable and thus results in a great loss to farmers. This disease is common in all cultivars of litchi. High temperatures (above 38 °C), low humidity (above 60%), and low soil moisture levels during fruit production lead to fruit cracking during the summer season as well as during fruit production. Sudden variations in day and night temperature, coupled with heavy irrigation during dry spells, increase cracking in litchi fruit. Rains at maturity and fluctuation in soil moisture and ambient temperature have a major impact on fruit splitting strength in litchi.

Integrated Management:

- Irrigate the orchard at 15 days intervals to maintain the proper moisture in the field during the summer months. Soil mulching with polyethylene sheets or locally available crop residues is helpful for maintaining the moisture.
- Foliar sprays of 1% borax at 15 days intervals should be given at pea stage.
- Spray GA3 @ 100 ppm, when fruits have attained 50% of their size.
- Fruit cracking can be controlled by the application of calcium nitrate (0.5%–1%), borax (0.4%–0.8%), Zn (0.4%), alone or in combinations.

16.4   CONCLUSIONS

Research into the sources, epidemiology, and disease management of litchi crops has considerable scope. The lack of a clear understanding of the causative agents of different diebacks and root rots with improved knowledge of pathogen biology would allow for more efficient disease control strategies to be established.

Possible disease prevention solutions cover a variety of strategies. We are likely to see a shift to the use of environment-friendly fungicides, combined with more focused chemical formulations, models for disease prediction, and improved application technology. Studies of natural defense systems in this crop can give rise to opportunities for defense elicitors to boost or induce antifungal compounds (Willingham et al., 2001), particularly concentration of nitrogen and calcium in fruit tissue. The role of these nutrients in the development of fruit disease in litchi should be investigated. While there has been very little success in biological control of plant diseases, this strategy, when combined with other control methods, should not be discounted.

KEYWORDS

- litchi
- diseases
- symptomatology
- epidemiology
- management

REFERENCES

Agnihothrudu, V. (1964). Annual Scientific Report, Tocklai Experiment Station, India, Indian Tea Assoc. for 1963: 64 (*Hypoxylon nummularium* inter alia; 3142).

Ann, P. J. and Ko, W. H. (1984). Blossom blight of litchi in Taiwan caused by *Peronophythora litchi*. *Plant Dis.*, 68: 826.

Awasthi, D. P., Sarkar, S., Mishra, N. K., and Kaiser, S. A. K. M. (2005). Disease situation of some major fruit crops in new alluvial plains of West Bengal. *Environ. Ecol.*, 23: 497–499.

Bhavakul, K., Topsol, M., Rakvitayasart, V., and Suwanketnikon, S. (1998). Studies on Phytophthora leaf blight of longan: symptoms, causal organism and chemical control. *Proceedings of a Seminar and Workshop on Longan Producing Technology*, Chiang Mai Phucome Hotel, Chiang Mai, Thailand, 14–15 September, 1998, pp 62–73.

Brown, B. I., Scott, K. J., and Mayer, D. G. (1984). Control of ripe fruit rots of guava, lychee and custard apple by postharvest prochloz dips. *Singapore J. Prim. Ind,* 12:40–49.

CAB International (2002). Crop Protection Compendium (2002 edition). Wallingford, UK: CAB International.

Chapman, R. L. (1984). An assessment of the current status of our knowledge of the Trentepohliaceae. In: Systematics of the Green Algae, D. E. G. Irvine and D. M. John (Eds). London: Academic Press, pp. 233–250.

Chen, C. C. (1961). A Species of Peronophythora Gen. Nov. Parasitic on Litchi Fruit in Taiwan. Special Publication. College Agricultural. National Taiwan University, 10: 1–37.

Chen, L. C., Lai, S. C., Lee, C. C., Chung, Y. W., and Ann, P. J. (1998a). Effect of environmental factors on mycelial growth of *Peronophythora litchii*. *Plant Pathol. Bull.*, 7: 128–133.

Chi, P. K., Pang, S. P., and Liu, R. (1984). On Downy blight of litchi (*Litchi chinensis Sonn.*). 1. The pathogen and its infection process. *Acta Phytopathol. Sin.*, 14: 113–119.

CMI (1989). CMI descriptions of pathogenic fungi and bacteria, set 98, Nos 971–980. *Mycopathologia,* 106: 183–211.

Coates, L. M., Sangchote, S., Johnson, G. I., and Sittigul, C. (2003). Diseases of lychee, longan and rambutan. Pp307–325 In: R. C. Ploetz (ed.) Diseases of Tropical Fruit Crops. CABI Publishing, Wallingford, UK, p. 527.

Cohen, M. (1955). Clitocybe rot of lychee tree. *Proc. Florida State Hortic. Soc.*, 68: 329–332.

Cohen, M. E. and Duncan, L. W. (1990). Nematode parasites of subtropical and tropical fruit trees. In: M. Luc, R. A. Sikora, and J. Bridge (Eds.) Plant Parasitic Nematodes in Subtropical and Tropical Agriculture, CAB International, Wallinford, UK, pp. 347–362.

Darvas, J.M. (1992). Armillaria Rot of Litchi (*Litchi chinensis*) in South Africa. *South African Association of Litchi Growers Year Book*, 4: 2–4.

DOA (2003a). Application for Market Access of Longan from Thailand to Australia. Department of Agriculture, Ministry of Agriculture and Cooperatives, Bangkok, May 2003.

DOA (2003b). Application for Market Access of Lychee from Thailand to Australia. Department of Agriculture, Ministry of Agriculture and Cooperatives, Bangkok, May 2003.

DPI (Department of Primary Industries) (1993). Diseases of Fruit Crops. D. Persley (ed.). Department of Primary Industry, Queensland, Brisbane, p. 178.

Gupta, J. H. (1992). Chemical control of algal rust of Litchi caused by *Cephaleuros virescens*. *Prog. Hortic.*, 24: 109–110.

Hall, G. (1989). *Pyrenophythora litchi*. CMI Descriptions of Pathogenic Fungi and Bacteria No. 974. Commonwealth Mycological Institute, Kew, UK.

Huang, H., Wang, C. P., and Xu, D. Y. (1983). On Peronophythora litchi. *Acta Mycolog. Sin.*, 2 (4): 201–206.

Jouebert, J. J. and Rijkenberg, F. H. J. (1971). Parasitic green algae. *Ann. Rev. Phytopathol.*, 9: 45–64.

Kao, C. W. and Leu, L. S. (1980). Sporangium germination of *Peronophythora litchi*, the causal organism of litchi downy blight. *Mycologia*, 72: 737–748.

Ko, W. H., Chang, H. S., Su, H. J., Chen, C. C., and Leu, L. S. (1978). Peronophythoraceae, a new family of Peronosporales. *Mycologia*, 70: 380–384.

Kumar, V. (2016). Diseases in litchi management. In: Litchi Global Perspectives. Drs Vishal Nath, M. Feza Ahmad, Hidayatullah Mir, Manoj Kundu, Sanjay Sahay, Shesh Dhar Pandey, Kuldeep Srivastava, Alemwati Pongener (Eds.), Excel India Publishers, New Delhi, pp.101–105.

Kumar, V., Anal, A. K. D., and Nath, V. (2014a). Prevalence of some threatening pests and disease of litchi (*Litchi chinensis* Sonn.) in Bihar state of India. *J. Appl. Hort.*, 16: 235–240.

Kumar, V., Kumar, A., and Nath, V. (2011). Emerging pests and diseases of litchi (*Litchi chinensis* Sonn.). *Pest. Manag. Hort. Ecosyst.*, 17: 11–13.

Kumar, V., Kumar, A., Nath, V., and Kumar, R. (2014b). New threats of insect pests and disease in litchi (*Litchi chinensis* Sonn.) in India. *Acta Hortic.*, 1029: 417–424. DOI:10.17660/ActaHortic.2014.1029.53.

Kumar, V., Purbey, S. K., and Anal A. K. D. (2016). Losses in litchi at various stages of supply chain and changes in fruit quality parameters. *Crop Prot.*, 79: 97–104. DOI:10.1016/j.cropro.2015.10.014.

Li, J. (1997). Diseases and pests and their control. In: Zhang, Z. (ed.). Litchi Pictorial Narration of Cultivation. Pomology Research Institute, Guangdong Academy of Agricultural Science, p. 189.

McLauchlan, R. L., Mitchell, G. E., Johanson, G. I., Nottingham, S. M., and Hammerton, K. M. (1992). Effect of disinfection dose of irradiation on the physiology of Taiso Lychee. *Postharvest Biol. Tec.*, 1: 273–281.

McMillan, R. T. (1994b). Diseases of *Litchi chinensis* in South florida. *Proc. Florida State Hortic. Soc.*, 107: 360–362.

Menzel. C. M., Watson, B. J., and Simpson D. R. (1988). The lychee in Australia. *Queensland Agric. J.*, January–February 114: 1988, 19–26.

Milne, D. I. M. devilliers, E. A., and Holtzhausen, L. C. (1971). Litchi tree decline caused by nematodes. *Phytophylactica*, 2: 37–44.

Mishra, A. K. and Pandey, B. K. (2001). Disease of Litchi and Their Management. T. S. Thind (Ed.) Diseases of Fruits and Vegetables and their Management, first Edition. Kalyani Publishers, pp 150–155.

Mishra, B., Prakash, O., and Mishra, A. P. (1973). A serious leaf spot disease of litchi caused by *Cephaleuros virescens*. *Indian J. Mycol. Plant Pathol.*, 3: 219–220.

Nakasone, H. Y. and Paull, R. E. (1988). Tropical fruits. Crop Production Science in Horticulture Series No. 7 CAB International, Wallingford, UK.

Nakasone, H. K. and Paull, R. E. (1998) Tropical Fruits. CAB International, Wallingford, 132–148.

NHB (2015). Indian Horticulture database 2014. National Horticulture Board, Ministry of Agriculture, Government of India, Gurgaon, India. http://nhb.gov.in/area-pro/NHB_Database_2015.pdf. Accessed 06 October 2016.

Ou, H. X., Deng, W. S., and Wu, X. T. (1999). Experiment of control of litchi downy mildew disease by using 80% mancozeb wetted powder. *China Fruits*, 3: 32.

Ou, H. X., Sun, G. C., Jiang, Y. M., and Zhu, X. R. (2001). Pathogenesis-related proteins in litchi after inoculation with Peronophthora litchii. pp. 439–442. In: H. B. Huang and C. Menzel (Eds). Proceedings of the First International Symposium on Litchi and Logan. Guangzhou, China June 2000. *ISHS Acta Hortic.*, 558: 446.

Pathak, V. N., and Desai, B. G. (1971). A new species of on *Nephelium litchi* Camb. *Sydowia*, 24: 73–74.

Ponmurugan, P., Saravanan, D., and Ramya, M. (2010). Culture and biochemical analysis of a tea algal pathogen, *Cephaleuros Parasiticus. J. Phycol.*, 46(5): 1017–1023.

Prasad, S. S. (1962). Two new leaf spot diseases of *Nephelium litchi* Camb. *Curr Sci*, 31: 293.

Prasad, S. S. and Bilgrami, R. S. (1973). Investigation on diseases of Litchi. III. Fruit rots and their control by post harvest technology. *Indian Phytopathol.*, 25: 523–527.

Prasad, S. S. and Bilgrami, R. S. (1974). Investigations on diseases of litchi: VI. Post harvest diseases of fruit in India. *Plant Dis. Rep.*, 58(12): 1134–1136.

Sharma, V. K., Srivastava. A. K., and Chouhan, J. S. (1972). A new host record for parasitic alga *Cephaleuros virescens* Kunze (Litchi). *Sci. Cult.* 38:39–40.

Shaw, G. C. and Kile, G. A. (1991). Armallaria Root Diseases. Agriculture hand Book No. 691, Forest Service, U.S. Department of Agriculture, Washington, DC.

Srivastava, N., Rawat, V. S., and Ahmad, M (2000). Distribution of plant parasitic and soil nematodes associated with litchi fruit trees in Doon valley (U.P.) India. *Ind. J. Nematol.*, 30(1): 100–101.

Vien, N. V., Benyon, F. H. L., Trung, H. M., Summerell, B. A., Van, N. K., and Burgress, L. W. (2001). First record of *Peronophythora litchi* on litchi fruits in Vietnam. *Australas Plant Pathol.*, 30: 287–288.

Visitpanich, J., Sittigul, C., and Chanbang, Y. (2000). Longan leaf blight and fruit drop. *House Agric. Mag.*, 24(1): 144–148.

Willingham, S. L., Pegg, K. G., Cooke, A. W., Coates, L. M., Landon, P. W. B., and Dean, J. R. (2001). Rootstock influences postharvest anthracnose development in 'Hass' avocado. *Aust. J. Agric. Res.* 52: 1017–1022.

Zentmyer, G. A. and Mitchell, D. J. (1985/86). *Phytophthora* diseases of fruit trees in the tropics. *Rev. Trop. Plant Pathol.* 2: 287–309.

CHAPTER 17

MAJOR DISEASES OF COORG MANDARIN (*CITRUS RETICULATA*) AND THEIR MANAGEMENT

V. VENKATARAVANAPPA* and PRITI S. SONAVANE

Central Horticultural Experiment Station, ICAR-IIHR, Chettalli 571248, Kodagu, Karnataka, India

Corresponding author. E-mail: venkatrajani@gmail.com

ABSTRACT

Coorg Mandarin is one of the renowned citrus cultivar grown in Southern India, particularly Coorg and Wynad districts of Karnataka and Kerala. The fruits are greenish to yellow in color, having tight skin with sweet in taste, whereas some cultivar are having loose skin and lesser in sweet taste. With respect to shelf life, the Coorg oranges have a longer shelf life as compared to other mandarins. Among the various abiotic and biotic factors, diseases like root rot of *Phytophthora*, Citrus greening, and CTV (Citrus tristeza virus) are the major limitation for its cultivation and production. Major diseases Coorg Mandarin are discussed below.

17.1 INTRODUCTION

Citrus plant includes "Citron/Malta" (*Citrus medica*), "Lime/Acid lime" (*Citrus aurantifolia*),"Key lime" (*Citrus aurantiifolia*) {A citrus hybrid (*C. micrantha* × *C. medica*)}, "Kaffir lime/Makrut lime/Mauritius papeda" (*Citrus hystrix*), "Rangpur lime is a hybrid cross between orange and citron (*Citrus limonia*)}, Seedless lime/Bearss lime/Tahiti lime" (*Citrus latifolia*), "Lemon" (Citrus limon), "Rough lemon" (*Citrus jambhiri* Lush), "Meyer lemon" {Hybrid citrus fruit native to China (*Citrus meyeri*)}, "Sweet orange"

(*Citrus sinensis*), "Mandrin orange" (*Citrus reticulata*), "Bitter orange/ Seville orange/Sour orange/Bigarade orange/Marmalade orange/Kamala orange" (*Citrus aurantium*), Grapefruit (*Citrus paradisi*), Ttrifoliate orange/ Poncirus trifoliata/Citrus trifoliate (*Citrus trifoliate*), Bergamot orange (*Citrus bergamia*), Satsuma mandarin/Satsuma orange (*Citrus unshiu*), Mousambi/ Musambi/Sweet lime, Sweet lemon/Sweet limetta (*Citrus limetta*), Papeda/ Ichang papeda (*Citrus ichangensis*), Cleopatra mandarin (*Citrus reshni*), Indian wild orange (*Citrus indica*), "Pomela" (*Citrus paradisi*). The below-described diseases are all prevailing species of citrus.

Coorg orange is one of the renowned citrus cultivar grown in Southern India, particularly Coorg and Wynad districts of Karnataka and Kerala. In Karnataka, the crop being cultivated in pepper and coffee-based planta-tions with an area of 24,000 hectare in Hassan Chikmagalur, and Kodagu districts. Due to the change in climate and diseases, the area of cultivation is drastically reduced to 2000 hectares. The fruits are greenish to yellow in color, having tight skin with sweet in taste, whereas Nagpur oranges are having loose skin and lesser in sweet taste. With respect to shelf life, the Coorg oranges have a longer shelf life as compared to other mandarins. The area of cultivation decreasing in Coorg mandarin due to the attack of many pests, shortage of quality planting material, improper nutrients and water, introduction of other oranges. Per plant yield had come down 10 kg which was more than 50–100 kg in the early period of crop introduction. The average fruit production was more than 45,000 tones. Among the various abiotic and biotic factors, diseases like root rot of *Phytophthora*, Citrus greening, and CTV (Citrus tristeza virus) are the major limitation for its cultivation and production (Fawcett, 1936). The crop flowers twice in a year, that is, monsoon which is harvested in June–July and main crops harvested in October–November. As the harvesting stage revolves around the monsoon and post monsoon season, the crop is highly vulnerable to different pests and diseases. Following are some important diseases and their management observed in Coorg mandarin.

Coorg oranges or mandarins	Root rot/foot rot/ gummosis of *Phytophthora*	*Phytophthora nicotianae* var. *parasitica, P. palmivora, citrophthora, P. hibernalis, P. syringae, P. cactorum*
Coorg oranges or mandarins	Powdery mildew	*Acrosporium tingitaninum*
Coorg oranges or mandarins	Scab	*Eisinoe fawcetti (I.S: Sphaceloma fawcetti)*

Coorg oranges or mandarins	Anthracnose or wither tip/die back	*Colletotrichum gloeosporioides*
Coorg oranges or mandarins	Dry root rot	*Macrophomina phaseoli, Fusarium* spp. *and Diplodia natalensis*
Coorg oranges or mandarins	Sooty Mould disease	*Capnodium citri*
Coorg oranges or mandarins	Canker	*Xanthomonas axonopodis* pv. *citri*
Coorg oranges or mandarins	Citrus Greening disease/ (Huanglongbing (HLB)	*Candidatus Liberibacter asiaticus*
Coorg oranges or mandarins	Tristeza	*Citrus tristeza virus*
Coorg oranges or mandarins	Citrus yellow mosaic badnavirus (CMBV)	Citrus yellow mosaic badnavirus (CMBV)
Coorg oranges or mandarins	*Citrus exocortis*	*Citrus exocortis viroid* (CEV)

17.2 FUNGAL DISEASES

17.2.1   ROOT ROT/FOOT ROT/GUMMOSIS

Introduction/Economic Importance:

Phytophthora causes root rot, foot rot, collar rot, stump rot, leaf fall, brown root rot, and gummosis in many citrus species worldwide. This type of symptom on different citrus species is widespread in Punjab and Assam. *Phytophthora* infections have been considered as a major disease in India (Fraser and Singh, 1966). Among them, lime is more susceptible than rough lemons sweet orange and grapefruit. In South India, *Phytophthora* root rot/ foot rot/gummosis disease is more prevailing in the Coorg mandarin and sweet orange.

Symptoms:

The oozing or gummosis disease is caused by multiple species of Phytoph-thora. The pathogen will infect roots and above the ground parts such as stem, trunk, branches, blossoms, and fruit. The pathogen also causes damping off disease in nursery at the seedling stage. Foot rot, initially the symptoms will occur at ground level near the crown region, the fungus produces dark lesions and it will extend to the bud union in tolerant rootstocks. Root rot will be occurring upon the infection of the cortex of fibrous roots (Uppal and Kamath, 1936). The infected roots slough off their cortex by leaving tissue of the inner fibrous root. The most important symptom is exudation gum from the bark, stem, and crown region of the tree (Figure 17.1a).

FIGURE 17.1A   Gum exudation at tree trunk.

The disease may cause tree decline, and in severe cases the tree become wilted and their death takes place (Graham and Menge, 1999; Figure 17.1b).

FIGURE 17.1B   Coorg mandarin plant showing the complete wilting symptoms under field condition.

The infected plants show yellowing and blighting symptoms, if the disease is more severe, the bark becomes completely rotting, the trunk becomes girdling and the tree ultimately dies. Before death, the plants bloom heavily and die before the fruits are matured (Kumbhare and Moghe, 1976). The disease is more severe during prolonged moisture availability, which in turn enhances the production of zoospores by the fungus and spread through rain splash into tree trunks (Timmer, 1977). The secondary infections normally occur through lesions created by Phytophthora.

Causal Organism: *Phytophthora nicotianae* var. *parasitica, P. palmivora, P. citrophthora, P. cinnamomi, P. hibernalis, P. citricola, P. syringae, P. cactorum*

Etiology:

Phytophthora palmviora is a cosmopolitan pathogen having a very wide host range in the tropics (Erwin and Ribero, 1996). The fungal hypha is irregular and up to 7 μm in dia. The sporangia are caducous, and ovoid or ellipsoid, pedicels are <5 μm in length. They are 40–60 × 25–35 μm in size. Chlamydospores are spherical to subspherical produced in terminal as well as intercalary and average 37 μm in diameter. The fungus is heterothallic producing oospores either crosses between two mating types (A1 and A2) are available or when they are crossed with the opposite type of several other species. Oospores are spherical or round in shape about 22–24 μm in dia and amphigynous sexual reproduction. The minimum temperatures for its growth are 11 °C, 27.5–30 °C, and 35 °C, respectively.

Phytophthora nicotianae produces noncaducous, ovoid, ellipsoid, pyriform to spherical shaped sporangia in a single papillum (Erwin and Ribero, 1996). The sporangia produced singly are in sympodia on stalks that range from 100 to 595 μm in length. The fungus survives by the production of resting spores such as chlamydospores about 13–16 μm in diameter. Most of the isolates are heterothallic. Antheridia is amphigynous, the oogonia are smooth spherical or oval in shape about 15–64 μm in diameter. Oospores are aplerotic and the minimum temperature required for its growth is 5–7 °C, 27–32 °C, and 37 °C, respectively.

Phytophthora cinnamomi produces unique coralloid aseptate mycelium. The sporangia are elliptical to nonpapillate, noncaducous, and ovoid in shape, are rarely produced in culture. The sporangia dimension range dramatically ranged from 11–123 μm × 11–63 μm depending upon the host they infect. The chlamydospores are terminal and intercalary about 31–50 mm in dia are produced profusely in culture and usually in botryose clusters (Naqui,

1988). The fungus is heterothallic, which produces oogonia about 21–58 μm in dia, antheridia is amphigynous, and oospores are plerotic in nature. The minimum temperatures for growth and infection are 5–15 °C, 20–32.5 °C, and 30–36 °C, respectively.

Phytophthora citricola produces noncaducous sporangia that are obclavate and obpyriform and obovoid in shape and flatten at one side (Erwin and Riberio, 1996). They are semipapillate and having a single apex or deeply bifurcated or irregularly shaped with three or four spices. The size of the sporangia ranges from 30 to 75 μm × 21 to 44 μm and produces chlamydospores rarely in natural condition. The fungus is homothallic. The sexual reproduction by paragynous and oogonia is 18–30 μm in dia and is plerotic and size is 16–30 μm in diameter. Its cardinal temperature for it growth is 25–28 °C and 31 °C.

P. citropthora produces capricious, noncaducous sporangia (Erwin and Ribireo, 1996) which are different in shape, namely spherical, avoid, obpyriform, obtuncate, ellipsoidal to extremely distinct, and the size is ranged from 23 to 90 μm × 18 to 60 μm. The sporangia is papillate, and having two and more papilla. The sporangiophores are branched unevenly from singly or some in sympodia having swellings at the branch points. The fungus will not produce the resting spores. The lowest temperature for growth is <5 °C, the optimum is between 24 and 25 °C, and the maximum is 32–33 °C.

Disease Cycle:

The fungus will survive for long period as a dormant mycelium or oospores in soil and plant debris. The secondary spread of sporangia and zoospores is by rain splash or irrigation water.

Epidemiology:

The tree trunk is in contact with water for a long time as waterlogged areas or due to flood irrigation and soil becomes heavy, these factors enhance the disease. The disease is more in black soils than in light soils, high moist conditions, and high rainfall areas.

The minimum temperature for growth is <5 °C, the optimum is between 24 °C and 25 °C, and the maximum is 30–35 °C with high relative humidity.

Management:

- Precautionary measures like providing the proper drainage.

- In budded plants, the bud union should be 45 cm above the ground level at the time of planting and avoid making any injuries to the plants near the ground level.
- The basin ring for irrigation must be made 50–60 cm away from the plant base.
- Application of *Trichoderma* spp. to soil along with farm yard manure before and after the monsoon.
- Cover the trunk with 10% Bordeaux paste at the height of 60 cm from the soil level premonsoon and postmonsoon.
- Scraping the diseased affected parts with a sharp knife, and paste with 10% Bordeaux or copper fungicides followed by 3–5 spraying with Fenamidone 10% + Mancozeb 50% WDG (1 g/L) or Fenamidone 4.44% + Fosetyl-Al 66.66% WDG (1 g/L) or Famoxadone 16.6% + Cymoxanil 22.1% SC (1.5g/L) of water at 15 days interval to contain the disease effectively.
- Drench the soil with Fenamidone 10% + Mancozeb 50% WDG (2 g/L) or Fenamidone 4.44% + Fosetyl-Al 66.66% WDG (2 g/L) or Famoxadone 16.6% + Cymoxanil 22.1% SC (2 g/L) of water along with 0.5% *Trichoderma* commercial formulation.

17.2.2 POWDERY MILDEW

Introduction/Economic Importance:

Powdery mildew is caused by fungi, which leads to a significant loss in terms of productivity and yield of the tree. The disease is more serious in mandarins, particularly nursery stock, in Asia, due to poorly ventilated orchards. There disease was also reported from other countries like Uganda, Israel, and the USA. In India, it is a serious problem in nurseries of South India especially Coorg, Nilgiris, palnis Wynad, and Shevaroy hills where mandarins are interplanted in coffee plantations.

Symptoms:

The fungus will produce cottony white powdery growth on leaves. Due to cottony or powdery growth on leaves, the photosynthetic activity gets reduced. Later, affected leaves will turn yellow and cause defoliation (Figure 17.2). In severe conditions, it also causes premature leaf and fruit drop and twig and branch dieback. The white powdery growth will be more on the lower surface of the leaves compared to the upper surface of the leaves (Baiswar et al., 2015).

FIGURE 17.2   Powdery mildew symptoms on Coorg mandarin leaves.

Causal Organism: *Acrosporium tingitaninum*

Etiology:

The powdery mildew disease is polycyclic in nature and produces a huge number of asexual spores on plant surfaces. The specialized hyphae are produced either singly or in chains called conidiophores. Conidiophores are epiphytic hyphae, emergence through stomata. At the crop, the fungus will produce sexual spores, such as ascospores, in a sac-like ascus fruiting body known as chasmothecium (cleistothecium). The chasmothecium spherical shape with no natural opening; asci and ascospores are released when the fruiting body wall is cracked (Sastra-Hidayat, 1992). Chasmothecia have a variety of appendages on the surface of the fruiting body. These appendages help in attaching the fruiting bodies to the host.

Disease Cycle:

The primary source of the inoculums of fungi is overwintering infected on plant debris crops or weeds containing chasmothecium or mycelial mats. In the spring, the ascospores are released from the cleistothecia, which are spread through splashing raindrops, wind, or insects. The condia will be dispersed more in dry conditions because the powdery mildew fungi require a wet leaf surface for infection.

Epidemiology:

For fungus for spore germination, the relative humidity of the air needs to be high. Therefore, the disease is more severe in crowded plantings, due to poor air circulation and shaded areas. The incidence will be more when the relative humidity of air is 80% and above.

Management:

- Immediately, after noticing the cottony growth on newly flash leaves, spraying of 0.50% Potassium bicarbonate or 0.3% Sulfex 80% WP or 0.05% Hexaconazole 5EC, or Carbendazium 50% WP, two times at 10–15 days interval to contain the disease effectively

17.2.3 SCAB DISEASE

Introduction/Economic Importance:

The scab disease is caused by *Elsinoë fawcettii* which belongs to the family Elsinoaceae. In India, it has been observed on mandarin fruits in Nilgiri, Palani, and B. R Hills of South India and also reduced market value of mandarin fruits due to scurfy appearance of the rind.

Symptoms:

The pathogen infects shoots, leaves, and fruits mandarin cultivars. The quality of juice and yield is greatly reduced in susceptible cultivars that are severely infected. The disease is more severe when the temperature is low and the humidity should be high. Small corky lesions will occur on the lower surface of the leaves. Initially, minute water-soaked lesions appear on young leaves, later turn into creamy-yellowish or bright-colored pustules (Sivanesan and Critchett, 1998). The central area becomes outgrowth is depressed and becomes dark grayish in color. Scab lesions are rough surfaced and dusky-colored and cracked and fissured (Figure 17.3).

On the infected fruits, raised pustules appear on the rind of the fruits, which are varied in shape, size, and color (Chung, 2010). They appear crater-like outgrowths all over the fruits. High humidity and rainfall is the main key factor for the development of scab.

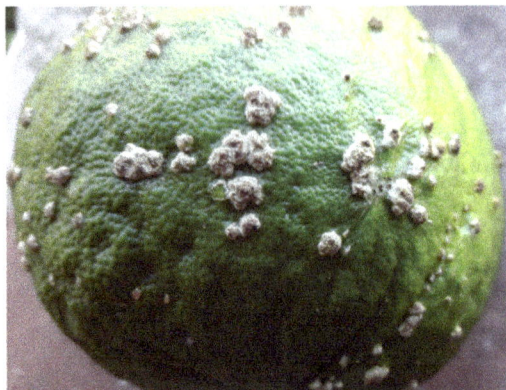

FIGURE 17.3   Scab of Coorg mandarin.

Causal Organism: *Eisinoe fawcetti (I.S: Sphaceloma fawcetti)*

Etiology:

The causal agent is *Elsinoe fawcetti*, Bitancourt and Jenkins. The conidia are colorless, nonseptate, ovoid to curved. Acervuli erupt from leaves or from the infected fruit which releases masses of clustered conidiosphores and bear conidia on top and sides (Jenkins, 1934).

Disease Cycle:

The spores are produced from the diseased fruits, leaves, and dead twigs. These spores are disseminated through rain splash or irrigation water. Fungus propagates primarily through conidia, which are produced from scab pustules. The pathological role of ascospores is very less.

    Three to four hours wetness is required on susceptible plant tissues for infection to occur. The young leaves are more prone to infection and become resistant just before they become full size. The pathogen will survive on diseased dried leaves and fallen twigs and fruits (Naqui, 2004).

Epidemiology:

The citrus scab is required frequency and duration of relative humidity, inoculum to induce conidial formation, germination, and infection. In general, scab incidence is more severe in low-lying areas with frequent wetting, compared with elevated areas. At optimum temperature (24–28 °C), a short wetting period of 2–3 h is required for infection to occur.

Management:

- Remove and destroy the infected parts from the orchards.
- Spraying of Propaconazole 10% EC (0.5 mL) or Hexaconazole 5% EC (0.5 mL/L) or Copper oxychloride 50% WP (2.5 gm/L of water) are helpful in containing the disease.

17.2.4 ANTHRACNOSE/WITHER TIP/TWIG BLIGHT/DIE BACK

Introduction/Economic Importance:

Colletotrichum gloeosporioides has been observed after *Phytophthora* infection in mandarin crops in Nagpur area of India (Kumbhare and Moghe, 1976) and caused many symptoms in mandarin plants. The symptoms are more pronounced in old and neglected orchards and cause reduction in yield.

Symptoms:

The pathogen will infect shoots, leaves, and tender fruits. The affected leaves show symptoms of circular spots, which are light tan in color with a purple margin. On the fruit lesions are gray to black spots about 1.5 mm in dia. The disease severely affects twigs (Figure 17.4) and starts to wither from the tip downwards (Kassahun et al., 2006). The drying gradually progresses downward; the leaves turn yellow and droop.

FIGURE 17.4 Anthracnose or wither tip/die back of Coorg mandarin.

Causal organism: *Colletotrichum gloeosporioides*

Etiology:

The fungus produces whitish to gray lawn of aerial mycelium with orange spore masses in acervuli. Conidia is hyaline, single-celled, (8–20 × 2.5–5 µm) may be cylindrical with rounded apex having a narrow truncated base. Conidiophores are produced in acervuli are irregular, light brown, and produce orange in spore mass. Acervuli produced on fallen infected leaves, branches, and fruits and conidia remain viable for long period under natural conditions (Simmonds, 1965).

Disease Cycle:

Acervuli and /ascomata of fungus occur normally on senescent tissues such as fruits, leaves, or twigs, although teleomorphs have no role in disease development. The fungus disseminate from diseased parts to healthy organs and sporulate on them (Fitzel, 1987). Heavy rain and wind are most favorable for conidial dispersal. On fruits, the infection can take place at any phase of the developmental stage; conidium germinates forming a germ tube and a terminal appressorium from which an infection peg emerges and penetrates the outer wax layer and cuticle of the fruit skin (Binyamini and Schiffmann-Nadel, 1972). Nevertheless, reported by Prusky et al. (1983), the presence of antifungal compounds in the unripe fruits (i.e., epicatechin) limits the growth of the infection peg, which remains in latency until fruit ripening. During ripening, levels of these compounds decrease by enzymatic action (i.e., laccase) (Guestsky et al., 2007). Then, latent infections become reactivated and the hyphae invade the flesh until most of the fruit is rotten. In the initial stages, acervuli are produced beneath the fruit surface. The conidia will be released, when the cuticle and epidermal cells are ruptured and later are dispersed by water.

Epidemiology:

The infection of the disease will be favored if the relative humidity is 80% and temperature is 18–26 °C.

Management:

- Prune the infected twigs and protect the wound portion with copper fungicide (copper oxychloride 50% WP or 10% Bordopaste) is most effective in checking the disease.

- Spraying of hexaconazole 5% EC (0.5 mL/L of water) or carbendazim 50% WP to contain the disease effectively.

17.2.5 DRY ROOT ROT DISEASE

Introduction/Economic Importance:

Dry root rot disease of citrus was first reported by Ramakrishnan in 1954 and identified causal agents as *Diplodia natalensis* and *Fusarium solani*. Later, Reddy and Paparao (1960) reported the root rot disease on Sathgudi sweet orange and acid lime from Anantharajpet of Andhra Pradesh. The incidence of the disease is reported to be increasing in acid lime and sweet orange growing areas especially the orchards that are in a dry state (Reddy et al., 1999; Gopal et al., 2000). The disease is spreading all over the country where ever the citrus is grown. The incidence of dry root-rot is 5%–50% and 10%–15% of the infected trees are being killed every year. The Jamberi root stocks and Sweet oranges are more susceptible to the disease as compared to Rangpur lime stock.

Symptoms:

The dry root rot disease is caused by complex pathogens. The affected plant becomes wilted at an early stage of its growth, resulting in death that may occur within a month. The infected plants show yellowing of foliage, drooping, and wilting. The most important symptoms are bark decay at early stages and dead wood underneath. The affected roots will emit a foul smell. The plants become excessive flowering, heavily bearing with small-sized fruit, and ultimately death of the plant (Mohammed, 2013).

Causal Organism: *Macrophomina phaseoli, Fusarium* spp. *and Diplodia natalensis*

Disease Cycle

M. phaseolina produces microsclerotia within the roots and stems of the mandarin plants. These resting spores (microsclerotia) are overwinter in the soil and dead crop residue as a primary source of inoculum in the spring for the spread of the disease.

Epidemiology:

M. phaseolina is a drought loving pathogen, which produces huge quantities of microsclerotia under harsh conditions such as low moisture and high temperature.

Management:

- If the plant starts wilting, try to locate infected roots and destroy them. The wound portion should be protected with 10% Bordeaux paste.
- Avoid ploughing deeply or digging that may cause damage to the primary roots.
- Mulch in the tree basins with dried leaves.
- Soil application of *Trichoderma* spp. along with farm yard manure before and after the monsoon.
- Soil drenching with 0.5% *Trichoderma vride* + 0.2% Chlorothalonil 75% WP or application of neem cake (10 kg) will reduce the disease intensity.
- The disease can be checked by drenching with 0.25% Chlorothalonil 75% WP or 0.1% Carbendazium 50% WP followed by drenching of 0.25% Mancozeb 75% WP.

17.2.6   *GANODERMA* ROOT ROT

Introduction/Economic Importance:

The Ganoderma was identified by Karsten (1881) with *Ganoderma lucidum* (W. Curt, Fries) and it is distributed worldwide in tropical and temperate regions of the world. The pathogen will survive in hot and humid conditions (Pilotti et al., 2004). Ganoderma species is a facultative parasite that can live as saprobes on rotted stumps and roots (Pilotti et al., 2004). Ganoderma root rot is more in old neglected orchards of different citrus species.

Symptoms:

- Whitish mycelium spread on the surface of the root bark and gradually moves down the trunk.
- On the affected portion of the stem, bracket like fungal fruiting bodies will appear at the ground level during rainy season (Figure 17.5).
- Lateral root/roots become wilted and die.

FIGURE 17.5 Bracket like fungal fruiting body has appeared at ground level, which is caused by *Ganoderma lucidum*.

Casual Organism: *Ganoderma lucidum*

Etiology:

Basidiocarps, the sexual structures of the fungus will grow from a living or dead trunk or branch of a tree and form a bracket at ground level. The fungus will produce two types of basidiocarps, based on the species: a laccate fruiting body having a shiny upper surface, or a nonlaccate fruiting body having a dull upper surface (Hapuarachchi et al., 2015; Smith and Sivasithamparam, 2000; Pilotti et al., 2004).

Disease Cycle:

The fungus is disseminated through basidiospores. These spores will land on the soil surface and germinate once the moisture is available. The fungus damages will be observed always at the ground closest to the soil surface and moves in the center of the trunk causing a cone-like shape seen in the infected trunk.

Management:

- Remove the dead plants or old stumps from the orchard.

- Growing green manure crops and incorporating into the soil in orchards.
- Collection of brackets found near the collars of other hosts and destroying them.
- The infested roots will be destroyed periodically.
- Soil drenching with 0.5% trichoderma viride + 0.3% copper oxychloride 50% WP.
- Apply sulfur 80% WP (0.5–1 kg) with the soil around the trench helps in containing the disease. Spray antifungal antibody like Aureofungin 46.15% SP (1.5 g in 5 L of water) or Vitavax (500 ppm) in the basin of the plant was found very effective.

17.2.7 SOOTY MOLD

Introduction/Economic Importance:

Sooty mold is also known as blotch, which is characterized by splotchy black coatings on different parts of the plants such as on leaves, some time on stems, and fruit. The black powder like substances consist of dark fungal threads of different ascomycetes fungi (*Capnodium* spp. and *Fumago* spp.). These fungi will grow normally on leaves, stem, and fruits, due to honeydew secreted by different sucking pests while feeding. The fungus is not harmful but it prevents the photosynthetic activity of the host.

Symptoms:

The sooty mold diseases are more destructive, if the plants are infected with aphids, mealy bug, and scale insects, due to these insects suck the sap from the host by puncturing the host leaf and suck the plant juices, while sucking the sap from the plants, they secrete the honeydew like substances, which enhance the growth of the sooty mold fungi. Sooty mold consists of black masses of spores, which are made up of individual hyphae having conidiophore and conidia (Williams, 2002).

Sooty mold is a non parasitic pathogen and does not penetrate the plant tissue, it grows superficially on the honeydew excretions. Sooty mold causes a certain degree of injury when its growth is high by prevention of sunlight to the leaf and fruit and making it unattractive (Williams, 2002). Fruit that are covered with black sooty mold will affect both quality as well as quantity. The affected leaves become shriveled under dry conditions.

Causal Organism: *Capnodium citri*

Sooty molds are of two types. The first one is deciduous growth on the leaves and remains till the leaf is lost. The second is persistent growth on twigs stem and woody plants.

Disease Cycle:

A densely populated orchard where there is low light intensity provides favorable conditions for sooty mold disease. Sugary substance secreted by the insects is stated to be a condition favorable for the development of sooty mold. Incidence of insects on the shoot is directly associated with disease severity.

Epidemiology:

High humidity and high incidence of insects proved to be congenial for the development of the disease.

Management:

- The fungi will be grown on sugary substances, which are secreted by different insects like aphids, mealy bug, and scale. This can be controlled by spraying of systemic insecticides. Spraying 1% sodium bicarbionate would effectively check the fungus.

17.3   BACTERIAL DISEASES

17.3.1   CITRUS CANKER

Introduction/Economic Importance:

In India, citrus is the most cultivated fruit crop after mango and banana. Canker is one of the major limitations for cultivation and production of crops in different parts of the country. This was first recorded in Punjab (Luthra and Sattar, 1942; Bedi, 1961). Later in different states of India such as Karnataka, Rajasthan, Madhya Pradesh, Andhra Pradesh, Tamil Nadu, Assam, and Uttar Pradesh. Subsequently, it was recorded in almost all districts of the country (Das, 2003).

Symptoms:

It is one of the most important bacterial diseases of citrus, which causes serious yield loss to the crop in terms of quality. The disease will occur in all parts of the tree (Figure 17.6a and b). Initial symptoms will appear on the newly emerged leaves as small, raised translucent spots, which gradually enlarge about 2–10 mm in size surrounded by a yellow halo. Later the spot becomes corky in texture (Timmer et al., 2005). Old lesions are irregular and corky with brown in color on leaves. These leaves become prematurely dropped. On the twigs, the water-soaked dark green lesions later turn brown. On the fruit, the lesions may vary from 1 to 10 mm dia, which are attached to a few millimeters into the rind. The infected fruits drop prematurely. The lesions are restricted to the rind only and affect the fruit quality. The infected fruit's marketability is seriously affected (Das, 2003).

FIGURE 17.6A   Canker growth on twigs.

FIGURE 17.6B   Canker growth on the leaves.

Causal Organism: *Xanthomonas axonopodis* pv. *citri*

Etiology:

The disease is caused by *Xanthomonas axonopodis* pv. *citri* is Gram-negative, a rod-shaped, non spore forming, aerobic bacteria and having polar flagellum. Colonies are usually yellow due to "xanthomonadin" pigment production in media. Culture medium is supplemented with glucose or other sugars, the bacteria will produce exopolysaccaride slime. The bacteria will grow very well at the temperature of 28–39 °C, respectively (Das, 2003).

Disease Cycle:

The pathogen will survive in the infected fruits and dead plant materials like fallen cankerous leaves and twigs as primary inoculums (Rao and Hingorani, 1963). The pathogen will survive on infected twigs and branches for many years (Chakravarti et al., 1966). The pathogen also survived in the leaves for more than 6 months (Vasudeva, 1958). The bacterium survives epiphytically at a low number on different citrus hosts (Goto, 1970, 1972; Leite and Mohan, 1984). The disease was disseminated through wind and rain. However, the bacteria will enter the host through wounds caused by feeding sucking.

Epidemiology:

Pathogen required a thin film of water for 20 minutes for infection and a temperature of 20–30 °C for disease development.

Management:

- Pruning and destroying the infected parts of the host before the monsoon. The wound portion is protected by 10% Bordo paste to prevent the spread of the disease.
- The disease will appear more on newly emerging leaves, immediately after noticed spray 2-bromo-2-nitro-1, 3-propanediol (bronopol) (0.1%, 0.5–1 g/L) or streptomycin sulfate 500–1000 ppm (0.5–1.5 g/L of water) in combination with copper oxychloride 50% WP @ 2.5 g/L of water to effectively check the disease.

17.3.2   CITRUS GREENING DISEASE/HUANGLONGBING (HLB)

Introduction/Economic Importance:

HLB, also called as citrus greening is caused by unculturable phloem-limited bacteria, which is one of the limitations for the production of citrus worldwide. The disease is known to be caused by three different species of *Ca. Liberibacter*. The isolate from Asia (*Ca. L. asiaticus*) is more severe and widespread in different parts of the country. Most of the citrus cultivars are susceptible to HLB bacteria. Normally, the symptoms moderate on lemon, sour orange, and grapefruit, however, the symptoms are more severe in mandarins and their hybrids and sweet orange. Although, trifoliate orange, pummelo, and lime are listed as the most tolerant cultivars. The infected leaves in the trees develop a blotchy mottled appearance (Figure 17.7).

FIGURE 17.7   Yellowing and mottling symptoms of Coorg mandarin caused by *Ca. Liberibacter asaticus*.

Symptoms:

Persistently affected trees display twig and limb dieback, premature dropping of fruit with tiny leaves that point upward. The HLB-infected fruits are very small, poorly colored, and contain aborted seeds (Bove, 2006). The infected fruits retain their green color till they mature, hence the common name "greening disease of citrus." Further, the fruit produced by infected trees is not suitable for either the fresh market or juice processing due to the significant increase in acidity and bitter taste. The main indicator plants are kagzi lime grapefruit and sweet orange for this disease. Symptoms alone may not be useful to diagnose a tree as infected with HLB. This is because it resembles other diseases, namely stubborn and tristeza disease and zinc deficiency.

Causal Organism: *Candidatus Liberibacter asiaticus (Fastidious Phloem limited Bacterium*, Gram-negative bacterium)

Etiology:

Candidatus Liberibacter is a genus of Gram-negative bacteria in the Rhizobiaceae family. There are three *Ca. Liberibacter'* species that have been identified as causal agents of HLB in different countries and climates of the world, primarily by 16s rDNA and other gene sequences: "*Ca.* L. *asiaticus, Ca.* L. *africanus* (Hocquellet et al., 1999)," and "*Ca.* L. *americanus*" (Coletta-Filho et al., 2005), respectively. These are vectored by diverse species of psyllids.

Transmission:

This disease is transmitted by two species of psyllid insect, namely the Asian citrus psyllids (*Diaphorina citri*) and the African citrus psyllid (*Trioza erytreae*; Dahiya et al., 1994). Both the species are transmitting the disease from infected trees to healthy plants. Usually the adult psyllids are readily observed on newly developed leaves. The female psyllids lay the eggs in unopened leaves and inside the buds. A single female psyllid can lay eggs up to 800 to 1000 in the entire lifespan (Halbert and Manjunath, 2004; Tsai and Liu, 2000). The factors which enhance the psyllid population are the availability of young growing shoots and temperature. The insect population was built on the early stage of the new flushes and reach to peak at the first summer new flush, which normally occurs May or June. During the summer months, psyllid populations will oscillate due to the availability of new flush. Summer temperatures above 30 °C may shorten the longevity of psyllid adults (<30 days) and lower their reproductive fitness. The acquisition times for *D. citri* are 15–30 min and up to 5 h. *D. citri,* the fourth and fifth instar larvae nymphs will acquire the pathogen, and adults will transmit the disease. Citrus greening is also transmitted through the grafting of infected budwood, by dodder, and, possibly, by infected seed (Xu et al., 1988).

Management:

- Growing healthy and certified disease-free planting materials.
- Spraying the crop with Imdiocloprid @ 0.5 mL/L or Thiamethoxam @ 0.2 g/L of water.
- Temporary suppression of the disease is possible spray 2-bromo-2-Nitro-1,3-propanediol (bronopol; 0.1%, 0.5–1 g/L) or streptomycin sulfate 500–1000 ppm (0.5–1.5 g/L of water) in combination with copper oxychloride 50% WP @ 2.5 g/L of water helps in effective management of the disease.
- Temporary suppression through injecting of the infected plants with Ladermycin @ 500 ppm or Tetracycline 500 ppm/1 L of water.
- Spray 0.5% zinc sulfate + 0.3 % neem oil/L of water at intervals of 40 days.
- Soil application of zinc-EDTA chelate (Zn-12%, @ 10 kg/ac) or zinc sulfate monohydrate (Zn-33%, @ 15 kg/ac) or zinc sulfate heptahydrate (Zn-21%, @ 10 kg/ac) could effectively manage the yellowing of leaves.
- Removal of HLB infected trees routinely in young plantings stage
- Yellow sticky card traps should be placed 18–20 (46–51 cm) inches above the ground to attract the adult pysllids.

17.4   VIRAL DISEASES

17.4.1   CITRUS TRISTEZA DISEASE

Introduction/Economic Importance:

Tristeza virus is first reported in China. Tristeza also known as "quick decline" is one of the major limitations for the production of citrus worldwide. This disease was first noticed in Italy and Florida on *Citrus aurantifolia* and *C. sinensis.* In India, the disease was spread to Andhra Pradesh, Karnataka, Madhya Pradesh, Tamil Nadu, Maharashtra, Punjab, Bihar, West Bengal, and Sikkim.

The virus affects all kinds of citrus species but mainly orange, grapefruit, and acid lime. The affected tree shows quick or chronic tree decline and the disease is more severe on sour orange root stocks propagated trees. The indicator plants used for detection the CTV of was Kagzi lime and Nasnaran.

Symptoms:

Depending on the citrus cultivars, virus strain, and scion-rootstock combination, it will cause different kinds of symptoms on citrus plants. In Asia, three types of CTV strains have been reported based on the symptoms they cause to the host, such as seedling yellows (CTV-SY), tristeza (CTV-T), stem pitting (CTV-SP), and a mild type. Any of these strains may exist together or single on the citrus plant.

Initially, the infected small branches and twigs of plants show dieback, yellowing of leaves and plants are bearing heavily with small fruits (Savita et al., 2011). Later, the infected plants showed severe chlorosis and mottling of leaves. A huge number of pits were seen on trunks and stems of plants when it was split open. The disease is primarily disseminated through the grafted bud-wood under in vitro and citrus aphid (*Toxoptera citricida*), which will transmit the virus into the healthy plants under natural conditions or field conditions (Gottwald et al., 2002).

Causal Organism: Citrus Tristeza Virus

Etiology:

Citrus tristeza virus (CTV) has a 19.3-kb ssRNA genome (Bar-Joseph et al., 1979; Karasev, 2000; Agranovsky, 1996; Dolja et al., 2006) belongs to family *Closteroviridae,* which contains viruses with mono-, bi-, and tripartite genomes. The genomic RNA of CTV was organized into 12 open reading frame (ORFs), which contains potentially 19 different final proteins (Karasev, 2000). Ten 3' genes are expressed at nested set of 3' co-terminal sgmRNAs (Hilf et al., 1995), which consist of ORFs (Pappu et al., 1994) plus 5 nonconserved genes.

Epidemiology:

CTV is transmitted vertically by vegetative propagation of citrus when grafting infected budwood onto a rootstock. The pathogen is also transmitted horizontally by several aphids' species in a semipersistent mode. Generally, acquisition of the virus by aphids requires 30 min to 24 h, and aphid can transmit for 1–2 days. It was found that *Aphis gossypii* and a few other aphid species will transmit CTV isolates, but the most efficient vector is *Toxoptera citricida* (William et al., 2015). A large number of aphid populations and the species predominant in each region are critical factors in CTV epidemiology, with the spatial and temporal spread patterns caused by *T. citricida* or *A. gossypii* being different. Owing to the semipersistent mode of transmission, chemical control of aphids may not be useful to prevent the CTV spread (Gottwald et al., 2002).

Management:

- Healthy and certified buds will be used for the production of disease-free planting materials.
- As such there is no control measure for the virus so, the best management is pruning infected plants or uprooting and burning to avoid further spread.
- The disease (virus) is spread by insect vector such as aphids, which are more confined to young shoots, if aphids are noticed on young shoot, immediately spray the crop with Imidachloprid 17.8% SL @ 0.25 mL/L of water or Acetamiprid 20 SP @0.3 g/L or Thiamethoxam 25% WG @ 0.25 g/L or water may effectively check the citrus aphids.

17.4.2   CITRUS YELLOW MOSAIC BADNAVIRUS (CMBV)

Introduction/Economic Importance:

The disease was first reported on sweet orange (*Citrus sinensis*) from southern India (Dakshinamurti and Reddy, 1975). However, the etiology of the disease was not known until the early 1990s. It was later studied in detail and proved to be badnavirus on citrus cultivar, Pummelo (*Citrus grandis* [L.] Osbeck), grown in Karnataka state in southern India (Ahlawat et al., 1996a, b; Ghosh et al., 2014). Later, the disease was recorded on citrus cultivars–Rangpur lime and acid lime grown in Andhra Pradesh (Baranwal et al., 2005; Sai Gopal et al., 2000). The losses attributed to this disease alone could be around 70% in some citrus species (Ahlawat et al., 1996a; Chung and Briansky, 2012) and currently appears it is restricted to only India.

Symptoms:

The infected tree shows the symptoms of mosaic and flecking along the veins of the leaves (Ahlawat et al., 1996a; Figure 17.8).

Casual Organism: Citrus yellow mosaic badnavirus (CMBV)

Transmission:

The Citrus yellow mosaic badnavirus is transmitted by mealybug (*Plano-coccus citri*) with an acquisition access period and inoculation access period of 24 hours, respectively. The virus was artificially transmitted through the mealybug. *Citrus decumana* plants showed mosaic symptoms similar to field symptoms after 20 days (Reddy et al., 2010). The virus is also transmitted by grafting, dodders, and mechanical inoculation to other citrus species (Ahlawat et al., 1985, 1996a, b).

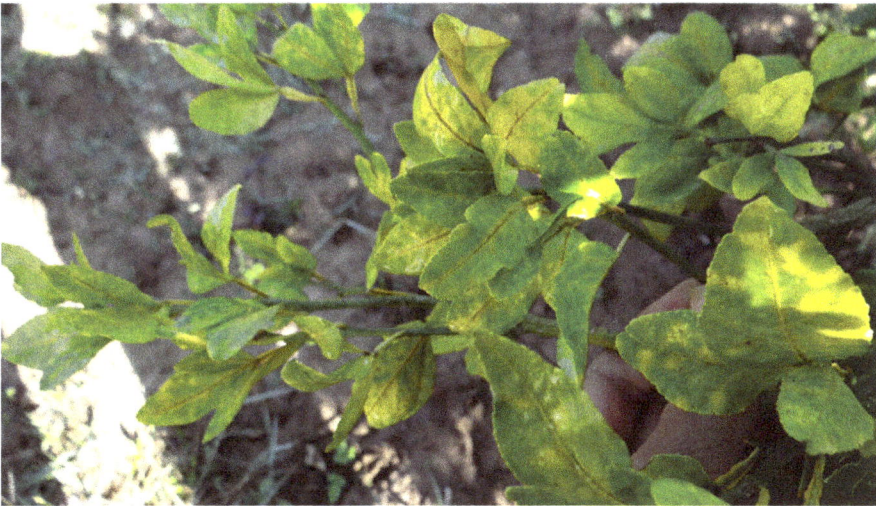

FIGURE 17.8   Mosaic and flecking along the veins of the leaves caused by citrus yellow mosaic badnavirus.

Host Range:

CMBV has been recorded mainly on oranges (*Citrus sinensis*), *Aegle marmelos*, *C. limonia*, *Fortunella* spp., mandarins (*C. reticulata*), lemons (*C. limon*), grapefruits (*C. paradisi*), and pummelos (*C. maxima*), Citrus spp., (*C. volkameriana, C. jambhiri, C. aurantium, and C. mitis*) were mostly susceptible, while *C. limettioides, C. aurantiifolia, and C. medica* were more rarely infected and gave milder symptoms (Ahlawat et al., 1996).

Etiology:

The causal agent of the disease is Citrus yellow mosaic badnavirus, is bacilliform in morphology, measuring about 130×30 nm in size (Ahlawat et al., 1996a, b). The viral genome consists of a single circular ds DNA of approximately 7.5 kb in size that contains six ORFs. All ORFs are present in the plus strand and encode different proteins (Bouhida et al., 1993; Futterer et al., 1997; Huang and Hartung, 2001). The intergenic region, which is approximately 700 nucleotides long, contains a putative promoter and initiator for transcription signal. The virus has been classified tentatively as a member of the badnavirus genus under the family Caulimoviridae (Huang and Hartung, 2001).

Transmission:

CMBV is transmitted vertically by clonal propagation of citrus when grafting infected budwood onto a rootstock. The virus transmits horizontally by mealybug (*Planococcus citri*) with an acquisition access period of 24 h followed by an inoculation access period of 24 hours (Reddy et al., 2010). The viruses are also transmitted through uncertified scion budding used for planting in nontraditional areas of citrus cultivation besides traditional areas. Due to ignorance and nonavailability of certified scion material in huge quantities, the private nurseries collect scion material from farmers' fields which are a source of latent infection. Once these infected scions are used in budding, the virus is spread under field conditions to healthy plants by different methods (Aparna et al., 2012).

Management:

- Healthy and disease-free bud-wood should be used for the production of disease-free planting materials.
- Tools used for grafting could be disinfected with 2% sodium hydroxide along with 2% formaldehyde to prevent mechanical transmission
- The disease (virus) is spread by mealybug, which is more confined to young shoots, if insect vector is noticed on a young shoot, immediately spray with Diamethoate 30% EC @ 1.5 mL/L or Imidachloprid 17.8% SL @ 0.25 mL/L of water or Thiamethoxam 25% WG @ 0.25 g/L or Acetamiprid 20 SP @ 0.3 g/L of water may effectively check the citrus mealybug.

17.5   VIROID DISEASE

17.5.1   *CITRUS EXOCORTIS*

Introduction/Economic Importance:

The *Citrus exocortis viroid* disease is probably one of the most ancient viroid disease (Bar-Joseph, 2003), has the widest reported naturally occurring host, and present worldwide wherever the citrus is grown. The citrus cultivars cultivated commercial are symptomless carriers and they may be stunted in growth. The causal agent of the disease is *Citrus exocortis viroid* (CEV), which was first recorded on trifoliate orange rootstock and described as a bark-shelling disorder (Fawcett and Klotz, 1948). Benton et al. (1949) recorded the disease in Australia in the early 1930s as "scaly butt" and found it to be transmissible.

Symptoms:

The viriod infected plants showed drying of outer bark, which is separate from the inner live-bark. The typical symptoms are given below on different citrus sp.

- Trifoliate orange (*Poncirus trifoliate*): bark scaling and stunting of citrus trees.
- Lemon (*Citrus lemon*): longitudinal shelling of bark and cracking occurs, leaves become blotchy mottle.
- *Citrus medica* (*Etrog citron*): leaf epinasty and rugosity and underside of the leaf tip veins show browning and stunting.
- Rangpur lime (*Citrus limonia*): older twigs show the symptoms of yellow blotches.

Casual Organism: *Citrus exocortis viroid* (CEVd)

Etiology:

CEVd is the causal agent of the exocortis disease on citrus. The pathogen has the widest host range among the Pospiviroids. It consists of 371–375 nucleotides and is a species within the genus Pospiviroid belonging to the Pospiviroidae family (Flores et al., 2000).

Transmission:

The viroid is highly contagious and mechanically transmitted. It is spread by leaf contact, contaminated planting, and cultivating equipments like tools and knives during clipping of fruit, pruning, or collecting of budwood (Garnsey and Jones, 1967, 1968).

The viriod is disseminated mostly as a result of cultural operations through contaminated knives, tools, and hands. However, in vegetatively propagated citrus and related species, the main means of CEVd spread is by bud-grafting and/or use of contaminated tools during grafting and pruning operations (Garnsey and Jones, 1967). The viriod replicates by using a host DNA-dependent RNA polymerase II involved in the synthesis of mRNA synthesis through an asymmetric rolling circle mechanism and accumulate in the nucleus (Kovalskaya and Hammond, 2014).

Management:

- The disease can be managed through viroid-free budwood.
- Tools used for grafting could be disinfected with 2% sodium hydroxide plus 2% formaldehyde to prevent the spread of viriod.

KEYWORDS

- **Coorg mandarin**
- **Diseases**
- **Symptomatology**
- **Epidemiology**
- **Management**

REFERENCES

Agranovsky, A. A. (1996). Principles of molecular organization, expression, and evolution of closteroviruses: over the barriers. *Adv. Virus Res.* **47:**119–158.

Ahlawat, Y. S., Pant, R. P., Lockhart, B. E., Srivastava, M., Chakraborty, N. K., and Varma, A. (1996a). Association of badnavirus with citrus mosaic disease in India. *Plant Dis.*, **80:** 590–592.

Ahlawat, Y. S., Pant, R. P., Shukla, A., and Lockhart, B. E. (1996b). Partial characterization of a badnavirus associated with citrus yellow mosaic in India. In: Proceedings of 13th Conference of the International Organization of Citrus Virologists (China; pp. 208–217).

Ahlawat, Y. S., Chenulu, V. V., Vishwanath, S. M., and Pandey, P. K. (1985). Studies on a mosaic disease of citrus. *Curr. Sci.,* **54**:873–874.

Aparna, G. S., Venkata Subbaiah, K., Saigopal, D. V. R., and Reddy, M. N. (2012). Citrus yellow mosaic: a transmissible virus of citrus species in India. *Bioscan,* **7**(2):283–287.

Baiswar, P., Ngachan, S. V., and Rymbai, H. (2015). *Erysiphe quercicola,* a powdery mildew fungus on Khasi mandarin in North East India. *Australas. Plant Dis. Notes,* **10**:30.

Baranwal, V. K., Singh, J., Ahlawat, Y. S., Gopal, K., and Charaya, M. U. (2005). Citrus yellow mosaic virus is associated with mosaic disease in Rangpur lime rootstock of citrus. *Curr. Sci.,* **89**:1596–1599.

Bar-Joseph, M. (2003). Natural history of viroids—historical aspects. In A. Hadidi, R. Flores, J. Randles W., Semancik J.S. (Eds.), Viroids (pp. 246–251). Collingwood: CSIRO.

Bar-Joseph, M., Garnsey, S. M., and Gonsalves, D. (1979). The closteroviruses: a distinct group of elongated plants viruses. *Adv. Virus Res.,* **25**:93–168.

Bedi, K. S. (1961). Some important observations on the citrus canker in Punjab. *Punjab Hort. J.,* **2**:89–91.

Bender, G. S., Menge, J. A., Ohr, H.D., and Burns, R. M. (1982). Dry root-rot of citrus: its meaning for the grower. *Citrograph,* **67**:249–254.

Benton, R. J., Bowman, F. T., Fraser, L., and Kebby, R. G. (1949). Stunting and scaly butt of citrus associated with Poncirus trifoliata rootstock. *Agr. Caz. NSW,* 60:521–526,

Binyamini, N. and Schiffmann-Nadel, M. (1972). Latent infection in avocado fruit due to *Collectotrichum gloeosporioides. Phytopathology,* **62**:592–594.

Bouhida, M., Lockhart, B. E., and Olszewski, N. E. (1993). An analysis of the complete sequence of a sugarcane bacilliform virus genome infectious to banana and rice. *J. Gen. Virol.,* 74(Pt 1), 15–22.

Bove, J. M. (2006). Huanglongbing: a destructive, newly-emerging, century-old disease of citurs. *J. Plant Pathol.,* **88**:7–37.

Chakravarti, B. P., S. Porwal, and Rangarajan M., (1966). Studies on citrus canker in Rajasthan. I. Disease incidence and survival of the Pathogen. *Labdev J. Sci. Tech.,* **4**: 262–265.

Chung, K. R. (2010). *Elsinoë fawcettii* and *Elsinoë australis*: the fungal pathogens causing citrus scab. *Mol. Plant Pathol.,* **12**:13.

Chung, K. R. and Briansky, R. H. (2012). Citrus diseases exotic to Florida: Citrus Yellow Mosaic. EDIS:#PP293.

Coletta-Filho, H. D., Takita, M. A., Targon, M. L. P. N., and Machado, M. A. (2005). Analysis of 16S rDNA sequences from citrus huanglongbing bacteria reveal a different "*Ca. liberibacter*" strain associated with citrus disease in Sao Paulo. *Plant Dis.,* **89**:848–852.

Dahiya, K. K. Lakra, R. K., Dahiya, A. S., and Singh, S. P. (1994) Bioefficacy of some insecticides against citrus psylla, *Diaphorina citri. Crop Res.,* **8**:137–140.

Dakshinamurti, V. and Reddy, G. S. (1975). Mosaic—a transmissible disorder of sweet oranges. *Indian Phytopathol.,* 28:398–399.

Das, A. K. (2003). Citrus canker—a review. *J. Appl. Hort.,* **5**(1):52–60.

Dolja, V. V., Kreuze, J. F., and Valkonen, J. P. T. (2006). Comparative and functional genomics of closteroviruses. *Virus Res.,* **117**:38–51.

Fawcett, H. S. (1936). Citrus diseases and their control. McGraw Hill Book Co. New York and London; p. 656.

Fawcett, H. S. and Klotz, L.J. (1948). Bark shelling of trifoliate orange Calif. *Citrogrph,* **48**:230.

Fitzel, R. D. (1987). Epidemiology of anthracnose disease of avocados. South African Avocado Growers' Association Yearbook **1**:113–116.

Flores, R., Randles, J. W., Bar-Joseph, M., and Diener, T. O. (2000). Viroids. In M. H.V. van Regenmortel, C. M. Fauquet, D. H. L. Bishop, E. B. Carstens, M. K. Estes, S. M. Lemon, *et al.* (Eds.), Virus taxonomy 7th Report of the International Committee on Taxonomy of Viruses (pp. 1009–1024). San Diego.

Fraser, L. R. and Singh, D. (1966). Root rot of Citrus in India. *Indian Horticult.*, **11**(1):15–17, 26.

Futterer, J., Rothnie, H. M., Hohn, T., and Potrykus, I. (1997). Rice tungro bacilliform virus open reading frames II and III are translated from polycistronic pregenomic RNA by leakyscanning. *J. Virol.*, **71**:7984–7989.

Garnsey, S. M. (1968). Exocortis virus can be spread by contaminated tools. *Citrus Ind.*, **49**:13–16.

Garnsey, S. M. and Jones, J. W. (1967). Mechanical transmission of exocortis virus with contaminated budding tools *Plant Dis. Rep.*, **51**:410–413

Ghosh, D. K., Bhose, S., Mukherjee, K., Aglave, B., Warghane, A. J., Motghare, M., Baranwal V. K., and Dhar A. K. (2014). Molecular characterization of citrus yellow mosaic badnavirus (CMBV) isolates revealed the presence of two distinct strains infecting citrus in India. *Phytoparasitica*, **42**:681–689.

Gopal, K., Subbi Reddy, G., Reddy, M. R. S., Chenchu Reddy, B., Ramakrishna Rao, A., Madhavi, M., and Venkataramana, K. T. (2000). Citrus decline in Andhra Pradesh—causes and their management. Paper presented in International Symposium on Citriculture held at Nagpur. November 23–27, 1999 at NRCC, Nagpur, India, pp. 107.

Goto, M. (1970). Studies on citrus canker III. Survival of *Xanthomonas citri* (Hasse) Dowson in soils and on the surface of weeds. *Bull. Fac. Agric. Shizouka Univ.*, **20**:21–29.

Goto, M. (1972). Survival of *Xanthomonas citri* in the bark tissues of citrus trees. *Can. J. Bot.*, **50**:2629–2635.

Gottwald, T. R., Abreu-Rodr´iguez, E., Yokomi, R. K., Stansly, P. A., Riley, and T. K. (2002). Effects of chemical control of aphid vectors and of cross-protection on increase and spread of citrus tristeza virus. *Proc. Int. Organ. Citrus Virol.*, **15**:117–30.

Gottwald, T., Polek, M., and Riley, K. (2002). History, present incidence, and spatial distribution of citrus tristeza vírus in the California Central Valley. In: Duran-Vila, N., Milne, R.G., da Graça, J.V., (Eds.), Proceedings of the 15th Conference of the International Organization of Citrus Virologists. IOCV, Riverside, CA, USA, pp. 83–94.

Graham, J. H. and Menge, J. L. (1999). *Phytophthora* induces disease In: Timmer, L. W., Garnesy, S. M., Graham, J. H., editors) "Compendium of Citrus diseases," American Phytopathological Society, St. Paul, MN, USA, pp. 12–15.

Guestsky, R., Kobiler, I., Avilla-Quezada, and G., Prusky, D. (2007). Metabolism of epicatechin by laccase of *Collectotrichum gloeosporioides*. Book of Abstracts of the VI World Avocado Congress, Vina del Mar, Chile, p. 98.

Halbert, S. E. and Manjunath, K. L. (2004). Asian citrus psyllids (Sternorrhyncha: Psyllidae) and greening disease of citrus: a literature review and assessment of risk in Florida. *Fla. Entomol.*, **87**:330–353.

Hapuarachchi, K. K., Wen, T.C., Deng, C. Y., Kang, J. C., and Hyde, K. D. (2015). Mycosphere Essays 1: Taxonomic Confusion in the *Ganoderma lucidum* Species Complex. Mycosphere **6 (5)**:542–559.

Hilf, M. E., Karasev, A.V., Pappu, H.R., Gumpf, D.J., Niblett, C.L., and Garnsey, S.M. (1995). Characterization of citrus tristeza virus subgenomic RNAs in infected tissue. *Virology*, **208**:576–82.

Hocquellet, A., Toorawa, P., Bove, J. M., and Garnier, M. (1999). Detection and identification of the two "*Candidatus liberobacter* species" associated with citrus huanglongbing by PCR amplification of ribosomal protein genes of the β operon. *Mol. Cell. Probes,* **13**:373–379.

Huang, Q. and Hartung, J. S. (2001). Cloning and sequence analysis of an infectious clone of Citrus yellow mosaic virus that can infect sweet orange via Agrobacteriu mediated inoculation. *J. Gen. Virol.*, **82**:2549–2558.

Jenkins, A. E. (1934). A species of *Sphaceloma* on Avocado. *Phytopathology*, **24**:84–85.

Karasev, A.V. (2000). Genetic diversity and evolution of closteroviruses. *Annu. Rev. Phytopathol.*, **38**:293–324.

Karsten, P. A. (1881). Enumeralio boletinearum et polypore arum fennicarum, systemate novo dispositarum. *Rev. Mycol.*, **3**:16–19.

Kassahun, T., Temam, H., and Sakuja, P. K. (2006). Management of fruit and leaf spot disease in Ethiopia. *Agricultura Tropica et Subtropica*, **39**:242–245.

Kovalskaya N. and Hammond, R .W. (2014). Molecular biology of viroid–host interactions and disease control strategies. *Plant Sci.,* **22**:48–60.

Kumbhare, G. B. and Moghe, P. G. (1976a). Leaf fall disease of Nagpur orange caused by *Phytophthora nicotianae* var. *parasitica* Waterhouse. *Curr Sci.*, **45**:561–562.

Kumbhare, G. B. and Moghe, P. G. (1976b). Leaf fall disease of Nagpur orange caused by *Phytophthora nicotianae var. paracitica. Curr. Sci.*, **45**(15):561–562.

Leite, R. P. and S. K. Mohan, (1984). Survival of *Xanthomonas campestris* pv. *citri* (Hasse) Dye in soil and in association with some gramineous plants. *Proc. Int. Soc. Citric.*, **2**:365–368.

Luthra, J. C. and A. Sattar, (1942). Citrus canker and its control in Punjab. *Punjab Fruit J.,* **6**(1):179–182.

Mohammed, Y. (2013). Pseudocercospora leaf and fruit spot disease of citrus: achievements and challenges in the citrus industry: a review. *Agri. Sci.,* **4**(7):324–328.

Naqui, SAMP. (2004). Diagnosis and management of certain important fungal diseases of citrus. In: Naqui, SAMP. (Ed.) *Diseases of Fruits and Vegetables (Vol I)*, Kluwer Academic Publisher, The Netherlands, pp. 247–290.

Naqvi, SAMH. (1988). Prevalence of *Phytophthora* species pathogenic to citrus in orange groves of Vidarbh, Mahrashtra, *Indian J. Myco. Plant Pathol.*, **18**:274–276.

Pilotti, C. A., Sanderson, F. R., Aitken, A. B., and Armstrong, W. (2004). Morphological variation and host range of two Ganoderma species from Papua New Guinea. *Mycopathologia*, **158**:251–265.

Prusky, D., Keen, N. T., and Eaks, I. (1983). Further evidence of the involvement of a preformed antifungal compound in the latency of *Collectotrichum gloeosporioides* on unripe avocado fruit. *Physiol. Plant Pathol.,* **22**:189–198.

Ramakrishnan, T. S. (1954). Common diseases of citrus in Madras State: Government of Madras Bulletin.

Rao, Y. P. and M. K. Hingorani, (1963). Survival of *Xanthomonas citri* (Hasse) Dowson in leaves and soil. *Indian Phytopath.,* **16**:362–364.

Reddy, B. V. B., Ahlawat, Y. S., and Pant, R. P. (2010).Transmission of citrus yellow mosaic virus and its detection in mealybugs (*Planococcus citri*) by dot-blot hybridization. *Crop Res.*, **39**(1/2/3):145–148.

Reddy, B., Govindarajulu, B., Aariff Khan, M.A., and Begum, H. (1999). Root rot diseases. The major cause of acidlime decline in Andhra Pradesh paper presented in International Symposium on Citriculture held at Nagpur, November 23–27, 1999 at NRCC, Nagpur, India, pp. 157.

Reddy, G. S. and Papa Rao, A. (1960). Problems of sweet orange decline in Andhra Pradesh. *Andhra Agri. J.,* **7:**175–188.

Ryvarden, L. (2004). Neotropical polypores Part 1. Synopsis Fungorum, 19: 1–229.

Sai Gopal, D. V. R., Sreenivasulu, M., Raghavendra Rao, G., Venkata Prasanna, T. M., and Subbaiah, K. V. (2000). Characterization and identification of a citrus mosaic badnavirus infecting acid lime (*Citrus aurantifolia*) in Andhra Pradesh. In: Proceedings of the International Symposium on Citriculture (HI TECH Citrus Management, Nagpur, India; pp. 845–852).

Sastra-Hidayat, A. R. (1992). Preliminary epidemiological study of powdery mildew (*Oidium tingitaninum*) on *Citrus sinensis.* Indonesian Center for Agricultural Library and Technology Dissemination, Pusat Perpustakaan dan Penyebaran Teknologi Pertanian (Indonesia) ICALTD Asian citrus rehabilitation conference. Malang (Indonesia).

Savita Virk, G. S., and Nagpal, A. (2011). *In vitro* selection of calli of *Citrus jambhiri* Lush. for tolerance to culture filtrate of *Phytophthora parasitica* and their regeneration. *Physiol. Mol. Biol. Plants,* **17:**41–47.

Simmonds, J. H. (1965). A study of the species of *Colletotrichum* causing ripe fruit rots in Queensland. *Q. J. Agricult. Animal Sci.,* **22:**437–459.

Sivanesan, A. and Critchett, C. (1998). *Elsinoë australis.* Commonwealth Mycological Institute (CMI), Description of Pathogenic Fungi and Bacteria (No. 440). CAB International, Wallingford, UK.

Smith, B. J. and Sivasithamparam, K. (2003). Morphological studies of Ganoderma (Ganodermataceae) from the Australian and Pacific regions. *Australas. System. Botany,* **16:**487–503.

Timmer, L. W. (1977). Preventive and curative trunk treatments for control of Phytophthora foot rot of citrus. *Phytopathology,* **67:**1149–1154.

Timmer, L. W., Roberts, P. D., and Chung K. R. (2005). Florida citrus pest management guide: Citrus scab. University of Florida, pp. 146.

Tsai, J. H., and Liu, Y. H. (2000). Biology of *Diaphorina citri* (Homoptera: Psyllidae) on four host plants. *J. Econ. Entomol.,* **93:**1721–1725.

Uppal, B. N. and Kamath, M. N. (1936). Gummosis on Citrus in Bombay. *Indian J. Agric. Sci.,* **6:**803–820.

Vasudeva, R. S. (1958). *Sci. Rep. Indian Agric. Res. Inst.,* New Delhi, pp. 93.

William, O. D., Moshe, Bar-Joseph, Stephen, M. G., and Pedro M. (2015). Citrus Tristeza Virus: Making an Ally from an Enemy. *Annu. Rev. Phytopathol.,* **53:**137–55.

Williams, L. (2002). Sooty Mold: Control sooty mold on plants by stopping it before it starts, Department of Horticulture, Okaloosa County Cooperative Extension Service, University of Florida (http://okaloosa.ifas.ufl.edu/sootymold.html).

Xu, C. F., Xia, Y. H., Li, K. B., and Ke, C. (1988). Further study of the transmission of citrus huanglongbing by a psyllid, *Diaphorina citri* Kuwayama. Pages 243–248 in: Proc. of the 10th Conf. Intl. Organ. Citrus Virol. Timmer, L.W., Garnsey, S.M. and Navarro, L., eds. IOCV, Riverside, CA.

CHAPTER 18

MANGO (*MANGIFERA INDICA* L.) DISEASES IN INDIA: PERSPECTIVE AND MANAGEMENT STRATEGIES

ASHISH KUMAR[1,*], K. K. SHARMA[2], and SHARAD BISEN[3]

[1]*Department of Plant Pathology, College of Agriculture, Jawahar Lal Nehru Krishi Viswavidyalay (JNKVV), Jabalpur, Madhya Pradesh, India*

[2]*Regional Research Station (PAU), Ballowal Saunkhri, Balachaur 144521, SBS Nagar, Punjab, India*

[3]*Department of Horticulture, College of Agriculture, Jawahar Lal Nehru Krishi Viswavidyalay (JNKVV), Jabalpur, Madhya Pradesh, India*

Corresponding author. E-mail: ashishashish2612@mail.com

ABSTRACT

Mango (*Mangifera indica*) belonging to family—*Anacardiaceae*—leads in the world's fifth most important fruit crop. Mango is grown in India, China, Mexico, Thailand, Pakistan, Philippines, Indonesia, Nigeria, and Brazil. In India, it is grown in Bihar, Bangladesh, Gujarat, Karnataka, Maharashtra, Rajasthan, Uttar Pradesh, West Bengal, and other states. Various diseases have been reported in different countries also in India mainly from fungal origin affecting the leaves, twigs, branches, stem, and roots. In many areas, diseases are the most important constraint to fruit production. Major diseases of mango are discussed in this chapter.

18.1 INTRODUCTION

Mango (*Mangifera indica*) the "King of Fruits" of the family Anacardiaceae and order Sapindales is considered as the national fruit of India grown in all the states. It is the most popular fruit and is liked by everyone in India and abroad for its flavor in addition to its dietetic and medicinal value. Besides the common consumption in mango juice, jelly, pickles, and so forth, it is also used for making some unique desserts from mango namely ice cream, milk shakes, squashes, and so forth. It has been grown in India since 500 decades. Approximately 27 species within the genus *Mangifera* produced edible fruits. Mangoes are considered to be native of the eastern part of India, Burma, and Andaman Islands belong to Southern Asian regions (Litz, 1997). The "Indo-Burma" region was proposed as its centre of origin by Vavilov. Mango occupied 35.4% area (2163.5 thousand ha) under total fruits cultivation even though it contributed only 21.4% (18527.0 thousand MT) of the total fruit production of India with 8.6 MT/ha productivity during 2014–2015. The state-wise survey showed that Uttar Pradesh and Andhra Pradesh were the leading states in mango production (23.5%) and area (14.6%) under cultivation, respectively in 2014–2015 (Indian Horticulture Database, 2014–15). In Punjab, mango is cultivated in whole submountainous districts and its cultivation has now being extended to canal irrigated arid areas of northern India. More than 1000 varieties of mangoes are found in India and near about 30 varieties are grown commercially for export business including Dashehri, Alphonso, Kesar, Totapuri, and so forth and India holds the first rank globally among the countries export the mango. Among them, Totapuri mangoes are on top in export which was estimated a value of USD 19504972, from January to April 2017 (www.exportgenius.in). Several abiotic (nutritional and physiological) and biotic factors (diseases) are responsible for low productivity (8.1 MT/ha) of mango in India regardless of its highest global production and maximum acreage under cultivation. Some diseases are being discussed here which are economically very important.

18.2   FUNGAL DISEASES

18.2.1   MANGO MALFORMATION

Introduction/Economic Importance:

Malformation is one of the most potential threats of mango production causing substantial yield loss (Hiffny et al., 1978) in India and other tropical and subtropical countries of the world growing mangoes (Crane and Campbell, 1994; Misra and Singh, 2002). The first report of the disease came in 1891 from district Darbhanga of Bihar state and was credited to Maries (Watt,

1891). The occurrence of the disease in our country was reported in Maharashtra, Haryana, Bihar, Punjab, and Andhra Pradesh including Uttar Pradesh where it causes huge damage. Approximately 50% disease incidence was documented in orchards of Northwestern part of India; although it was not recorded in coastal areas. Although, research is being done for several decades but none of the management strategy or method was found effective till now (Pant, 2000; Bains and Pant, 2003; Misra and Singh, 2002).

Symptoms:

Different groups of workers illustrated discrete forms of symptoms as seedlings exhibit bunchy top and malformation of vegetative and floral parts (Varma, 1983).

Bunchy Top Phase:

Symptoms of this phase start in the nursery where shoots become small and thick, appear bunchy with undersized rudimentary leaves. The affected young branches remain stunted which results in the bunchy top look of the affected plant.

Vegetative Malformation:

These types of symptoms are quite common on young saplings (Nirvan, 1953) and are characterized by small shoot branches with tiny scaly leaves which provide a bunchy look to the shoot apex. The affected seedlings develop abnormal vegetative growths with very short internodes. Several vegetative buds sprout and the seedling remains stunted, hence as apical dominance is lost (Kanwar and Nijjar, 1979). The seedlings infected in the later stages may restart to grow normally beyond the diseased portion but early infected seedlings remain stunted and die subsequently (Kumar and Beniwal, 1992a).

Floral Malformation:

This is the malformation of panicles where the florets get converted into vegetative buds. Rachises become thick and remain undersized. Malformed panicles are greener and heavy branching gives it the witches' broom appearance having numerous unopened and predominantly male flowers (Singh et al., 1961; Hiffny et al., 1978). The female parts of affected bisexual florets get extremely enlarged which become unable to fertilize with the least viable pollens (Mallik, 1963; Shawky et al., 1980).

FIGURE 18.1   Vegetative malformation symptoms.

FIGURE 18.2   Floral malformation symptoms.

Casual Organism: *Fusarium moliliforme* var. *subglutinans*

Pathogen:

Single or bicelled microconidia develop from polyphialides which are oval to fusiform in shape but bicelled or tricelled, falcate macroconidia are hard to produce. The production of chlamydospores is not reported.

Survival and Spread:

- Dispersal of the pathogen at long distances occurs mainly through infected propagating material or infected pruning equipment. The spread of this pathogen is slow within diseased orchards.
- The feeding injuries caused by mango bud mite (*Aceria mangiferae*) enable fungus to initiate and spread infection from plant to plant within orchards which provide indirect evidence of mite association with disease.

Favorable Conditions:

The Northwestern part of India is reported for the high incidence of disease with 10–15 °C mean temperature and humid weather at flowering which also leads to disease development.

Management:

- New mango plantings should be established with noninfected nursery stock.
- Avoid establishing nurseries in those orchards either having the disease or its previous history. Furthermore, scion material should not be obtained from affected orchards. An assay based on PCR technique for identifying *F. mangiferae* has been reported and is being developed for use in a quarantine diagnostic kit (Ploetz et al., 2002; Iqbal et al., 2006).
- In infected orchards, symptomatic tissues should be removed from trees and burned. By doing this for two or three consecutive seasons, MMD can be reduced to insignificant levels. De-blossoming between 20th January and 25th February was found effective to rejuvenate new and more panicles in the same season which was also efficacious along with spray of NAA at 200 ppm (Singh and Dhillon, 1986b; Tripathi and Ram, 1998).

- According to Rai and Singh (1967) and Yadav (1972), the disease was also managed when diseased parts were pruned and sprayed with diazinon.
- Spray of NAA at 200 ppm in first week of October and 500 ppm etherel during February at bud initiation stage was found very effective to reduce the severity of floral malformation (Singh and Dhillon, 1986a).
- The inefficacy of carbendazim was documented by Kumar and Beniwal (1992) when it was injected either alone in the trunk or applied through soil or in combination with cultural practices.

18.2.2 POWDERY MILDEW

Introduction/Economic Importance:

This disease affects nearly all cultivars of mango which has also become globally a serious problem in mango production. In Uttar Pradesh, Maharashtra, and Karnataka states of the country, the disease appears in a severe form generally from December to March. In India, the disease is widespread causing 20%–90% losses in different parts which become a serious risk to mango cultivation. The pathogen attacks and causes infection on flower, leaf, and young fruits during the winter season under dry weather. The characteristic symptom of the disease is the white superficial powdery fungal growth on leaves, stalk of panicles, flowers, and young fruits. The affected flowers and fruits drop prematurely reducing the crop load considerably or might even prevent the fruit set. Rains or mists accompanied by cooler nights during flowering are congenial for the disease spread.

Symptoms:

Symptoms of the disease appear on foliage, florets, flower shoots, and younger fruits as a creamish-white floury coating of fungal growth. Affected parts including inflorescence and fruits turn brownish in color and finally become dry. The crop becomes highly prone for infection when it is in the full blooming phase. The diseased flower may be crushed easily with one hand. Infection of the pathogen usually produces aborted flowers. The fruits are small if produced which subsequently drop-off. Fruit infection after its setting leads to the development of purplish-brown lesions or blotches causing corky fruit with cracking as it increases in size. The pathogen generally infects and parasitizes the lower surface of young and new leaf alongside the veins; however, upper surface of the leaves may also get infected of the varieties which are susceptible. If shoots get infected early, it results in distortion, curling, and size reduction of panicles and newly developed leaves. Diseased portion finally becomes brown and necrotic.

Casual Organism: *Oidium mangiferae* Berthet

Pathogen:

Causal organism of the disease is *O. mangiferae*, Berthet. Initially, the perfect stage was attributed to *Erysiphe cichoracearum* but on the basis of shape of haustoria and conidial germination, it was changed to *Erysiphe polygoni* (Palti et al., 1974). Daily temperature ranges from 10 °C to 31 °C coupled with 60%–90% relative humidity are congenial for the spread of powdery mildew. Pathogen persists on diseased plants and on fallen plant stubble comparatively for a short duration in the form of conidia. Seasonal flushes of new growth of vegetative and reproductive parts produce an enormous population of *O. mangiferae* which results in its rapid build-up. The air-borne conidia are disseminated for long distances.

Survival and Spread:

The off-season survival of the pathogen occurs in quiescent buds. The overwintering fungus under a favorable environment for its development in spring season starts to produce and release the spores causing new infection. Spores produced on newly infected portions of the plant cause secondary spread of the disease.

Favorable Conditions:

Low temperature in nights with hot and humid weather is favorable for the development of the disease.

Management:

- Wettable sulfur at 0.3% was found effective for the management of the disease when sprayed before its appearance twice: once before the bud opening and second after the fruit set. The better control was also recorded with other fungicides including benomyl and piparazin (Palti et al., 1974).
- Two to three applications with the dust of fine sulfur may also decrease the disease.

- The spray before flowering and after fruit set (peanut stage) with Karathane at 0.1% or Cosan at 0.1% has shown the efficacy to suppress the disease.
- Spray with fungicide wettable sulfur (Gupta and Yadava, 1984), Tridemorph and Dinocap (Sharma et al., 1993) are also documented for their effectiveness for managing the disease.
- Planting of resistant varieties like Neelum, Janardhan pasand, Banglora, Torapari-khurd, and Zardalu is always the best strategy.

18.2.3   ANTHRACNOSE

Introduction/Economic Importance:

At present, anthracnose is known as a disease of great importance which globally affects mangoes in the field as well as after harvest (Ploetz and Prakash, 1997). Among all mango growing countries where high relative humidity succeeds during the season, it becomes a major limiting factor in production as initial infection generally starts in the field before harvesting. However, postharvest infection is highly destructive and of great concern causing global economic (qualitative and quantitative) loss. The quality deterioration directly reduces the market value of the fruit. Sattar and Malik (1939) credited for the first report of anthracnose from Punjab where they reported its incidence in several districts of the state. Presently, it is prevalent in all regions of India where mango is grown.

Symptoms:

Leaves, petioles, shoots, inflorescence, and fruits may get infected and exhibit symptoms of the disease. Small, irregular-shaped, brown to black, necrotic spots appear on both surfaces of the leaf which become extensive necrotic lesions as they enlarge in size and may be dropped out in dry weather conditions. Panicles show small and dark brownish-black lesions which may merge with an increase in size causing the death of blossoms which results in huge yield reduction. Petioles, shoot, and its branches may also be attacked and develop characteristic black lesions which expand on fruits, foliage, and florets. Leaf curl symptoms are generally observed on severely infected leaves.

Conidia are developed on young and old lesions of the disease. Latent infection occurs on older leaves without any lesion development and the pathogen rests inactive till the senescence of leaf tissue. The fungus restarts to grow and later fruiting bodies are developed on lesions of the necrotic tissues. In few instances, conidia are disseminated through wind or rain splashes and attack the young shoots resulting in twig die-back under a favorable set of environmental conditions (Ploetz et al., 1994). Depressed, dark, and brownish-black spots develop on infected ripened fruits at pre- or post-harvest stage causing premature fruit drop. Generally, the fruit lesions may unite and finally penetrate deeper into the skin causing extensive fruit rot. Latent infection on unripe green fruits remains invisible and they look healthy at the time of harvest which produces vital symptoms upon ripening. Another type of symptom in the form of linearly arranged necrotic regions on the fruit as "tear stain" which may or may not be related with cracks in the epidermis that extend deep into the pulp of the fruit.

Postharvest symptoms express as circular brown-black lesions with irregular margins on fruit. After initial latent infection on green fruits, it becomes evident on ripe fruits as darker, sunken, and round lesions which enlarge quickly. They may even cover the entire fruit surface in extremely severe cases. The lesions on severely infected fruits bigger more than 2 cm in size are very frequent which may unite and cover the wide areas of the fruit producing a distinctive tear-stain symptom (Arauz, 2000). Generally, anthracnose lesions are limited to fruit skin but they may go beyond the peel but the pathogen penetrates deeper in fruit pulp under advanced stages of disease development and produces asexual fruiting structures (acervuli) where ample masses of orange-pinkish conidia are developed.

Casual Organism: *Colletotrichum gloeosporioides*

Pathogen:

Mango anthracnose is caused by *Colletotrichum gloeosporioides* (Penz.) Sacc. imperfect (conidial) stage of the ascomycetous fungus *Glomerella cingulata* (Stonem) Splaud and Schrenk.

Survival and Spread:

The pathogen survives mainly in the previous season infected leaves, defoliated branches, mummified flowers, and flower brackets and spread through rain splash or wind driven rain water. Under favorable conditions, conidia are dispersed and invade young twigs causing twig dieback in twigs.

Favorable Conditions:

High relative humidity, frequent rains, and temperature ranging from 24 °C to 32 °C favored the disease development but dry weather permits the growth of fruits which is unfavorable to develop the disease. Mango usually starts flowering during this dry season in majority of the tropical regions. The florae and newly developed fruits may get infected and it becomes worst if mango blooms early prior to commencement of dry season although, the final stage of fruit development occurs in the rainy season. Hence, high disease incidence is common on those fruits that develop in the rainy season.

Management:

- Planting should be done with wide spacing to avoid over-crowding in orchards that will prevent a high incidence of the disease.
- For effective management, growers must adopt proper orchard sanitation practices. The collection and burning of dropped fruit and tree debris have been suggested (Lim and Khoo, 1985). Infected fallen leaf residue, shoots, or twigs should be pruned and destroyed by burning.
- Tree pruning of infected parts and removal of fall down plant debris from the orchard on yearly basis decrease the inoculum.
- Intercropping with nonhost trees for mango anthracnose will prevent the epidemics.
- On susceptible cultivars under a favorable environment, the applications of fungicidal sprays at appropriate dose and time play an important role in disease management. Spraying of suggested fungicides has to be initiated at the emergence of the panicles which is continued at the commended intervals till the fruits attain 2 inches long size approximately.
- Application of Bavistin (carbendazim) at 0.1% or Kavach (chlorothalonil) at 0.2% spray which is continued till harvesting with 14 days interval is also suggested.
- To manage the disease at postharvest stage during storage, it is found that prestorage hot water treatment of the fruits at 50–55 °C for 15 min or dipping the fruits in benomyl at 500 ppm or Thiabendazole at 1000 ppm for 5 min or exposing the fruits to gases like ammonia and sulfur dioxide gave good results.
- Anthracnose at postharvest stage can be managed if started from field or treatment after harvest although, a combination of both preharvest and postharvest management practices gave better results against the disease.

- The chemical, prochloraz was reported to show its efficacy as a protective spray or to eradicate the pathogen (Estrada et al., 1996). A protective coating of mancozeb or copper should be maintained up to harvest.
- Two sprays of Strobilurin fungicides at flowering and 10 days before harvest is beneficial in reducing anthracnose infection in fruit.
- Infection of the blossom may be controlled with two sprays of Bavistin (carbendazim) at the rate of 0.1% at an interval of 15 days throughout flowering.
- The copper fungicides like Blitox-0 (copper-oxychloride) at 0.3% spray are advocated to manage the foliar infection.
- All or majority of mango cultivars grown commercially are vulnerable to anthracnose although, these cultivars exhibit variable reactions with mild to severe symptoms of the disease from one location to another. Presently, there is no cultivar available which was found to be resistant hence; selected protectant fungicides are sprayed under moist weather because more chance of infection is likely to occur under these environmental conditions (Dodd et al., 1997).
- A number of reports are available on the pathogen of mango anthracnose to show resistance against the Benzimidazole group of fungicides but it was not yet reported against prochloraz even widespread use and 13 years have been completed of its registration in Taiwan. Moreover, the resistance was not recorded in field screening of pathogenic isolates over a wide area and a large population in Taiwan under different concentrations and treatment rates (Kuo, 2001).
- It has also been reported for tropical environments if mature foliage is sprayed with potassium nitrate, the flowering can be achieved by a few weeks in advance (Nunez-Elisea, 1985).
- Removal of dried panicles and diseased withered fruits is tedious and timewasting. Moreover, covering the fruits as a management alternative with some physiological drawbacks has also been studied which may simply be unnoticed under the severity of disease (Hoffman et al., 1997).

18.2.4   DIE-BACK

Introduction/Economic Importance:

The orchards affected with die-back of mango caused by *Lasiodiplodia theobromae* (Syn: *Botryodiplodia theobromae*) may destroy within a few days or weeks which has been documented from various countries world-wide including India, Pakistan, Brazil, USA, and so forth. The fungus is a soil-borne pathogen which initiates the infection through wounds and it may affect all parts of the host at any stage. So, die-back has become one of the serious diseases causing significant losses affecting mango industries and marketing. In Indian states, it is predominantly found in Uttar Pradesh, Maharashtra, Punjab, Rajasthan, Haryana, Delhi, Gujarat, Tamil Nadu, and Orissa. The disease can be noticed in conspicuous form during the month of October–November.

Symptoms:

Initially, disease is noticed as a general symptom of wilting on a mango tree with twig die-back. Later on, disease symptoms as curling and drying of leaves, twigs, and branches followed by complete defoliation of tree as the disease progresses. At this stage, trees appear as scorched by fire and vascular regions of the stems and branches become discolored. The die-back symptoms become visible on bark as dark and discolored lesions. The lesions progress and start withering of young green twigs or branches resulting in the death of these affected portions which shrivel and finally fall-off. This may be accompanied by the exudation of gum. A characteristic brown discoloration with or without yellow gummy material may be seen in vascular tissues on the longitudinal splitting of old branches.

Under favorable conditions, diseased tree shows characteristic die-back symptom as the death of twigs and branches from the top which progresses downwards including reproductive structures. Affected branches sequential drying in one by one manner under in severe infection of disease and ultimately causing death of the mango trees. It is quite difficult to manage the disease after the appearance of extensive die-back symptoms although; a variable range of symptoms may be seen on mango varieties with their susceptible reaction for the fungus (Khanzada et al., 2015; Naqvi and Perveen, 2015).

In vivo work conducted by Kazmi et al. (2005), Paolinelli-Alfonso et al. (2016) stated that die-back pathogen shows aggressiveness to colonize host tissues under abiotic stress (heat or drought) conditions to the host.

Casual Organism: *Lasiodiplodia theobromae* (Syn: *Botryodiplodia theobromae*)

Pathogen:

Survival of the pathogen, *L. theobromae* occurs on the dead twigs of the tree as well as fallen on the ground. Invasion by the fungus starts from the tips of twigs and branches of the host tree producing die-back and blight symptoms. The pathogen survives during the off-season in the form of pycnidial bodies on diseased wood. These fruiting bodies produce and discharge an enormous number of bi-celled, brown, striated conidia which are disseminated by air and rain splashes. Infection progresses with the landing of these conidia on newly cut or injured woody portion. Upon conidial germination on invaded wood tissue, it starts to damage the vascular system of the plant.

Survival and Spread:

Pathogen survives on plant debris such as previous season infected leaves, diseased twigs providing primary inoculums which spread through rain splash or wind driven rainwater.

Favorable Conditions:

It can be seen very commonly during the month of October and November. High humidity and moist conditions favor disease development. Prevalence of high temperatures reduces the plant vitality and thus predisposes them to attack of the fungus.

Management:

- The disease can be managed by adopting some precautionary methods for example, select the scion from disease-free and healthy tree and disinfect the budding/grafting knife, and so forth. After grafting plant should be kept under a comparatively dry environment and gradually expose them to full sunlight.
- Pruning of diseased twigs and branches should be done including 3–4 inches of the healthy portion below the point of infection. After this, cut ends should be sprayed with Bordeaux mixture or Blitox-50 (copper oxychloride) at 0.3% or can be pasted these cut ends with the paste of above-listed chemicals.

18.2.5   PHOMA BLIGHT

Introduction/Economic Importance:

This blight disease on mango is relatively new and it was initially reported at the Central Institute of Subtropical Horticulture, Lucknow. Later on, it has been noticed from the other regions of Lucknow where mango is grown and now it has become a disease of economic importance.

Symptoms:

Only old leaves show visible symptoms which initiate as angular to irregular, small, yellow to light-brown colored lesions dispersed on the leaf. Lesions turn cinnamon brown and become nearly irregular as they expand which develop dark borders with a grayish centre. As advanced stages of infection such lesions are merged to form large patches ranges from 3.5 to 13 cm in size which occurs the complete leaf withering and defoliation.

Casual Organism: *Phoma glomerata* (Cords) Woll. Hochapf.
Fungus nowadays called as *Peyronellaea glomerata* and formerly known as *Phoma glomerata* thereby the common name originated.

Survival and Spread:

It is a ubiquitous and widespread fungus that survives in soil and on various dead and living plant materials (seeds, foliar plant parts, fruits) normally without causing symptoms. It can also be observed inside the houses on woody and cemented material or on oil paint surfaces and paper. However, in some hosts at certain environmental condition, it caused disease.

Favorable Conditions:

Rainy seasons favor/moist conditions at a temperature ranging from 26 °C to 37 °C favored the development of the disease.

Management:

- For the management of this disease, spray the plant with Blitox 50 (copper oxychloride) at 0.3% rate as soon as the disease appears followed by successive spray(s) at 20 days interval.
- The disease is also managed by the spraying of Benomyl at 0.2% as soon as the disease appears followed by another spray of Miltox (copper oxychloride + zineb) at 0.3% at 20 days interval.

18.2.6   STEM END ROT

Introduction/Economic Importance:

Stem end rot is an important postharvest disease of ripe mango fruits. Besides India, its occurrence is also reported from Sri Lanka, Burma, USA, Phillippines, and Mauritius. In markets of India, 4%–6% annual loss is usually reported due to spoilage of fruits. The disease adversely affects the export of fruit because of its appearance during transit and storage. Species of *Lasiodiplodia*, better known as diplodia, like other members of the family *Botryosphaeriaceae*, are pathogens with worldwide distribution in tropical and subtropical regions and cause different types of diseases.

Symptom:

It is easy to recognize the disease by its characteristic symptoms. In the initial stage of infection, the epicarp of the fruit nearby the pedicel becomes dark. Within some hours of infection, the infected area expands forming a black circular patch which under humid conditions extends rapidly and within two or three days the whole fruit turns black. The pulp in the diseased fruit softens and turns brownish. The infected fruits lose ascorbic acid rapidly with reduced amount of nonreducing sugars in fruit pulp. At the portion of stem-end, black rot develops in the form of a dark brown ring. This rot continues in the direction of the opposite end and it causes dark streaking of the xylem tissues which differentiates stem end rot from anthracnose.

Casual Organism: *Diplodia natalensis* (Syn: *Lasiodiplodia* theobromae, *Botryodiplodia theobromae*)

Pathogen:

The pathogen, *Diplodia natalensis* (P. Evans) is the causal organism of the stem end rot disease of mango. The pathogen is capable of producing pectolytic and cellulolytic enzymes which help in the production of soft rot symptoms. This pathogen goes inside through mechanical injuries made on the stem or skin. The fungus develops from the pedicel forming black circular lesions around it.

Survival and Spread:

Pathogens survive in plant debris such as in previous season infected leaves, diseased twigs which is the source of primary inoculums and spread through rain splash or wind driven rainwater.

Favorable Conditions:

Humid condition under high relative humidity is favorable for developing the disease.

Management:

- Preventive measures such as the selection of scion from healthy trees, sterilization of the budding knife, keeping the grafted tree in a rather dry environmental condition followed by exposing gradually into full sunlight have shown their effectiveness in decreasing the disease.
- Pruning of diseased twigs and branches should be done including 3–4 inches of the healthy portion below the point of infection. After this, cut ends should be sprayed with Bordeaux mixture or Blitox-50 (copper oxychloride) at 0.3%.
- Both types of treatment that is, preharvest spray and postharvest dip have been found successful. Besides these, certain handling practices help in reducing the disease. Preharvest spray of carbendazim (0.1%) can be given. The fruits should be harvested during a dry and sunny day and immediately after harvest, the fruits should be covered and brought to the ripening houses. Growers should be careful to prevent shaping off of the pedicel and injuries should be avoided during handling. The bare superficial area of the pedicel or stem end scar can be coated with Chaubatia paint. Fruits can also be given postharvest dip with borax at 43 °C for 3 minutes.

18.2.7　SOOTY MOULD

Introduction/Economic Importance:

The sooty mold is quite common in those orchards infested with mealy bugs, hoppers, and scale insects which are not managed effectively. It is present wherever honeydew or sugary substances secreted by the above-mentioned insects exist. The old orchards with a dense canopy and diffused intensity of light are related with severe disease. This disease adversely affects the marketability of the fruit as the fruit looks almost blackened.

Symptoms:

The characteristic field symptom of the disease can be seen as a superficial black and smooth coating of fungal growth on the leaf lamina. Under severe conditions, the mold covers the complete surface of leaves and twigs or branches of the tree which results in the affected tree appearing black. The severity of the disease is related with honeydew which is secreted by the above-listed insects and this sticky secretion adheres on the surface of the leaves providing the required medium for fungi to grow. The sooty mold fungus is primarily saprobic which does not take its nutrition from the host and it causes indirect damage by affecting the photosynthesis due to blockage of stomata of the leaf. The sooty mold may also be observed on mango fruits of late varieties under the presence of high humidity during harvesting.

Casual Organism: *Meliola mangiferae* Earle; *Capnodium mangiferae* Cke. and Brown; *Capnodium ramosum* Cke. and *Tricospermum acerinum* (Syd.) Speg.

Pathogen:

Sooty mold of mango is caused by *Meliola mangiferae* Earle; *Capnodium mangiferae* Cke. and Brown; *Capnodium ramosum* Cke. and *Tricospermum acerinum* (Syd.) Speg. The rating to be used for sooty mold is based on the percentage leaf area affected on the selected shoot for sampling. The number of sooty mold infected shoots in respect of the severity scale-out of shoots of the selected tree should be recorded.

Survival and Spread:

The fungus spreads through wind-borne spores and the brutality of sooty mold infection is determined by honey dew secretions secreted by scale insects providing the essential medium for fungi to grow there.

Favorable Conditions:

High humidity and moist condition favor disease development.

Management:

- For preventing the spread of disease, affected twigs or branches should be pruned and destroyed quickly.
- Application by spraying of wettable sulfur (0.2%) along with insecticides has been found effective to control the disease and honeydew secreting insects. Spray of starch at 2% is also noticed for its effectiveness against sooty mold.

18.3   BACTERIAL DISEASES

18.3.1   BACTERIAL CANKER

Introduction/Economic Importance:

The canker of mango caused by *Xanthomonas campestris* pv. *mangiferae indicae* is predominantly found to occur in the states namely Uttar Pradesh, Bihar, Madhya Pradesh, Delhi, Haryana, Maharashtra, Karnataka, Tamil Nadu, Andhra Pradesh, Kerala, and possibly in some other regions where mango is grown. Besides causing disease on varieties of mango, the bacterium is found to be pathogenic and able to infect wild mango and cashew-nut including weeds. The fruit drop ranging from 10% to 70% was reported to cause 10%–85% of yield loss; however, 5%–100% of fruit rot is reported during storage. A number of cultivars namely Dashehari, Amrapali, Mallika, Totapuri, and Langra are found susceptible to diseases which are commercially grown.

Symptoms:

The bacteria attack on aerial parts including foliage, petioles, twigs or branches, and fruits and initial symptoms are visible as water-soaked irregular lesions which subsequently turn into characteristic cankers. The initial leaf lesions are light-yellow which later increase and turn dark brown-black. Later on, lesions become angular, raised producing cankers which are bordered with yellow halos. Several of such lesions merge producing irregular cankerous patchy areas which result in leaf yellowing and defoliation under severe conditions of the disease. The mature lesions on twigs and branches appear dark-brown and somewhat raised forming longitudinal cracks without bacterial ooze. The infected fruit also develops dark brown-black cankerous lesions which release contagious gluey ooze of bacterial cells upon bursting.

Casual Organism: *Xanthomonas campestris* pv. *mangiferae indicae*

Pathogen

The disease is caused by a bacterium *Xanthomonas campestris* pv. *mangiferae indicae*. The bacterium survives on diseased plant parts and weeds where it multiplies at an early stage. Transmission of the bacterium takes place from infected weeds and diseased plant parts lying on the orchard ground to the younger twigs through water splash during rains. *Xanthomonas* is well known phytopathogenic, rod-shaped, and Gram-negative bacterium. In addition to the Mangifera, numerous other genera belonging to Anacardiaceae family were also reported as the host of *xanthomonads*. Bacteria affect a variety of hosts, namely, citrus, cotton, beans, rice, and so forth.

Survival and Spread:

The bacterium survives for a long duration in affected diseased tissues of the plant.

Favorable Conditions:

Spring season favors for the development of the disease.

Management:

- The precautionary methods like planting of seedlings from certified sources and orchard hygiene with monitoring and inspection for infection and its development help to manage the disease.
- Management of bacterial canker disease by copper-based fungicides has been reported by several workers (Brodrick, 1971; Singh, 1978).
- Carbendazim is also reported to control the disease (Ramkishun, 1981, 1985).
- The canker is managed by the spraying of streptocycline or Agri-mycin-100 at 100 ppm at 10-days intervals as soon as initial symptoms appear.

18.4 ALGAE DISEASE

18.4.1 RED RUST

Introduction/Economic Importance:

Red rust disease was caused by an alga which was found to occur differently in mango growing areas. It is typically a harmless, generally, minor disease characterized by negligible leaf spots but sometimes may cause economic injury to the plant. The algae initially attack leaves causing reduced photosynthesis and leaf defoliation affecting plant vigor.

Symptom:

The brick-red colored rusty spots primarily on foliage are the characteristic symptoms of this disease; although these are initially velvety and greenish-gray colored. These types of spots may occasionally develop on petioles and bark of new twigs. Sometimes, these small and slightly raised spots may merge and form bigger and irregular spots.

Fruiting bodies of the algae are developed under moist weather. The new infection is initiated by zoospores produced in sporangia. The cracks on the stem provide the route to enter the zoospore causing stem infection. The affected areas develop cracks which can be scaled off. Severely infected plants due to disease show thickening of the bark with stunted twigs and sparse foliage which finally dry out.

Casual Organism: *Cephaleuros virescens* Kunze (alga)

Pathogen:

An alga, *Cephaleuros virescens* Kunze is the causal organism of this disease and which has the widest host range among the species of genus Cephaleuros.

Survival and Spread:

The causal organism survives and multiplies in disease spots of red rust on the leaf and stem or in fallen plant refuse.

Favorable Conditions:

Environmental conditions like hot weather and regular rains favor the growth and spread of pathogens whereas poor nutrition to plants, improper soil drainage, poor aeration, and other stresses predispose the host to algal infection. Dense canopy and wet humid environment also favor disease development.

Management:

- Trees should be thinned to improve ventilation and sunshine penetration.
- Serious outbreaks can be controlled with 2–3 sprays of copper fungi-cides (copper oxychloride) at 0.3% is found effective to manage the disease.

KEYWORDS

- **mango**
- **diseases**
- **symptomatology**
- **epidemiology**
- **management**

REFERENCES

Arauz, L. F. (2000). Mango anthracnose: Economic impact and current options for integrated management. *Plant Dis.*, 84: 600–611.

Bains, G. and Pant, R. C. (2003). Mango malformation: Etiology and preventive measures. *Physiol. Mol. Biol. Plants*, 9: 41–61.

Brodrick, H. T. (1971). Mango diseases. *Fmg. S. Afr.*, 1: 29–32.

Crane, J. H. and Campbell, C. W. (1994). Fact Sheet HS-2, Horticultural Science Department, Florida Cooperative Extention Service. Institute of Food and Agricultural Sciences, University of Florida, USA.

Dodd, J. C., Prusky, D., and Jeffries, P. (1997). Field diseases. In: The Mango: Botany, Production and Uses, Litz, R.E. (Ed.). CAB International, Wallingford, UK, pp. 257–280.

Estrada, A. B., Jeffries, P., and Dodd, J. C. (1996). Field evaluation of a predictive model to control anthracnose disease of mango in the Philippines. *Plant Dis.,* 45: 294–301.

Gupta, J. H. and Yadava, A. S. (1984). Chemical control of powdery mildew of mango. *Indian J. Mycol. Plant Pathol.,* 14(3): 297–298.

Hiffny, H. A. A., El-Barkouki, M., and El-Banna, G. S. 1978. Morphological and physiological aspects of the floral malformation of mangoes. *Egypt. J. Hort.,* 5: 43–53.

Hoffman, P. J., Smith, L.G. Joyce, D.C. Johnson, G.I., and Meiburg, G. F. (1997). Bagging of mango (*Mangifera indica* cv. `Keitt'*) fruit influences fruit quality and mineral composition. *Postharvest Biol. Tech.,* 12: 83–91.

Indian Status of Mango. National Mango Database retrieved on 02/05/2018 from http:// mangifera.res.in/indianstatus.php.

Iqbal, Z., Rahman, U. M., Dasti, A. A., Saleem A., and Zafar, Y. (2006). RAPD of fusarium islates causing mango malformation disease in Pakistan. *World J. Microb. Biotechnol.,* 22: 1161–1167.

Kanwar, J. S. and Nijjar, G.S. (1979). Some observations on mango malformation. Proceedings of the Research Report Mango Workers Meeting, May 2–5, Goa, India, pp. 415–419.

Kazmi M., Fateh, F., Majeed, K., Kashkhely, A. M., Hussain, I., Ahmad, I., and Jabeen, A. (2005). Incidence and etiology of mango sudden death phenomenon in Pakistan. *Pak. J. Phytopathol.,* 17: 154–158.

Khanzada, M. A., Lodhi, A. M., Rajput, A. Q., Syed, R. N., and Shahzad S. (2015). Response of different mango cultivars to mango decline pathogen, *Lasiodiplodia theobromae Pat. Int. J. Biol. Biotechnol.*, 12: 643–647.

Kumar, J. and Beniwal, S. P. S. (1992). mango malformation disease. In: Plant Diseases of International Importance, Diseases of Fruit Crops, Kumar, J., H.S. Chaube, U.S. Singh and A.N. Mukhopadhyay (Eds.). Vol. III. Prentice Hall, New York, p. 456.

Kuo, K. C. (2001). Sensitivity of mango anthracnose pathogen, *Colletotrichum gloeosporoides*, to the fungicide Prochloraz in Taiwan. *Proc. Natl. Sci. Counc.*, Repub. China B,, 25: 174–179.

Lim, T. K. and Khoo, K.C. (1985). Diseases and Disorders of Mango in Malaysia. Tropical Press, Kuala Lumpur, Malaysia.

Litz, R. E. (1997). The Mango: Botany, Production and Uses. CAB International, Wallingford.

Mallik, P. C. (1963). Mango malformation: symptoms, causes and cure. *Punjab Hort. J.*, 3: 292–299.

Misra, A. K. and Singh, V.K. (2002). *Fusarium subglutinans* (*F. moniliforme* var. *subglutinans*) in relation to mango malformation. *Indian J. Plant Pathol.*, 20: 81–83.

Naqvi, S. A. H. and Perveen, R. (2015). Mango quick decline manifestation on various cultivars at plants of particular age in the vicinity of district Multan. *Pak. J. Phytopathol.*, 27: 31–39.

NHB (2018). Horticultural Statistics at a Glance 2017 retrieved on 03/05/2018 from http://nhb.gov.in/statistics/Publication/Horticulture%20At%20a%20Glance%202017%20for%20net%20uplod%20(2).pdf.

Nirvan, R. S. (1953). Bunchy top of young mango seedlings. *Sci. Cult.*, 18: 335–336.

Nunez-Elisea, R. (1985). Flowering and fruit set of a monebryonic and a polyembronic mango as influenced by potassium nitrate sprays and shoot decapitation. *Proc. Fla. State Hortic. Soc.*, 98: 179–183.

Palti, J., Pinkas, Y., and Matheieda, C. (1974). Powdery mildew of mango. *Plant Dis. Rep.*, 58: 45–49.

Paolinelli-Alfonso, M., Villalobos-Escobedo, J. M., Rolshausen, P., Herrera-Estrella, A., Galindo-Sánchez, C., López-Hernández, J. F., and Hernandez-Martinez, R. (2016). Global transcriptional analysis suggests *Lasiodiplodia theobromae* pathogenicity factors involved in modulation of grapevine defensive response. *BMC Genom.*, 17: 615, doi: 10.1186/s12864–016-2952–3.

Ploetz, R.C. (1994). Anthracnose. In: Compendium of Tropical Fruit Diseases, R.C. Ploetz (Ed.). APS Press, St Paul, MN, USA, pp. 35–36.

Ploetz, R. C. and Prakask, O. (1997). Foliar, Floral and Soil Borne Diseases. CAB International, Wallingford, UK, pp. 281–326.

Ploetz, R., Zheng, Q. I., Vazquez, A., and Abdel Sattar, M. A., (2002). Current status and impact of mango malformation in Egypt. *Int. J. Pest Manage.*, 48: 279–285.

Rai, B. K. and Singh, N. (1967). An observation on recovery from malformation in mango sapling. *Curr. Sci.*, 36: 525–526.

Ram-Kishun (1981). Studies on bacterial canker of mango. Paper presented in Third Int. Symp. on Plant Pathol., New Delhi, p. 15 (abstr.).

Ram-Kishun (1985). Stem injection of chemicals for the control of bacterial canker of mango. IInd Int. Symp. on Mango, Bangalore, 20–24 May, p. 60 (abstr.).

Sharma, I. M., Raj, H., and Sharma, N. K. (1993b). Field evaluation of some new systemic fungicides against powdery mildew of mango in Himachal Pradesh. *Pestology*, XVII: 25–27.

Shawky, I., Zidan, Z., El-Tomi, A., and Dashan, D. I. (1980). Flowering malformation in relation to vegetative growth of Taimour mango. *Egypt. J. Hort.*, 7: 1–8.

Singh, L. S., Singh, S.M., and Nirvan, R. S. (1961). Studies on mango malformation review, symptoms, extent, intensity and cause. *Hort. Adv.*, 5: 197–207.

Singh, R. N. (1978). Mango. ICAR, New Delhi. p. 91.

Singh, Z. and Dhillon, B. S. (1986a). Effect of plant growth regulators on floral malformation, flowering productivity and fruit quality of mango (*Mangifera indica* L.). *Acta Hort.,* 175: 315–319.

Singh, Z. and Dhillon, B. S. (1986b). Effect of foliar application of boron on vegetative and panicle growth, sex expression, fruit retention and physiochemical characters of fruits of mango (*Mangifera indica* L.) cv. Dushehari. *Trop. Agric.,* 64: 305–308.

Tripathi, P. S. and Ram, S. (1998). Effect of deblossoming on panicle regeneration and control of malformation in mango. *Prog. Hort.,* 30: 40–43.

Watt, G., (1891). A Dictionary of Economic Products of India. Vol. 5, Govt. Printing Press, Calcutta, pp. 149.

www.exportgenius.in . Mango Export from India, Jan to April 2017: Analysis Report, retrieved on 03/05/2018 from https://www.exportgenius.in/blog/mango-export-from-india-jan-to-april-2017-analysis-report-37.php.

Yadav, T. D. (1972). Role of mango bud mite, *Aceria mangiferae*, in mango malformation. *Acta Hort.,* 24: 238–238.

CHAPTER 19

STATUS AND STRATEGIES FOR MANAGING MANGO (*MANGIFERA INDICA* L.) DISEASES

PEZANGULIE CHAKRUNO, SUSANTA BANIK*, and KAVI SUMI

Department of Plant Pathology, SASRD, Nagaland University, Medziphema 797106, Nagaland, India

Corresponding author. E-mail: susanta.iari@gmail.com

ABSTRACT

Mangoes (*Mangifera indica* L.) are universally considered as one of the choicest fruits in tropical and subtropical areas of the world and mangoes belong to the family *Anacardiaceae*. It is grown in at least 87 countries but nowhere in the world it is so greatly value as in India where 50% of total fruits grown in India is only mango. Mango is affected by a number of diseases at all stages of the development right from nursery to post harvest including storage and transit. Mango plant parts, namely, stems, branch twig, root, petiole, flower, and fruit are affected by various pathogens and causes huge losses to mango production. This chapter deals with important diseases of mango and their management strategies.

19.1 INTRODUCTION

Mango (*Mangifera indica*) belongs to the family *Anacardiaceae*, it is one of the most widely consumed fruit with ever-increasing demand (Nasir et al., 2014). In worldwide, it is considered as the third most important tropical fruit (Pérez-Rodríguez et al., 2017). They are extensively cultivated for commercial purposes in tropical and subtropical regions like Thailand, Indonesia, China, India, and Mexico (Mukerjee and Litz, 2009; Kumari et al., 2017), and India ranks first mango production worldwide (Laxmi et al., 2011). During 2015–2016, India produced 18,643,000 MT from an area of 22.9 lakhs ha (National Horticulture Board, 2017). Because of its delicious taste, high nutritive value, superior flavor, medicinal value, and wide range of diversity as well as great religion and historical significance, it is also referred to as the "King of the fruits" (Purseglove, 1972). Mango despite its unique qualities and esteemed status in the world, like any other crop it is reported to be susceptible to many pathogenic diseases and biotic stress which thereby disturbs the normal physiological functions during growth and development (Kumari et al., 2017; Rani et al., 2013).

List of Diseases in Mango

Sr. No.	Name of the Disease	Causal Organism
Major Diseases		
1.	Anthracnose	*Colletotrichum gloeosporioides* (Penz.)
2.	Mango malformation	*Fusarium mangiferae, Fusarium moniliforme*, Eriophyid mite and physiological disorder
3.	Powdery mildew	*Oidium mangiferae*
Minor diseases		
4.	Red rust of mango	*Cephaleuros virescens*
5.	Sooty mold	*Meliola mangiferae*
6.	Blight of mango	*Macrophomina mangiferae*

Some major diseases causing severe economic losses are discussed as following:

19.2   ANTHRACNOSE DISEASE

Introduction/Economic Importance:

In the worldwide scenario, anthracnose is categorized as the most destructive disease of mango both at field and postharvest phase (Ploetz and Prakash, 1997). Anthracnose disease severely endangers the production of mangoes especially in those regions having high rainfall or high humidity during the cropping season (Akem, 2006). Kumari et al. (2017) reported that anthracnose can cause losses up to 60% during the rainy season. Anthracnose disease of mango is caused by a fungal pathogen *Colletotrichum gloeosporoides* (Diedhiou et al., 2014).

Symptoms:

An anthracnose symptom in mango first appears on leaf, petioles, panicles, fruits, and twigs. The lesions on the leaves start as small, irregular-shaped dark spots which later coalesce to form a large necrotic patch mostly along the margin. On the affected panicles, dark brown to black spots appear killing the inflorescence stalk as well as the individual flowers thereby reducing the yield (Nelson, 2008). Affected fruits develop dark brown to black decay sunken lesions accompanied by pink slimy spore masses (acervuli) in the advanced stage of the disease and fruits may drop prematurely. Sometimes the lesions penetrate deep into the fruit resulting in extensive fruit rot (Jeffries et al., 1990; Awa et al., 2012). The lesions mostly develop on the young, tender tissue resulting in the formation of a large number of conidia which is observed in all ages of lesions. Under favorable conditions conidia are disseminated and invade young twigs causing twig die-back (Ploetz et al., 1996).

Causal Organism: *Colletotrichum gloeosporioides* (Penz.)

Colletotrichum gloeosporioides (Penz.) cause anthracnose disease in mango; it is also reported to cause blossom blight, leaf blight, and die-back in mango (Arauz, 2000). In India, anthracnose disease of mango caused by *C. gloeosporoides* was first reported by McRae (1924).

Disease Cycle:

Fitzell and Peak (1984) reported that conidia served as the primary source of inoculum in the dispersal and establishment of anthracnose disease. *C. gloeosporioides* conidia are produced numerous on the leaves, twigs, fruits, and panicles (Arauz and González-Lobo, 1986). The conidia that are rain splashed to other nearby leaves and panicles may serve as secondary infection, thus the disease is polycyclic in these organs. Sometimes even young developing fruits are also infected and some aggressive isolates can cause preharvest fruit loss (Gantotti and Davis, 1993.). The pathogen overwinters between seasons on the infected defoliated branches and mature leaves (Nelson, 2008).

Epidemiology:

C. gloeosporioides is an asexual facultative parasite which belongs to the family phyllachoraceae (ascomycota; Kumari et al., 2017). Prakash (1996) reported that relative humidity of more than 95% enhances the progress of disease development in wounded tissues and ripe fruits. Wet, warm, and humid environment favors the spread of anthracnose disease uniformly and effectively (Farr et al., 2006). *C. gloeosporioides* requires an optimum temperature of 25–28 °C, a pH range of 5.8–6.5, and high humidity for conidial germination and appressorium formation. In the dry season, the pathogen remains inactive but with the return of favorable environmental conditions it retains its active stage.

Integrated Disease Management:

Cultural Method:

- Field sanitation practices involving the collection and incineration of the fallen fruit and other plant debris (Lim and Khoo, 1985; Johnson and Hofman, 2009).
- Wrapping of developing fruits in paper bags (Hofman et al., 1997).
- Advancing the mango flowering by several weeks with foliar application of potassium nitrate on mature foliage greatly reduces the severity of anthracnose disease (Nunez-Elisea, 1985)

Chemical Method:

- McMillan (1984) reported that foliar spray of benomyl and thiophanate methyl @ 1.8 g/L gave good control of anthracnose disease on mango.
- Systemic fungicides benzimidazoles, benomyl, and carbendazim were reported to provide excellent anthracnose control under field conditions (Akem, 2006).
- Singh et al. (2008) reported that carbendazim was found to be effective providing 71.41% disease control under field conditions followed by hexaconazole and difenconazole in which the disease control 66.62% and 57.14%, respectively.
- Chowdhury and Rahim (2009) reported that the spray application of Dithane M-45 along with the practice of other cultural methods gave good control over anthracnose disease.
- Application of thiophanate methyl (systemic fungicides) before the onset of monsoon was reported to be effective against the disease (Diedhiou et al., 2014).

19.3   MANGO MALFORMATION

Introduction/Economic Importance:

Mango malformation disease is reported to be a major constraint in the production of mango in India and other mango growing countries of the world (Crane and Campbell, 1994: Ploetz, 2001). Economic losses due to this disease are 50%–60% every year and it may extend upto 100% in severe cases (Misra et al., 2000). In India, this disease was first observed in 1891 and since then several reports have been made of this disease from countries in Asia, America, and Africa (Marasas et al., 2006). Malformation of mango adversely affects the inflorescences and vegetative portions of the mango plant. The vegetative phase of the disease impedes canopy development and the floral phase greatly reduces the fruit yield; malformed inflorescences do not bear fruit, due to which high economic losses can occur (Freeman et al., 2014).

Symptoms:

Major characteristics of mango malformation are abnormal development of vegetative shoots and inflorescences (Yadav and Beg, 2015) and distortion of their meristem tissues (Ploetz, 2001).

The seedlings produce small shootlets bearing small scaly leaves with a bunch-like appearance on the shoot apices. The appearance of multibranching of shoot apex with scaly leaves is known as "Bunchy Top," also referred to as "Witch's Broom" (Ploetz, 2004; Gaur and Chakrabarti, 2009).

In the vegetative phase of the disease, mostly 3–4 months old nursery seedlings are the most vulnerable, especially where seedlings are grown beneath the canopies of affected trees (Youssef et al., 2007). The axillary or apical buds produce misshaped shoots, having shortened internodes with brittle leaves which are significantly smaller than those of healthy plants. The leaves appear narrow and bend back toward the stem, the whole seedlings become stunted and die (Kumar and Beniwal, 1992).

The symptoms of floral malformation are characterized by abnormal branching of the inflorescence and thickened panicles producing large number of flowers as compared to healthy ones (Marasas et al., 2006). The affected panicles also appear much greener and heavier possessing numerous flowers that remain unopened (Schlosser, 1971).

Causal Organism: *Fusarium moniliforme* var. *subglutinans*

In spite of the recognition of the disease in 1891, the causal factors of this malady are still not yet clarified and are being strongly disputed over a long time (Bains and Pant, 2003; Kumar et al., 2011). There have been many reports about its etiology, claiming it to be acarological, viral, fungal, and physiological abnormalities (Kumar et al., 1993; Rani et al., 2013). The cause of malformation has been controversial but many studies have proven that *F. mangiferae* is the pathogen responsible for mango malformation disease and Koch's postulates have been completed successfully with this fungus in various countries worldwide (Briz et al., 2002; Freeman et al., 1999; Iqbal et al., 2010; Kumar et al., 1993).

Disease Cycle:

Grafting is reported to spread the malformation disease to new areas (Kumar et al., 1993), but the dissemination of the pathogen from tree to tree and within trees in orchards and nursery is poorly understood (Ploetz, 2004). Reports have been made that macro- and micro-conidia of *F. mangiferae* are most likely the infective propagules since they are the only propagules that are produced by the fungus and form profusely on the different malformed tissues (Freeman et al., 2004). Aerial dispersal serves as the primary mechanism for the dispersal of conidia of *F. mangiferae* and the infection sites appear at the apical and lateral bud areas (Gamliel-Atinsky et al., 2009; Youssef et al., 2007). *Aceria mangiferae*, the mango bud mite acts as a vector in carrying conidia of *F. mangiferae* on its body, assisting the fungal dissemination and infection (Doreste, 1984; Gamliel-Atinsky et al., 2009; Crookes and Rijkenberg, 1985).

Epidemiology:

Severity of mango malformation has been reported to be variable in variety, individual tree of the same variety, and cycle to cycle (Nath et al., 1987), further variation in the occurrence and severity of the disease was reported to correlate with the ambient temperature at flowering (Majumdar and Sinha, 1970). Factors like temperature, age of the tree, time, and so forth govern the susceptibility of mango varieties to malformation, early-blooming varieties are more susceptible to the disease than late-blooming varieties (Khurana and Gupta, 1973). In another report made by Chakrabarti and Kumar (1998) that seasonal variation in the incidence of mango malformation is due to host metabolites and mangiferin content and environmental parameters.

Integrated Disease Management:

Cultural Method:

- Use of pathogen-free nursery planting materials (Ploetz, 2001), removal and destruction of infected material (Ploetz et al., 2002).
- Sirohi et al. (2009) have reported that malformed panicles should be pruned at panicle emergence (January month in case of cultivar Dasheri) for effective suppression of the disease. Further continuation of this practice for 2 or 3 consecutive years can reduce the disease incidence (Muhammad et al., 1999; Ploetz, 2001).
- Combination of pruning, use of insecticides, fungicides, and growth regulators may control the mango malformation disease (Ploetz, 2001; Varma et al., 1974).
- Increase in nitrogen dose application has been reported to reduce the panicle malformation (Minessey et al., 1971)

Bio-control Method:

- Spraying of a bioactive metabolite of *Streptomyces aureofaciens* has been reported to show reduction of mango malformation disease (Wafaa et al., 2014).

Chemical Method:

- Clipping of malformed branches at 45 cm followed by a spray of benomyl @ 0.2% (Iqbal et al., 2011).
- Prochloraz spray @ 0.1% have been reported to have the potential to reduce disease in the field (Freeman et al., 2015).

19.4   POWDERY MILDEW

Introduction/Economic Importance:

Plant pathogen, namely fungus, bacteria, nematodes, and viruses are all reported to cause infectious disease in mango. Irrespective of ecological and geographical differences, powdery mildew is reported as the most wide-spread, common, and damaging among all the diseases of mango affecting almost every cultivar (Palti et al., 1974; Prakash, 2004). Heavy economic loss befalls the mango growers due to powdery mildew disease, which heavily affects the inflorescences leading to leaf shedding and premature fruit drop (Misra, 2002). In India, the presence of powdery mildew is known before 1874, it is also considered to be the oldest known disease of mango in the world (Nelson, 2008). This disease was first recorded in the year 1914 from Brazil and Berthet has given a full description of the fungus in detail (Jones, 1923).

Symptoms:

Superficial white powdery mass on the inflorescence leaves and young tender fruits are the most conspicuous symptoms of powdery mildew disease on mango (Naqvi et al., 2014). Generally, the symptom begins to appear when new flushes start emerging and grow vigorously (Gupta, 1976). First, the disease appears on under the leaves surface, as small water-soaked lesion, which turns to white powdery growth and causes necrosis of the leaves with further development of the disease (Khalid and Alam, 2000). Sometimes, these necrotic lesions coalesce forming a large dead patch on the leaf, frequently resulting in distortion and curling (Burchill, 1978). Infected younger leaves drop prematurely, while purplish-brown spot develops on the infected matured leaves. Usually, the affected fruits are malformed and off colored sometimes followed by die-back symptoms (Singh, 2000). Premature drop of infected fruits occurs. White powdery growth appears on mature fruits producing purplish-brown blotch or corky surfaces (Gupta, 1989; Joubert et al., 1993).

Causal Organism: *Oidium mangiferae*

Powdery mildew disease on mango is caused by a fungus *O. mangiferae* Berthet (Johnson, 1994; Akhtar et al., 1999; Naqvi et al., 2014), recently named as *Pseudoidium anacardii* (F. Noack) U. Braun and R.T.A. Cook 2012 (Nasir et al., 2014).

Disease Cycle:

O. mangiferae Bert. is a biotrophic parasite which needs host plant tissues to live and develop (Naqvi, 2014). Under unfavorable conditions, the pathogen survives as dormant mycelium or haustoria in the detached older leaves or when it is still attached to the host plant, it may hide in the malformed green panicles or in the buds. With the return of the favorable conditions, the dormant mycelium or haustoria gets activated which then starts to produce conidia in great number causing both fresh and secondary infection (Nasir et al., 2014; Rawal and Saxana, 1997). With the slightest wind, the powdery conidial masses are dispersed causing a fresh infection on the young healthy tissue (Gupta, 1988). Within 5–7 hours of contact with the host tissue the conidia start to germinate, in 2 days mycelium is produced and within nine days the pathogen life cycle is completed (Wagle, 1928; Bose, 1953; Sinha et al., 2002).

Epidemiology:

Factors like the host plant, tissue specificity, environmental conditions, dormant mycelium, and asexual form of the fungus greatly influence the epidemiology and infection process of powdery mildew of mango (Nasir et al., 2014). Low night temperatures and warm humid weather accompanied by high wind velocity greatly favor the spread and establishment of powdery mildew (Singh, 2000). Gupta (1989) reported that atmospheric temperature plays a vital role in the establishment and development of the disease, the maximum infection was recorded at 26 °C and 1,000% relative humidity. And further concluded those temperature ranges of 11 °C–14 °C (maximum) to 30 °C–40 °C (minimum) with a humidity range of 65%–78% were the most favorable for the establishment of the disease.

Integrated Disease Management:

Cultural Method:

- From the early stage of infection, the potential inoculum of the pathogen should be checked (Joubert, 1991) by removal/pruning of infected plant parts like malformed panicles, leaves, and close monitoring of the mango orchard (Prakash and Raoof, 1994).

Chemical Method:

- Raheel et al. (2008) reported that Anpower 5ME (Hexaconazole) and Score 250EC (Diphenoconazole) showed the best result in reducing the incidence of powdery mildew disease in mango. Thind et al. (2005) also reported that triazole fungicides showed better result over nontriazole fungicides.
- Use of sulfur as dust or sprays reasonably reduces the incidence of powdery mildew (Chavan et al., 2009).
- Systemic fungicides, namely carbendazim, benomyl, and thiophante-methyl are equally effective for the control of powdery mildew in

mango from florescence to opening of flowers (Chavan et al., 2009; Gupta and Dang, 1984; McMillan, 1973).

19.5   GRAY BLIGHT OR *PESTALOTIA* LEAF SPOT

Introduction/Economic Importance:

In India, many research workers reported the gray blight disease in mango, namely Tandon et al. (1955), Sarkar (1960), Prakash and Srivastava (1987) and Vala et al. (1985), and Verma et al. (1991). The pathogen of this disease is considered as a week parasite capable of infecting injured host plant tissues and healthy fruits in contact with the diseased ones.

Symptoms:

The main characteristic symptom of grey blight disease is the appearance of brown spots on the leaf lamina. The spots may develop at the tips or margins, but in rare cases, are distributed irregularly on the entire leaf. Initially, the spots appear as minute brown spots, later their size increases and becomes dark brown in color. Some spots may enlarge considerably forming large lesions, in which the central region turns grayish white or light olive gray encircled by a halo of tan color. At this stage, black spots of acervuli start to appear and become visible to the naked eye in the central region which are more on the upper surface of the leaf and never extend beyond the midrib. Similar symptoms also appear on mature green fruit, with grayish-white centre having numerous black dots of acervuli. The infected fruits drop off when the disease advances to the stalk end. In 3–4 months, the infected portions get detached from the leaf. Defoliation follows in case of severe infection (Sarkar, 1960).

Causal Organism: Gray blight disease in mango caused by *Pestalotiopsis mangiferae* (P. Henn) Stey was reported by Prakash and Srivastava (1987). In addition, fungal pathogens like *P. moniifolia* Guba, *P. versicolor* (Speg.) Stey (Prakash and Misra, 1987), *Penicillium glandicola* (Cast) Stey (Ullasa and Rawal, 1985) were also reported from various mango growing regions of India.

Epidemiology:

Sarkar (1960) reported this disease can thrive at temperatures between 10 °C and 35 °C, and maximum growth occurs at 20–25 °C and the spore germination is best in a temperature range of 10–34 °C with an optimum of 30 °C. The maximum growth and sporulation are observed on host extract and Richard's media, and the optimum temperature for growth and sporulation was 30 °C (Pandey and Mohammad, 1974–75).

Integrated Disease Management:

Tandon et al. (1955) reported that dusting the leaves with zinc sulfate controls the gray blight disease.

Combination of wettable sulfur (0.2%) and zineb (0.2%) after heavy rains followed by wettable sulfur (0.2%) before flowering, carbandezim (0.3%) at pea stage, and zineb (0.2%) before maturation of stone reduces the incidence (Jadeja and Vaishnav, 1984).

Devi and Devi (2017) reported that in vivo experiment, thiophanate methyl (0.5%) resulted best in reducing disease intensity to 19.40% which was closely followed by carbendazim (15.40%), garlic (17.20%), and *T. harzianum* (16.76%) as compared to control.

19.6   SOOTY MOLD DISEASE

Introduction/Economic Importance:

Sooty mold or sooty blotch disease in mango is a common disease in orchards. It is commonly found wherever honey dew or sugary substance secreting insects, such as mango hopper, coccids, mealy bugs, and scales are found (Kulkarni and Kulkarni, 1978; Singh and Singh, 1972; Peethambharan and Aravindakshan, 1975; Sarwar, 2015). Prakash and Srivastava (1987) reported that sooty mold disease appeared in epidemic form in Bulandshahar district of Uttar Pradesh during August 1985 and in certain areas. The fungi responsible for this particular disease are nonpathogenic, they neither enter the host tissue nor draw nutrients from the host plant. They draw their nutrition from the sweet honey dew and not from the host. The damage to the host is indirect, interfering with the normal functioning of the plant as they cut off the effective leaf area for photosynthesis.

Symptoms:

The presence of a black velvety, thin membranous coating growth that is, sooty mold on the leaf lamina is the characteristic symptom of sooty mold disease. The amount of honey dew the insects secrete determines the severity and intensity of this disease. In case of high severity, the tree completely turns black due to a heavy infestation of mold on the entire surface of leaves and twigs. Under dry conditions, the affected leaves curl and shrivel. The insects honey dew secretions to the leaf surface which provide the medium for fungal growth. The fungus does not directly cause damage but instead disturbs the photosynthetic activity of the leaf by blocking the stomata opening (Sarwar, 2015). In the flowering period, the blossoms are infected by the fungus affecting the fruit setting and sometimes lead to immature fruit drop. The disease also appears on mature fruits of late mango varieties (Prakash, 1988).

Causal Organisms: *Capnodium mangiferae*

Several fungi have been reported to cause sooty mold disease in mango. These are *Meliola mangiferae* Earle Butler and Bisby (1931), *Capnodium mangiferae* Cke. and Brown (Vaheeduddin, 1953), *Tripospermum acerinum* (Syd.) Speg. (Das and Mohanty, 1972).

Epidemiology and Disease Cycle:

A densely populated orchard where there is low light intensity provides favorable conditions for sooty mold disease. Trees exposed to the eastern side (sunlight) have less incidence while trees in center of the orchard. Sugary substance secreted by the insects is stated to be a condition favorable for the development of sooty mold. Incidence of insects on the shoot is directly associated with disease severity. High humidity and high incidence of insects proved to be congenial for the high incidence of the disease (Singh and Singh, 1972).

Integrated Disease Management:

- Indian oil formulation (tree spray oil) 1% at 15 days interval could control sooty mold effectively (Prakash, 1988; Misra and Prakash, 1993).
- Spraying of Wettasul + Metacid + Gum Accacia (0.2% + 0.1% + 0.3%) were reported effective against the disease in UP (Prakash, 1985).
- Singh and Singh (1972) reported that spraying of Elosal (900 g/450 L) at 10–15 days interval proved a quite effective control over sooty mold disease.
- Variety Alphanso is also reported as resistant (Peethambharan and Aravindakshan, 1975), Sharma and Badiyala (1991) also reported Alphanso and Malika as most resistant against sooty mold disease.
- Spraying of 2% starch is found effective and spray of Antracol 80 WP is also reported to be equally effective against the disease (Hassan et al., 2002; Haggag, 2010).

19.7   RED RUST DISEASE

Introduction/Economic Importance:

Red rust disease in mango is caused by an alga that reduces the activity of photosynthesis leading to shedding of leaves which lowers vitality of the mango plant; this disease is mostly reported from places where the plant density in orchards is higher (Sarwar, 2015). Many workers have reported the disease from all mango growing regions of India (Chowdhury, 1975; Prakash and Raoof, 1985). Occurrence of red rust disease is very common in Tarai regions of Nepal and other humid regions of India. From India, it has been reported from Karnataka, Uttar Pradesh, and Bihar.

Symptoms:

The nature of alga is parasitic and epiphytic; it attacks the bark, twigs, and foliage of the mango plant. In severe cases the bark thickens, twigs remain enlarged and stunted and foliage becomes sparse and it ultimately dries up. Appearance of red rusty fructification of the alga on the surface of the infected young twigs, leaves, veins, and petiole are the characteristic symptoms of the red rust disease. Greenish gray, velvety spots initially appear on the infected leaves which later turn reddish brown. Spots are usually circular to irregular in shape, slightly elevated, and sometimes coalesce to form larger and irregular spots. Algal spot is slightly elevated and usually 2 mm in diameter, though in some cases it may be as much as 1 cm. At the later stage of the infection after shedding of spores, the remnants of the algal matrix get attached to the leaf surface resulting in a creamy white mark at the original rust spot (Singh, 1960; Sarwar, 2015).

Causal Organism:

The red rust disease in mango is caused by an alga *Cephaleuros virescens* (Syn. C. *parasiticus* Karst, C. *mycoidea* Karst; Prakash and Misra, 1992).

Pathogen:

Cephaleuros virescens has long unbranched reddish erect, stout, brown, filaments measuring 52.05–235.96 × 10.41–15.61 µm, emerging through the thalloides disk, 5–12 stalked. The sporangia are subspherical to oval, purple brown in color measuring 19.08–27.76 ×15.61–20.82 µm (Mishra et al., 1973).

At the early stage of disease development, on the affected portion of leaf, a thick and compact mat of alga thallus beneath the cuticle can be observed. The thallus is disc-like and is composed of symmetrically arranged cells radiating dichotomously from the center to the periphery. The parasite is sandwiched between the cuticle and the epidermis and in some cases, it may extend up to the palisade layer (Agnihothrudu, 1964).

C. parasiticus produces a sporangium type asexual reproductive structure along with a few septate, orange colored sporangiophores, and setae arising toward the periphery of the thallus disc after rupturing the cuticle.

C. parasiticus filaments are very thin, erect, and green in color measuring 50.31–23 mm and 240.82 mm. Due to the presence of β-carotene the filament cells of this alga appear reddish-brown in color. The fertile filaments bear clusters of 5–6 sporangia at the apex. Sporangiophores are usually 7–8 septate, distributed at an equal distance. After the alga grows for 15 days it starts to produce sporangia, which are oval to elliptical measuring 22.12 mm × 18.18 mm. The mature sporangia get detached from the filaments which are easily dispersed by rainwater and wind (Chapman, 1984; Ponmurugan et al., 2010).

Disease Cycle:

The fruiting bodies of the alga are formed in a humid atmosphere. The fresh infection is initiated by zoospores formed by the sporangia. Initial entry into the host is achieved through inflicted wounds or washed into stomata during rains (Sarwar, 2015).

Epidemiology:

Thrimurty et al. (1981) reported that the disease infection increases during the rainy season, the rainwater splash serves as a mode of spread for alga from one infection site/plant to another. Temperature of 30 °C maximum and 25 °C minimum with high relative humidity and wind velocity and frequent moderate rains are favorable for the growth, development, and rapid dispersal of *Cephaleuros virescens* (Prakash and Misra, 1988).

Integrated Disease Management:

- Coates et al. (2009) reported that spraying of copper oxychloride (0.3%) two to three times was found to be effective in managing the disease. With the onset of rain, it is followed by two fungicide sprays at 15 days intervals during the month of July–August.

- Prakash and Singh (1979) recommended the application of Bordeaux mixture (2:2:250) followed by copper oxychloride. Spraying of Difolatan is also reported to be effective in checking the red rust disease in mango (Gupta et al., 1980).

19.8   BLACK TIP OR NECROSIS DISEASE/DISORDER

Introduction/Economic Importance:

Black tip disease in mango is confined to India. The disease has been reported from regions of Uttar Pradesh, Punjab, West Bengal, and Bihar in orchards located in close proximity of a brick kiln. The disease has not been reported from South India.

Symptoms:

The symptoms first appear as a small etiolated area at the distal end of the fruit, that gradually spreads which later turns nearly black and completely covers the tip. During the etiolating process, isolated grayish spots appear that turn dark brown in color, these spots enlarge and merge into a continuous necrotic area. At the later stage, the neighboring parenchyma also turns brown and the deposits are also noticed in the ducts. The browning and deposits gradually spread throughout the mesocarp and the affected cells disintegrate and turn into dead tissue. The necrosis may extend to the endocarp in severe cases.

The disease is most commonly found in the orchard in the vicinity of brick-kilns and the disorder is more common if the fruiting orchards fall in direction of the wind from the brick-kilns side. Gases like carbon dioxide, sulfur dioxide, and acetylene which are formed by brick-kilns fume are reported to cause the symptoms by polluting the air around the orchard.

Causal Organism: Gases like carbon dioxide, sulfur dioxide, and acetylene.

Integrated Disease Management:

- Sen (1943) recommended the orchard should be located not less than 1.6 km away from the brick kilns.
- Reddy and Kapoor (1965) reported that spraying of borax (0.6%) at an interval of 10–14 days is effective in controlling the black tip disease in mango.

- Nauriyal et al. (1972) reported that spraying of or caustic soda (0.8%) at an interval of 10–14 days is effective in controlling black tip disease in mango.

KEYWORDS

- mango
- diseases
- symptomatology
- epidemiology
- management

REFERENCES

Agnihothrudu, V. (1964). Annual Scientific Report, Tocklai Experiment Station, India, Indian Tea Assoc. for 1963: 64 (*Hypoxylon nummularium* inter alia; 3142).

Akem, C. N. (2006). Mango anthracnose disease: Present status and future research priorities. *Plant Pathol J.* 5(3): 266–273.

Akhtar, K. P., Khan, I. A., Kazmi, M. R., Hussain, R. I., and Fatima, B. (1999). Preventive control of powdery mildew disease of mango. *J. Agric. Marine Sci.* 4: 23–28.

Arauz, L. F. (2000). Mango anthracnose: Economic impact and current options for integrated management. *Plant Dis.* 84(6): 600–611.

Arauz, L. F. and González-Lobo, M. (1986). Fuentes de inóculo de algunas enfermedades de poscosecha del mango. *Agron. Costarric.* 10(1/2): 217–220.

Awa, O. C., Samuel, O., Oworu, O. O., and Sosanya, O. (2012). First report of fruit anthracnose in mango caused by *Colletotrichum gloeosporioides* in Southwestern Nigeria. *Int. J. Sci.. Technol. Res.* 1(4): 30–34.

Bains, G. and Pant, R. C. (2003). Mango malformation: Etiology and preventive measures. Physiol. *Mol. Biol. Plants.* 9: 41–61.

Bose, S. K. (1953). Powdery mildew of mango in Kumaon and its control. In: *Proceedings of Indian Science Congress*, Published by Kitabistan, Allahabad, India, pp. 186.

Burchill, R. T. (1978). Powdery Mildew of Tree Crops. In: The Powdery Mildews, edited by D. M. Spencer. London: Academic Press, pp. 473–493.

Butler, E. J. and Bisby, G. R. (1931). The fungi of India. *Sci. Monograph No. 1.* The Imperial Council of Agric. Res., India, Calcutta.

Chapman, R. L. (1984). An assessment of the current state of our knowledge of the Trentepohliaceae. In: Systematics of the Green Algae, pp. 233–250. Edited by D. E. G. Irvine and D. M. John. London: Academic Press.

Chavan, R. A., Deshmukh, V. D., Tawade, S. V., and Deshmukh, J. D. (2009). Efficacy of fungicides for managing powdery mildew of mango. *Int. J. Plant Prot.* 2: 71–72.

Chowdhuri, S. R. (1975). Occurrence of red rust in Madhya Pradesh. *Indian Phytopath.* 28: 261–262.

Chowdhury, M. N. A. and Rahim, M. A. (2009). Integrated crop management to control anthracnose (*Colletotrichum gloeosporioides*) of mango. *Agric Rural Dev.* 7(1–2): 115–120.

Coates, L., Akem, C., Cooke, T., Dann, E., and Young, A. (2009). Mango. In: Diseases of Fruit Crops in Australia, edited by T. Cooke, D. M. Persley, and S. House. CSIRO Publishing, Collingwood VIC, Australia. pp. 157–173.

Crane, J. H., Campbell, C. W. (1994). The Mango. The Horticultural Science Department, Florida Co-operative Extension Service, Institute of Food and Agricultural Science, University of Florida, p. 24.

Crookes, C. A. and Rijkenberg, F. H. J. (1985). A literature review of the distribution, symptomatology, cause and control of mango blossom malformation. *S. Afr. Mango Grow. Assoc. Res. Rept.* 5: 15–24.

Das, S. R. and Mohanty, N. N. (1972). A note on some fungi from Orissa. *Curr. Sci.* 41: 430.

Devi, M. I. and Devi, R. K. T. (2017). Management of gray leaf spot of mango with fungicides, botanicals and bioagents *in vitro* and *in vivo* condition in Manipur. *J. Mycopathol. Res.* 54(4): 527–530.

Diedhiou, P. M., Diallo, Y., Faye, R. Mbengue, A. A., and Sene, A. (2014). Efficacy of different fungicides against mango anthracnose in Senegalese Soudanian agroclimate. *Am. J. Plant Sci.* 5: 2224–2229.

Doreste, S. E. (1984). Información sobre el eriófido del mango, *Eriophyes mangiferae* (Sayed), en Venezuela. Rev. *Facultad Agrociencia* (Universidad Central de Venezuela). 13: 91–100.

Farr, D. F., Aime, M. C., Rossman, A. Y., and Palm, M. E. (2006). Species of *Colletotrichum* on agavaceae. *Mycol. Res.* 110: 1395–1408.

Fitzell, R. D., and Peak, C. M. (1984). The epidemiology of anthracnose disease of mango: Inoculum sources, spore production and dispersal. *Ann. Appl. Biol.* 104: 53–59.

Freeman, S., Gamliel-Atinsky, E., Maymon, M., Shtienberg, D., Youssef, S., and Levin A. G. (2015). Mango malformation disease: etiology, epidemiology and management. *Acta Horticulturae.* 1075: 207–214.

Freeman, S., Maimon, M., and Pinkas, Y. (1999). Use of GUS transformants of *Fusarium subglutinans* for determining etiology of mango malformation disease. *Phytopathology.* 89: 456–461.

Freeman, S., Shtienberg, D. Maymon, M., Levin, G. L., and Ploetz, R. C. (2014). New sights in mango malformation disease epidemiology, lead to a new integrated management strategy for subtropical environments. *Plant Dis.* 98(11): 1456–1466.

Gamliel-Atinsky, E., Sztejnberg, A., Maymon, M., Shtienberg, D., and Freeman, S. (2009). Inoculum availability and conidial dispersal patterns of *Fusarium mangiferae*, the causal agent of mango malformation disease. *Phytopathology.* 99: 160–166.

Gantotti, B. V. and Davis, M. J. (1993). Pectic zymogram analysis for characterizing genetic diversity of the mango anthracnose pathogen. *Acta Hortic.* 341: 353–359.

Gaur, V. P. and Chakrabarti, D. K. (2009). Incidence of malformation in mango (*Mangifera indica*) nurseries in eastern Uttar Pradesh. *Ind. J. Agric. Sci.* 79: 160–162.

Gupta, J. H. (1976). Reaction of mango varieties to powdery mildew (*Oidium mangiferae*) in Uttar Pradesh. *Prog. Hortic.* 8: 63–64.

Gupta, J. H. (1988). Dispersal of *Oidium mangiferae* Berthet causing powdery mildew of mango. *Prog. Hortic.* 20: 341–342.

Gupta, J. H. (1989) Perpetuation and epidemiology of powdery mildew of mango. *Actahorticulturae, Leuven.* 231: 528–533.

Gupta, J. H., Srivastava, V. P., and Pandey, L. C. (1980). Control of red rust disease of mango with fungicides. In: *Symposium on Plant Disease Problems.* November 10–12, 1980 held at PAU, Ludhiana, pp. 61.

Gupta, P. C. and Dang, J. K. (1980). Occurrence and control of powdery mildew of mango in Haryana. *Indian Phytopathol.* 33: 631–632.

Haggag, W. (2010). Mango diseases in Egypt. *Agricult. Biol. J North Am.* 1(3): 285–328.

Hassan, B., Ian, B., and Stuart, S. (2002). Pests, diseases and disorders of mangoes in the northern territory: an illustrated field guide. The Northern Territory Horticultural Association. p. 95.

Hofman, P. J., Smith, L. G., Joyce, D. C., Johnson, G. I., and Meiburg, G. F. (1997). Bagging Iqbal, Z., Akhtar, N., Ghazanfar, M. U., Shehzad, S. M., Ahmad, S., Asif, M., Yasin, M., Pervez, M. A., Dasti, A. A. and Saleem, A. (2011). Management of mango malformation through physical alteration and chemical spray. *African J. Agric. Res.* 6(7): 1897–1901.

Iqbal, Z., Pervez, M. A., Saleem, B.A., Ahmad, A., Dasti, A. A., and Saleem, A. (2010). Potential of *Fusarium mangiferae* as an etiological agent of mango malformation. *Pak. J. Bot.* 42(1): 409–415.

Jadeja, K. B. and Vaishnav, M. U. (1984). Effective spray schedule for control of anthracnose and leaf blight of mango. *Indian J. Plant. Prot.* 12: 93–96.

Jeffries, P., Dodd, J. C., Jeger, M. J. and Plumbley, R. A. (1990). The biology and control of *Colletotrichum* species on tropical fruit. *Plant Pathol.* 39: 353–366.

Johnson, G. I. (1994). Powdery Mildew. In: Compendium of Tropical Fruit Diseases, edited by R. C. Ploetz. America Psychopathological Society, St. Paul, MN, USA. ISBN: 9780890541623, pp. 38–39.

Johnson, G. I., and Hofman, P. J. (2009). Postharvest technology and quarantine treatments. In: The Mango: Botany, Production and Uses. 2nd edition, edited by R. E. Litz. Centre for Agriculture and Biosciences International (CABI) Wallingford, Oxfordshire, England. pp. 529–605..

Jones, B. H. R. (1923). Mycological Work in Egypt during the Period 1920e1922, vol. 49. Ministry of Agriculture, Egypt. Technical Series Bulletin, pp 1–129.

Joubert, M. H. (1991). Implications of Epidemiological Studies on Strategies for Control of Powdery Mildew and Anthracnose. In: S. Afr. Mango Growers' Association Yearbk. 11: 26–28.

Joubert, M. H., Manicom, B. Q., and Wingfield, M. J. (1993). Powdery mildew of mango in South Africa: A review. *Phytophylactica* 25: 59–63.

Khalid, P. A. and Alam, S. S. (2000). Powdery mildew of mango: A review. *Pakistan J. Biol. Sci.* 3(7): 1119–1122.

Khurana, A. D. and Gupta, O. P. (1973). Mango malformation in India. *Pesticides.* 7: 12–13.

Kulkarni, O. K. and Kulkarni, U. K. (1978). Physiology of mango leaves infected by *Capnodium ramosum* Cooke II. Mineral contents. *Biovigyanam.* 4(2): 173–174.

Kumar, J., Beniwal, S. P. S. (1992). Mango Malformation. In: Plant Diseases of International Importance, Diseases of Fruit Crops, edited by J. Kumar, H. S. Chaube, U. S. Singh, A. N. Mukhopadhyay. Vol. 3, Prentice Hall, New York, p. 456.

Kumar, J., Singh, U. S., and Beniwal, S. P. S. (1993). Mango malformation: one hundred years of research. *Annu. Rev. Phytopathol.* 31: 217–232.

Kumar, J., Singh, U. S., and Beniwal, S. P. S. (1993). Mango malformation: One hundred years of research. *Ann. Rev. Phytopathol.* 3: 217–32.

Kumar, P., Misra, A. K., and Modi, D. R. (2011). Current status of mango malformation in India. *Asian J. Plant Sci.* 10: 1–23.

Kumari, P., Rakesh, and Singh, R. (2017). Anthracnose of mango incited by *Colletotrichum gloeosporioides*: A comprehensive review. *Int. J. Pure Appl. Biosci.* 5(1): 48–56.

Lakshmi, B. K. M., Reddy, P. N., and Prasad, R. D. (2011). Cross-infection of *Colletotrichum gloeosporioides* Penz. Isolates causing anthracnose in subtropical fruit crops. *Tropi. Agri. Res.* 22: 183–193.

Lim, T. K. and Khoo, K. C. (1985). Diseases and disorders of mango in Malaysia. Tropical Press, Kuala Lumpur, Malaysia, p. 114.

Majumder, P. K., Sinha, G. C., and Singh, R. N. (1970). The effect of exogenous application of alpha naphthyl acetic acid on mango (*Mangifera indica* L.) Malformation. *Indian J. Hort.* 27(3–4): 130–131.

Marasas, W. F. O., Ploetz, R. C., Wingfield, M. J., Wingfield, B. D., and Steenkamp, E. T. (2006). Mango malformation disease and the associated Fusarium species. *Phytopathology.* 96: 667–672.

McMillan Jr., R. T. (1973). Control of anthracnose, powdery mildew of mango with systemic and non-systemic fungicides. *Trop. Agric.* (Trinidad). 50: 245–248.

McMillan, R. T. (1984). Control of mango anthracnose with foliar sprays. *Proc. Fla. State Hort. Soc.* 97:344–345.

McRae, W. (1924). Economic botany part-III. *Mycol. Ann. Rep.* (1922–23). Board of Scientific Advice, India.

Minessey, F. A., Biely, M. P., and El-Fahal, A. (1971). Effect of iron chilates in correcting malformation of terminal bud growth in mango. *Sudan Agric. J.* 6:71–74.

Mishra, B., Prakash, O., and Mishra, A. P. (1973). A serious leaf spot disease of litchi caused by *Cephaleuros virescens*. *Indian J. Mycol. Plant Pathol.* 3:219–220.

Misra, A. K. (2002). Integrated disease management in mango. Manual of Summer School on Recent Advances in Production, Protection and Post Harvest Management of Subtropical Fruits, pp. 1–8.

Misra, A. K. and Prakash, O. (1993). Host range and efficacy of different chemicals for the control of sooty mould of mango. *Natl. Acad. Sci.* 63(B)II: 233–235.

Misra, A. K., Negi, S. S., Rajan, S., and Kumar, R. (2000). Cultivar Elaichi a new source of resistance to mango malformation. Proceedings of the Indian Phytopathological Soceity-Golden Jubilee, Vol. II, (IPSGJ'00), Indian Phytopathological Society, New Delhi, pp. 750–751.

Muhammad, F., Ibrahim, M., and Pervez, M. A. (1999). Some physiological aspects of mango malformation. *Int. J. Agri. Biol.* 1: 94–96.

Mukerjee, S. K. and Litz, R. E. (2009). Introduction, botany and importance. In: The Mango: Botany Production and Uses. 2nd edition, edited by R. E. Litz. CAB International, Wallingford Oxon, UK, pp. 1–18.

Naqvi, S. A. H., Perveen, R., Manzoor, S. A., Umar, H. M. I., Iqbal, M. T., Liaquat, F., Majid, T., and Irshad, A. (2014). Evaluation of various mango varieties against the infection dynamics of powdery mildew (*Oidium mangiferae* Bert.). *Am. J. Plant Sci.* 5: 2372–2377.

Nasir, M., Mughal, S. M., Mukhtar, T., and Awan, M. Z. (2014). Powdery mildew of mango: A review of ecology, biology, epidemiology and management. *Crop Protect.* 64. 19–26.

Nath, R., Kamal R. S. Wanshi, and I. P. Sachan, (1987). Studies on mango malformation. *Indian. J. Mycol. Plant Pathol.* 17: 29–33.

National Horticulture Board Database. (2017). Ministry of Agriculture, Government of India 85, Institutional Area, Sector-18, Gurgaon, pp. 122 015.

Nauriyal, J. P., Chadha, K. L., and Rajput, M. S. (1972). *Acta Horticult.* 24: 215–216.

Nelson, S. C. (2008). Mango anthracnose (*Colletotrichum gloeosporioides*). UH–CTAHR. (College of tropical agriculture and human resources). http://www.ctahr.hawaii.edu/freepubs.

Nelson, S. C. (2008). Mango powdery mildew. In: College of Tropical Agriculture and Human Resources. Department of Plant and Environmental Protection Sciences. University of Hawaii at Manoa, Plant Disease, pp. 46.

Nuñez-Elisea, R. (1985). Flowering and fruit set of a monembryonic and a polyembryonic mango as influenced by potassium nitrate sprays and shoot decapitation. *Proc. Fla. State Hortic. Soc.* 98:179–183.

of mango (*Mangifera indica* cv. 'Keitt') fruit influences fruit quality and mineral composition. *Postharv. Biol. Technol.* 12: 83–91.

Palti, J., Pinkas, Y., and Chorin, M. (1974). Powdery mildew of mango. *Plant Dis. Rep.* 58: 45–49.

Pandey, S. C. and Mohammad, A. (1974–75). Studies on fruit rot of mango III. Brown spot caused by *Pestalotia mangiferae* Bult. *Proc. Bihar. Acad. Sci.* 22–23: 89–95.

Peethambharan, G. K. and Aravindakshan, M. (1975). Varietal susceptibility of mangoes to sooty mould. *Agric. Res. J. Kerala* 16(2): 260–261.

Pérez-Rodríguez, A., Monteón-Ojeda, A., Mora-Aguilera, J. A., and Hernández-Castro E. (2017). Epidemiology and strategies for chemical management of powdery mildew in mango. *Pesq. Agropec. Bras., Brasília.* 52(9): 715–723

Ploetz, R. C. (2004). The major diseases of mango: strategies and potential for sustainable management. *Acta Hort.* 645: 137–150.

Ploetz, R. C. and Prakash, O. (1997). Foliar, floral and soil borne diseases. In Mango: Botany, Production and Uses. CAB International, Wallingford, UK, pp. 281–326.

Ploetz, R. C., Benscher, D., Vazquez, A., Colls, A., Hagel, J., and Schaffer, B. (1996). Mango decline: Research in Florida on an apparently wide-spread disease complex. *Proc. Int. Mango Sympos.* 5th. *Acta. Horti.* 45: 547–553.

Ploetz, R., Zheng, Q. I., Vazquez, A., and Sattar, A. M. A., (2002). Current status and impact of mango malformation in Egypt. *Int. J. Pest Manag.* 48: 279–85.

Ploetz, R.C. (2001). Malformation: a unique and important disease of mango, *Mangifera indica* L. In: Fusarium: Paul E. Nelson Memorial Symposium (St Paul: APS Press), edited by B. A. Summerell, J. F. Leslie, D. Backhouse, W. L. Bryden and L. W. Burgess, pp. 233–247.

Ponmurugan, P., Saravanan, D., and Ramya, M. (2010). Culture And Biochemical Analysis Of A Tea Algal Pathogen, *Cephaleuros parasiticus. J. Phycol.* 46 (5): 1017–1023.

Prakash, O and Singh, U. N. (1979). Fungicidal control of red rust of mango. *Indian J. Mycol. Plant Pathol.* 9: 175–176.

Prakash, O. (1985). Management of mango diseases and their, control. *Ann-Rept., Cent. Inst, Hort. Northern Plains,* Lucknow. pp. 31–35.

Prakash, O. (1988). Sooty mould disease of mango and its control. *Int. J. Trop. Plant Dis.* 9: 277–80.

Prakash, O. and Misra, A. K. (1988). Growth of red rust C. *virescens* Kunz of mango during the unusual drought year 1987 in Kakori and Malihabad mango belt of India. *XXI Congresse Brasileiro de Fitopatologia Bras.* 13(2): 121.

Prakash, O. and Misra, A. K. (1992). Important diseases of mango and their effect on production. *Biol. Memoirs.* 18(1,2).

Prakash, O. and Raoof, M. A. (1985). Blossom blight disease of mango. *Indian J. Plant Path.* 3: 271–272.

Prakash, O. and Srivastava, K. C. (1987). Mango Diseases and their Management—A World Review. Today and Tomorrow's Printers & Publishers, New Delhi. pp 175.

Prakash, O. (2004). Diseases and Disorders of Mango and Their Management. In: Naqvi, S.A.M.H. (ed.), Diseases of Fruits and Vegetables. 1: 511–620.

Prakash, O. and Raoof, M. A. (1994). Studies on powdery mildew (*Oidium mangiferae*) disease of mango: distribution, perpetuation, losses and chemical control. *Biol. Mem.* 20: 31–45.

Purseglove, J. W. (1972). Mangoes west of India. *Acta Hortic.* 24: 107–174.

Raheel, M., Anwar, S. A., Javed, N., Ilyas, M. B., Iqbal, M., and Zia, A. (2008). Management of powdery mildew of mango by foliar spray fungicides. *Pak. J. Phytopathol.* 21(1): 173–174.

Rani, V., Ansari, M. W., Shukla, A., Tuteja, N., and Bains, G. (2013). Fused lobed anther and hooked stigma affect pollination, fertilization and fruit set in mango: a scanning electronmicroscopy study. *Plant Signal Behav.* 8(3): 231671.

Rawal, R. D. and Saxana, A. K. (1997) Diseases of Dryland Horticulture and Their Management. In: Silver Jubilee National Symposium Arid Horticulture, *HISAR Proceedings*, HSH/CCS, Hisar, HAU, pp 5–7.

Reddy, D. B. and Kapoor, S. P. (1965). Black tip in mango. *Intensive Agric.*, Delhi. pp. 9–10.

Sarkar, A. (1960). Leaf spot disease of *Mangifera indica* L. caused by *Pestalotia mangiferae* Bult, Loydia. 23: 1–7.

Sarwar, M. (2015). Practices for Integrated Control of Mango (*Mangifera indica* L.) Diseases to Protect in Preharvest as Well as Postharvest Phases. *Biosci. Bioeng.* 1(3): 57–62.

Sarwar, M. (2015). Practices for integrated control of mango (*Mangifera indica* L.) diseases to protect in preharvest as well as postharvest phases. *Biosci. Bioeng.* 1(3): 57–62.

Sattar, A. and Malik, S. A. (1939). Some studies on anthracnose of mango caused by *Glomerella cingulata* (Stonem.) Spauld. Sch. (*Colletotrichum gloeosporioides* Penz.). *Indian J. Agri. Sci.* 1: 512–521.

Schlosser, E. (1971). Mango malformation: Symptoms, occurrence and varietal susceptibility. FAO, *Plant Prot. Bull.* 19: 12–14.

Sen, P. M. K. (1943). Further studies on black tip of mango. *Indian J. Agric. Sci.* 13:300–333.

Sharma, I. M. and Badiyala, S. D. (1991). Relative occurrence of sooty mould on mango varieties in Himachal Pradesh. *Plant Dis. Res.* 6: 1.

Singh, A., Verma, K. S., and Mohan, C. (2008). Evaluation of fungicides against *Colletotrichum gloeosporioides* causing anthracnose of guava. *Plant Dis. Res.* 23: 91–92.

Singh, L. B. (1960). In: *The ma'lgo.* Leonerd Hill Book Ltd., London. pp. 438.

Singh, R. S. (2000). Powdery mildew of mango. In: Diseases of Fruit Crops. Oxford and IBH Publishing Co, New Delhi, pp. 157–159.

Singh, S. P. and Singh, A. K. (1972). Studies on sooty mould of mango (*Mangifera indica* L.) Bihar. *Proc. Third Intern. Symp. Subtropical Tropical Horticulture,* Bangalore. p 121.

Sinha, P., Prajneshu, R., and Varma, A., (2002). Growth models for powdery mildew development of mango. *Ann. Plant Prot. Sci.* 10: 84–87.

Sirohi, S. C, Prakash, S., Rana, P., and Singh, R. (2009). Response of mango malformation to severity of malformed panicle bearing shoot pruning. *Indian J. Hort.* 66: 393–395.

Tandon, R. N., Sisodia, U. S., and Bilgrami, K. S. (1955). Pathological studies of *Pestalotia mangiferae. Proc. Indian Acad. Sci.* 42: 219–255.

Thrimurthy, V. S., Kashyap, R., Singh, J., Garg, K., Bhattachargee, S., and Shastri, M. B. (1981). Studies on red rust of mango. *Third International Symposium on Plant Pathology,* Dec. 14–18, New Delhi. pp. 41.

Ullasa, B. A. and Rawal, R. D. (1985). Occurrence of a new post-harvest disease of mango due to *Pestalotia grandicola.* (Abs.) *Second Int. Symp. Mango,* Bangalore. pp. 65.

Vaheeduddin, S. (1953). *The Mango.* 121(11): 23.

Vala, D. G., Solanki. K. U., Desai, V. D., and Joshi, H. U. (1985). Diseases of mango occurring in Gujarat State (Abs.). *Second Int. Symp. Mango,* Bangalore. pp. 64–65.

Verma, K. S., Cheema, S. S., Kang, M. S., and Sharma, A. K. (1991). Hitherto unrecorded disease problems of mango from Punjab. *Plant Dis. Res.* 6: 2.

Wafaa, M. H, Shabaan, A. M., and Saleh, A. E. M. (2014). Biological management of mango malformation using antifungal compound from *Streptomyces aureofaciens. Int. J. Pharm. Sci. Rev. Res.* 29(2): 283–291.

Wagle, P. V. (1928). Studies in the shedding of mango flowers and fruits, Part 1. Memoirs of the Department of Agriculture. *Indian Bot. Sci.* 8: 219–249.

Yadav, R. K. and Beg. M. Z. (2015). Malformation on mango by *Fusarium mangiferae.* in the Eastern UP of India. *Indian J. Sci. Res.* 6(1): 53–55.

Youssef, S. A., Maymon, M., Zveibil, A., Klein-Gueta, D., Sztejnberg, A., Shalaby, A. A., and Freeman, S. (2007). Epidemiological aspects of mango malformation disease caused by *Fusarium mangiferae* and source of infection in seedlings cultivated in orchards in Egypt. *Plant Pathol.* 56: 257–263.

CHAPTER 20

MUSKMELON (*CUCUMIS MELO* L.): MAJOR DISEASES AND THEIR MANAGEMENT

S. K. MAHESHWARI*, B. R. CHOUDHARY, A. K. VERMA, and SHAKTI KHAJURIA

ICAR-Central Institute for Arid Horticulture, Bikaner, Rajasthan, India

Corresponding author. E-mail: maheshwariskciah@gmail.com

ABSTRACT

Muskmelon or Kharbooja (*Cucumis melo* L.), is an important member of the Cucurbitaceae family. It is native of tropical Africa and grown for its delicious fruit. Most often, it is ambiguously chunked in both vegetable and fruit. But, botanically, legally, and culinarily it is a fruit while biological and cultural similarity to other cucurbitaceous vegetable it is considered as vegetable. In India, it is cultivated in Uttar Pradesh, Andhra Pradesh, Punjab, Madhya Pradesh, and Karnataka during summer and rainy months. Meanwhile, for off-season production, it is grown as *Diara* cultivation in the river basin of the Ganga, Yamuna, Gomti, and other tributaries of these rivers during the winter and spring months. The ripe fruits of muskmelon have sweet and juicy pulp. Muskmelon is severely affected by many fungal as well as bacterial diseases, which cause heavy damage to the crop every year. *Alternaria* leaf blight, powdery mildew, downy mildew, Fusarium wilt, anthracnose, and bacterial wilt are important maladies of this crop. To reduce yield losses by the disease, a complete knowledge of its disease control approaches is required. Major diseases of muskmelon are discussed in this chapter.

20.1   INTRODUCTION

India ranks second in vegetables production in the world next to China. Development of improved vegetable varieties/hybrids/technologies through systematic research followed by their adoption by the farmers have helped in tremendous increase in area (10.1 million ha) and production (169.06 million tons) of vegetables in India during 2015–2016. The success or failure of crop production relies on interaction among varieties/genotypes, inputs used, environment, and management factors. Muskmelon (*Cucumis melo* L.) usually called as *kharbuja* in India is one of the major cucurbits cultivated as a "dessert crop" round the world. This is cultivated in both riverbed and irrigated conditions for supplying to local as well as distant markets of other states. Area of muskmelon crop was 45,000 ha with production of 935,000 MT in India during 2015–2016 (Horticulture database). The Cucumis genus comprises nearly 40 species (Whitaker and Davis, 1962). It probably originated in North Africa and Asia. It is native of tropical Africa more specifically in the eastern region, south of Sahara Desert (Robinson and Whitaker, 1974). A good amount of diversity also exists in India, Persia, Southern Russia, and China; hence, they are considered secondary center of origin (Chadha and Lal, 1993). Fruits are nutritious as they contain 0.6% protein, 0.2% fat, 3.5% carbohydrates, 32 mg Ca, 14 mg P, 1.4 mg Fe, 16 mg carotene, and 26 mg vitamin C per 100 g fresh weight of fruit (Aykroyd, 1963). Seed kernels are tasty and nutritious, hence can be eaten as they are rich in oils and energy. Wide morphological variation is found in fruit morphology and composition, and that is why, *C. melo* is recognized the most diverse species of the genus *Cucumis* (Bates and Robinson, 1995). The cultivated varieties of this species are eaten raw, pickled, or cooked. Fruits are fleshy with many seeded pepo, very variable is size and shape, globular, oblong, flat round, smooth or furrowed ring glabrous and smooth to rough and reticulate, pale to deep yellow, yellow brown or yellowish green and whitish. Seeds are flattened, white or cream, smooth, 5–15 cm in length. It is obtaining popularity among the farmers owing to its short duration and high-yield potential along with more nutritive and industrial value (Reddy et al., 2017). Nowadays, low-tunnel cultivation is becoming popular among the farmers for early and off-season production to fetch higher price in the market.

Plant diseases caused by fungi, bacteria, viruses, phytoplasmas, and nematodes reduce crop yields (Khoury and Makkouk, 2010). Major diseases and post-harvest fruit rots can infect the vegetables. These diseases are major challenges in the cultivation of vegetable crops in India. Muskmelon is severely affected by many fungal as well as bacterial diseases, which cause heavy damage to the crop every year. *Alternaria* leaf blight, powdery mildew, downy mildew, Fusarium wilt, anthracnose, and bacterial wilt are important maladies of this crop. To reduce yield losses by the disease, a complete knowledge of its disease control approaches is required. Attempts have been made for management of muskmelon following diseases in arid and semi-arid regions.

20.2 FUNGAL DISEASES

20.2.1 ALTERNARIA LEAF BLIGHT DISEASE

Introduction/Economic Importance:

Alternaria leaf blight incited by *Alternaria cucumerina* is an important disease in Rajasthan and also an important disease of muskmelon cultivation states of India. It was observed in the field of Rajasthan as well as Uttar Pradesh. It is a serious problem in the production of muskmelon under hot arid conditions of Rajasthan as well as in semi-arid regions. In India, the disease was first reported by Prasada and Khandelwal (1970) on watermelon from Jobner (Rajasthan). Yield losses due to the foliage loss are variable. Its disease severity varies depending up on the prevailed weather condition.

Symptoms:

Symptoms primarily observe on leaves as circular and light brown to dark brown spots of different size which later enlarge in a concentric pattern and margins appear. These spots coalesce to form larger areas on leaves.

Casual Organism: *Alternaria cucumerina* (Ell. & Ev.) Elliot

Taxonomic Position:

Division—Eumycota
   Subdivision—Deuteromycotina
      Class—Hyphomycetes
         Subclass—Sporomycetidae
            Order—Moniliales
               Family—Dematiaceae
                  Genus—*Alternaria*
                     Species—*Cucumerina*

Disease Cycle:

The fungus perpetuates as mycelium or spores in infected plant debris. Perennial weed/collateral hosts contribute primary inoculums. The initial infection starts from the lower leaves, where numerous conidia are produced. These are disseminated by wind, water, rain splash and become secondary source for disease dissemination. Conidia germinate in the presence of moisture at temperature nearly 25 °C–30 °C giving rise to germ tubes, which enter the host tissue.

Epidemiology:

A humid weather influences the disease to a great extent. The disease ranged in trace to moderate form depending on the environmental factors during rainy season. Availability of high humidity (80%–95%) coupled with a temperature of about 25 °C–28 °C with bright sunlight and dew or rain was conducive for fungal spore germination at which maximum disease development is obtained.

Integrated Disease Management:

Combining control of cultural and chemical methods can be effective against this disease for integrated disease management strategies.

Cultural Control:

- Collect affected plant residues and destroy them.
- Sowing certified seed in the field.
- Deep ploughing in the field.

Chemical Control:

- Sprayings of Indofil M-45 (0.2%) at regular interval was very effective against this disease.
- Pushpa et al. (1999) reported the benefits of Bavistin to manage the seed-borne infection of Alternaria species in pumpkin, cucumber, muskmelon, and watermelon.
- Spraying of mancozeb @0.25% proved the most effective fungicide against this disease in Watermelon under field condition (Maheshwari et al., 2015).

Biological Control:

- Application of bioagents (*Triochoderma viride* and *Triochoderma viride harzianum*) @ 5% was also effective against this disease.

20.2.2 POWDERY MILDEW

Introduction/Economic Importance:

This disease affects muskmelon leaves. The disease causes maximum damage in the warm and dry areas where moisture is present as dew. In India, the disease is prevalent in almost all the states and causes losses by reducing its yield and also quality (Gupta et al., 2001).

Symptoms:

The symptoms showed as small floury patches on leaves first on the lower surface followed by upper surface. Fungal growth on these spots results in production of powdery growth. Severely affected leaves lose their dark green color and become pale yellow green to brown and are shriveled and premature defoliation, drying, and collapse. Fruits remain undersized.

Causal Organism: *Sphaerotheca fuliginea*

Integrated Disease Management:

Cultural Control:

- All infected crop debris as well as collateral hosts should be removed and burnt. Deep ploughing may burry the pathogen where it may be killed by soil microflora of the pathogen present in below soil layers, after ploughing, may be brought to the soil surface and inactivated by solar radiation.
- Crop sequence with appropriate crops may be the most effective management practices in eliminating soil borne pathogens. Introduction of resistant crop in crop sequence can be helpful in reduction of the disease.

Chemical Control:

- Two to three sprays of wettable sulfur @ 0.2% at 15 days interval was effective.
- Three sprays of hexaconazole @0.05% were found effective.

- Spray the crop with Karathane @0.1% at 2 times at an interval of 10–15 days just after the appearance of disease.

Host Resistance:

- Sowing of moderately resistant varieties such as Arka Rajhans, Punjab Hybrid F1 and Kashi Madhu to avoid the incidence of powdery mildew.

20.2.3 *DOWNY MILDEW*

Introduction/Economic Importance:

Downy mildew is considered as one of the serious problems. In India, it is present in tropical and subtropical areas. It can cause reduction due to early infection of downy mildew in crop yield up to 60% (Gupta et al., 2001).

Symptoms:

Symptoms first appear on topside of older leaves in the form of small green to yellowish spots or water-soaked lesions. Its center turns brown and finally dies. The lesion develops a downy growth on lower part of the light yellow lesions present on the top of the leaf. The downy growth on the underside of the lesions is frequently speckled with dark purple to black sporangia (spore sacks) that can be observed with a hand lens.

Causal Organism: *Pseudoperonospora cubense* (Berk & Curt) Rostow.

The fungus belongs to class Oomycetes, Order—Peronosporales and family—Peronosporaceae.

Disease Cycle:

The pathogen produces sporangia. They are easily transferred to healthy plant tissue by air or rain. Once they reach on a susceptible host, they germinate and can directly infect the leaf within an hour. Natural pores serve as a primary point of its entry, which results in multiple infections on the leaf.

Epidemiology:

Cool, wet, and humid conditions are congenial for Downy mildew disease. Optimum sporangia production occurs between 15 °C and 20 °C. It requires high humidity for at least 6 hours.

Integrated Disease Management:

Cultural Control:

- Promote air movement inside the crop canopy.
- Excess overhead irrigation should be avoided.
- Weeds in the field should be kept under control.
- Avoid the spread of disease by maintaining proper hygiene. Field workers should wash their hands before moving from one field to another.

Chemical Control:

- Mancozeb (2 g/L) spray was effective against this disease.
- Chlorothalonil (kavach) @ 2 g/L can be used as a protective application at 7 days interval.
- Two sprays of mancozeb (0.1%) and metalaxyl (0.05%) mixture starting with the onset of conducive weather at 10 days interval give good control (Gupta et al., 2001).

Host Resistance:

- Grow moderately resistant variety "Kashi Madhu" against downy mildew.

20.2.4   FUSARIUM WILT

Introduction/Economic Importance:

This disease causes heavy losses in various cucurbits such as cucumber, bottle gourd, and muskmelon. Wilt is the most devastating disease of cucurbitaceous crops of all over the world (Radhakrishnan and Sen, 1985). The disease is the most destructive at places wherever soil temperatures are high enough for the pathogen to grow but unsuitable for the plants. Yield losses up to 80% have been reported in the worst affected areas (Gupta et al., 2001).

Symptoms:

Flowering stage is the most critical stage of the crop for appearance of the disease. The leaves are accompanied by yellowing and marginal necrosis. Older plants may first exhibit temporary wilting only during the heat of mid-day but will die within a few days. The infection results in lesion formation on the collar region and infected areas appear brown and water soaked. Wilting begins sudden after disease initiation. As a result of softening of the tissue, the plants shriveled and after that rapid plant mortality. The older plants wither and die during the growing season.

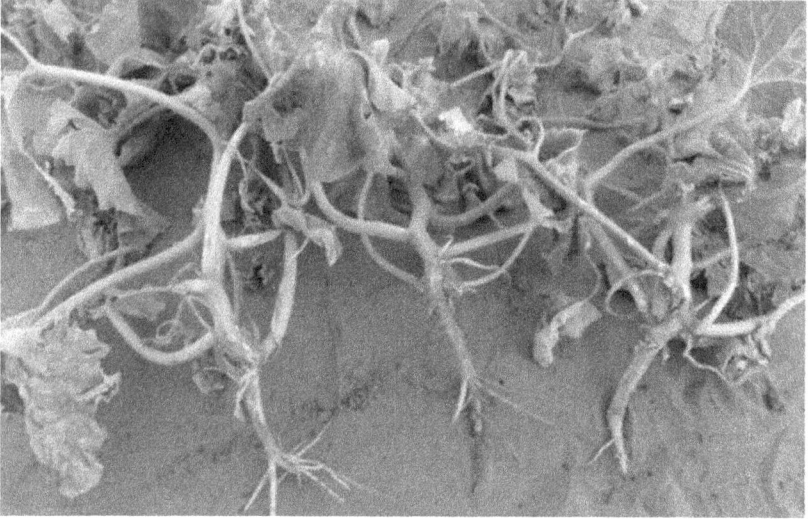

FIGURE 20.1 *Fusarium* wilt in muskmelon.

Causal Organism:

Wilt is caused by *Fusarium acuminatum* (Ellis and Everh) which was identified I.D. No. 9409.14 dated: 29-05-2014 from ITCC, Division of Plant Pathology, ICAR-IARI, New Delhi. Wilt in muskmelon is also caused by another fungus *Fusarium oxysporum* f. sp. melonis. It produces microconidia, macroconidia, and chlamydospores.

Taxonomic Position:

Division—Eumycota
　　Subdivision—Deuteromycotina
　　　　Class—Hyphomycetes
　　　　　　Subclass—Sporomycetidae
　　　　　　　　Order—Tuberculariales
　　　　　　　　　　Family—Tuberculariaceae
　　　　　　　　　　　　Genus—*Fusarium*
　　　　　　　　　　　　　　Species—*acuminatum*

Epidemiology:

A large population favored by continuous cropping of susceptible cultivars is required for serious disease outbreaks. Saprophytic activity of this disease is highest 60% moisture holding capacity of the soil, at pH 5.0–6.0 and temperature of 25 to 35 °C. Mostly symptoms are observed seriously up to 22 °C.

Integrated Disease Management:

Cultural Control:

- All infected plant should be collected and burnt.
- Crop rotation with garlic, radish, and onion has been found to reduce the disease in melons and cucumber (Sen, 1986).
- Deep ploughing should be practiced during summer.
- Growing the crop with nonhost crop.
- Certain soil amendments which may directly reduce the *Fusarium* population in soil or promote beneficial microflora which in turn destroy the soil inoculums, can also be used to control the disease.

Host Resistance:

- In muskmelon, the Indian cultivars (Durgapura Madhu and Punjab Sunehri) are resistant to *F. oxysporum* and *F. solani* (Radhakrishnan and Sen, 1985).
- Muskmelon cultivar "Pusa Madhuras" was found moderately resistant against this disease at New Delhi.
- Mark et al. (2005) found Hannahs Choice F1 as resistance source against Fusarium race 2 as well as powdery mildew.

Chemical Control:

- Drench the soil around roots with carbendazim (0.1%).
- Treat the seeds with Captan or Thiram or Bavistin @ 2 g/kg seed before sowing.
- Biological Control: Wilt in cucurbits can be suppressed by *Pseudo-monads* isolates.

20.2.5   FRUIT ROT/ROOT ROT

Introduction/Economic Importance:

It is a common disease of almost all cucurbits in India. It continues in transit, storage, and market. The disease is the most destructive at places where development of pathogen is very congenial.

Symptoms:

Symptoms develop on the surface in contact with the soil initially. Water-soaked lesions girdle the stem, later extending upwards and downwards. The rotting of infected tissues occur and plants collapse. The fungus causes rotting of fruits and entire fruit eventually rots.

Causal Organism: *Pythium aphanidermatum* and *P. butleri*

Taxonomic Position:

    Subdivision—Mastigomycotina
        Class—Oomycetes
            Order—Peronosporales
                Family—Pythiaceae
                    Genus—*Pythium*

Epidemiology:

High disease incidence is favored by high temperature and rainfall, also by excessive irrigation in hot weather. Fruit injuries caused by soil particles and insect bites help in fungal penetrations.

Integrated Disease Management:

- Avoid flood irrigation and raise the crop on drip system.
- Treat the seeds with thiram or Bavistin @ 2 g/kg seed. Apply *Trichoderma* @ 5 kg/ha in soil before sowing.
- Spray the crop with Bavistin @ 0.1% or copper oxychloride @ 0.2% which can be repeated if needed.
- Dipping or spraying the melons with 250 ppm imazalil in water followed by waxing leaves a residue of 0.5–2.0 ppm. is effective (Aharoni et al., 1992).

20.2.6   ANTHRACNOSE DISEASE

Introduction/Economic Importance:

Anthracnose was recorded in 1867 on gourds in Italy. The disease occurs principally in all humid regions of India. Cucurbit crops are the worst affected crops. Early infection may lead to premature fruit dropping and malformed fruits (Gupta et al., 2001). Anthracnose at severe condition can cause total crop failure or loss. The disease is the most destructive on fruits. Rapid defoliation of vines reduces fruit yield to a great extent. Fruit infection results in unsalable melons.

Symptoms:

The symptoms appear on stems, leaves, and fruit. Leaf symptoms include irregularly shaped, small water-soaked, or yellowish spots, dark brown lesions. Anthracnose-infected plants drop down, collapse, and die. Later these spots coalesce and cover the veins with entire area. Stem infections result in sunken, tan-colored cankers. Young infected fruits may drop. Fruit lesions are round and sunken. On mature fruits, symptoms appear as small, circular, slightly sunken, water-soaked lesions.

Taxonomic Position:

Division—Eumycota
   Subdivision—Deuteromycotina
      Class—Coelomycetes
         Order—Melanconiales
            Family—Melanconiaceae
               Genus—*Colletotrichum*
                  Species—*orbiculare*

Disease Cycle:

It is externally seed borne. In subtropical regions, it survives on the crop and weeds of the cucurbits. Humid or rainy season is essential for sporulation, dissemination, and penetration.

Epidemiology:

Humid or rainy weather is essential for infection, dissemination, and sporulation of the pathogen. Fungus occurs best between 22 °C and 27 °C.

Integrated Disease Management:

Cultural Control:

- Implementation of crop rotation with 2 years.
- Destruction of infected plant residues are cultural measures.
- Use of healthy seeds.
- Proper drainage.

Chemical Control:

- Seed treatment with thiram or carbendazim (2 gm/kg seed) is recommended. Sprays of zineb/ mancozeb (0.2%) at 7 days interval was found effective.
- Spraying of carbendazim (0.1%) or Difolatan (0.2%) can control the disease considerably (Bhardwaj and Sharma, 1999).
- Spray with carbendazim+ mancozeb (1 g/L each) when first fruit reaches size of a quarter at 15 days interval.
- Seed treatment with PGPR alone or as seed treatment + soil drenching at transplanting is useful to protect the crop (Gupta et al., 2001).

20.3   BACTERIAL DISEASES

20.3.1   BACTERIAL WILT

Introduction/Economic Importance:

Disease is also a serious malady in some growing areas of muskmelon. The bacterial pathogen multiplies within the vascular system of infected plants, causing collapse of vines. Yield losses can range from 10 to 20% in a disease-favorable season.

Symptoms:

Initial symptoms of bacterial wilt include flagging of leaves on one or more vines as dull green patches that rapidly increase in size. Symptom development proceeds rapidly; entire plants may collapse and die within a day or few days. Later infected leaves on the runner turn brown, wither, and die. The bacteria spread from the infected runner to the main stem and then to other runners. The entire plant soon wilts, shrivels, and dies. A diagnostic test for bacterial wilt can be performed in the field by cutting a wilted vine close to the main stem, rejoining the cut surfaces, then slowly drawing the sections apart. Bacterial wilt can be positively diagnosed if a thin strand of slime extends between the two sections.

Causal Organism: *Erwinia tracheiphila.*

The pathogen is transmitted by cucumber beetle vectors.

Disease Cycle:

Cucumber beetle is the primary inoculums of these bacteria. The bacteria overwinter in the digestive tracts of hibernating, adult striped, and spotted cucumber beetles. During spring, when beetles feed the plants, it introduces the bacteria into the leaves of cucurbits. The pathogen is not seed transmitted. Secondary inoculum is bacteria produced in infected plants and also transmitted from infected plants to healthy plants by cucumber beetles.

Epidemiology:

A temperature range of 25 °C–30 °C and high relative humidity is optimum for the growth of bacteria. The virulence of bacteria depends upon their isolate. Presence of enough dew from any resource is critical to allow the bacteria to enter the inner leaf tissue through feeding wounds for infection of cucurbits plants.

Integrated Disease Management:

Cultural Control:

- Destruction of weeds and infected plant debris.
- Use of healthy seeds.

Chemical Control: Its control is related to cucumber beetles control.

- Applications may be started as soon as seedling emerges or sometimes even before that. Frequent applications, especially in the seedling stage at 8- to 15-day intervals, are helpful. Systemic insecticide applied in the soil and it will help to manage for some time.
- Plants under kitchen garden can be saved against insects by covering with net for several weeks until flowering occurs. Use adapted, partially tolerant cucumber varieties as per availability is also a good option for prevention of the disease.

- Use adapted, partially resistant/tolerant cucumber varieties as per availability is also a good option for prevention of the disease.

KEYWORDS

- **muskmelon**
- **diseases**
- **symptomatology**
- **epidemiology**
- **management**

REFERENCES

Aharoni, Y., Copel, A., Davidson, H. and Barkai, R.G. (1992). Fungicide application in water and in wax for decay control in melons. *N.Z.J. Crop. Hortic. Sci.* 20: 177–179.

Aykroyd, W.R. (1963). ICMR Special Report. Series No. 42.

Bates, D.M. and Robinson, R.W. (1995). Cucumbers, melons and watermelons. In: *Evolution of Crop Plants* (Eds. Smartt J. and Simmonds N. W.). 2nd edn. Longman Scientific, Essex, pp. 89–96.

Bhardwaj, S.S. and Sharma, I.M. (1999). Diseases of minor fruits. In: *Diseases of Horticultural Crops* (Eds. Verma, L. R. and Sharma, R. C.). Indus Publ. Comp., New Delhi, pp. 541–562.

Chadha, M.L. and Lal, T. (1993). Improvement of cucurbits. In: *Advances in Horticulture.* Vol. 5. Vegetable Crops. (Eds. Chadha, K.L. and Kalloo, G.). Malhotra Publishing House, New Delhi, India.

Gupta, A.K., Sharma, R.C. and Sharma, S. (2001). Fungal diseases of cucurbits. In: *Diseases of Vegetable Crops* (Eds. Gupta, V. K. and Paul, Y. S.). Kalyani Publisher, New Delhi, pp. 71–86.

Khoury, W.E. and Makkouk, K. (2010). Integrated plant disease management in developing countries. *J. Plant Pathol.* 92(4): 35–42.

Maheshwari, S.K, Choudhary, B.R., Singh, D., Sharma, B.D. and Sharma, S.K. (2015). Management of Alternaria leaf blight of mateera (watermelon) by fungicides, bio-agents and botanical. *Indian J. Arid Hortic.* 10: 64–67.

Mark, J.H., Henry, M.M. and Moly, M.J. (2005). Hannahs choice F1: a new muskmelon hybrid with resistant to powdery mildew, Fusarium race 2 and potyviruses. *Horticulture Sci.* 40 (2): 492–493.

Prasada, R. and Khandelwal, G.L. (1970). Taxonomy, physiology and control of *Alternaria* leaf spot of watermelon (*Cucumis lanatus*). *Indian Phytopathol.* 23: 32–36.

Pushpa, K., Borkar, G.M., Patil, D.V. and Kamble, P. (1999). Studies on seed-borne pathogens of pumpkin, cucumber, watermelon and muskmelon. *J. Soils Crops.* 9: 234–238.

Radhakrishnan, P. and Sen, B. (1985). Efficacy of different methods of inoculation of *Fusarium oxysporum* and *F. solani* for inducing wilt in muskmelon. *Indian Phytopathol.* 38: 70–73.

Reddy, B.P.K., Hameedunnisa, B., Sunil, N. and Thirupathi Reddy, M. (2017). Correlation and path coefficient analysis in muskmelon (*Cucumis Melo* L.). *Int. J. Current Microbiol. Appl. Sci.* 6(6): 2261–2276.

Robinson, R.W. and Whitaker, T.W. (1974). *Cucumis.* In: *Handbook of Genetics* (ed. R.C. King). Vol. 2. Plenum, New York, USA, pp. 145–150.

Sen, B. (1986). Cultural management of soil borne diseases. In: *Vistas in Plant Pathology* (Eds. Varma, A. and Verma, J.P.). Malhotra Publishing House, New Delhi, pp. 367–381.

Whitaker, T.W. and Davis, G.N. (1962). Cucurbits. *International Science*, New York. p. 250.

CHAPTER 21

IMPORTANT DISEASES OF PAPAYA (*CARICA PAPAYA* L.) AND THEIR MANAGEMENT

B. K. PRAJAPATI[1,*] and N. M. GOHEL[2]

[1]*Agricultural Research Station, Sardarkrushinagar Dantiwada Agricultural University, District Banaskantha 385 535, Gujarat, India*

[2]*Department of Plant Pathology, B. A. College of Agriculture, Anand Agricultural University, Anand, Gujarat, India*

Corresponding author. E-mail: bindesh_prajapati@yahoo.in

ABSTRACT

Papaya (*Carica papaya* L.) belongs to Caricaceae family and is commercially cultivated all over the country. It has high nutritional value (vitamins A and C, calcium, carbohydrates, etc.) and great commercial potential. Its ease of propagation from seed, popularity, and versatility, and relatively short period from planting to production have made it an important food in many countries. Papaya crops have a narrow genetic base that may be partly responsible for the susceptibility of papaya to a wide range of diseases. Major diseases of muskmelon are discussed in this chapter.

21.1   INTRODUCTION

Papaya (*Carica papaya* L.) is the finest fruit commonly grown under tropical and subtropical climates. The fruit has very thin skin and thus rough handling contributes to heavy losses due to a variety of fungi and bacteria rots. Various diseases such as powdery mildew, *Phytophthora* root rot, anthracnose, stem end rot, black spot disease, and papaya ring spot and papaya leaf curl have affected commercial papaya production.

Diseases:

Stem rot/Foot rot: *Pythium aphanidermatum*
Powdery mildew: *Oidium caricae*
Papaya ring spot: *Papaya ring spot virus*
Leaf curl: *Papaya leaf curl virus*
Anthracnose: *Colletotrichum gloeosporioides*

21.2 *PYTHIUM* FOOT ROT/ROOT ROT

Pythium Foot rot/Basal stem rot/Root rot is only a severe issue when the soil reaches a higher temperature. The disease is very popular in both papaya-growing regions of the world and India.

Symptoms:

Symptoms initially appeared as water-soaked lesions on near-floor stem area. Such lesions coalesce and girdle the heart; tissue rotting that turns dark brown into black. Fruits get shriveled and malformed (Anonymous, 2012).

Casual Organism: *Pythium aphanidermatum*

Pathogen:

As oospores, hyphae, and/or sporangia, *Pythium aphanidermatum* overwinters in soil. Oospore produces and penetrates germ tube into the plant. It produces sporangia, producing motile, biflagellate zoospores that can swim, encyst, and germinate to the host plant. If it infects a seedling's roots, the mycelium can expand in the plant system and release digestive enzymes to break down plant cell walls to absorb the nutrients from the pathogen (Anonymous, 2018a).

Scientific Classification: (Kirk et al., 2008)

Kingdom:	Fungi
Phylum:	Oomycota
Class:	Oomycetes
Order:	Peronosporales
Family:	Pythiaceae
Genus:	*Pythium*

Disease Cycle and Epidemiology:

The pathogen is inhabitant of soil and able to live in rotting organic matter. Stem rot caused by *P. aphanidermatum* is common in plants of two years of age. During the monsoon, the disease emerges and the frequency slowly increases with the rainfall rate. Optimum 36 °C temperature, high humidity, heavy rainfall, and water-logged environment are ideal conditions for disease development (Anonymous, 2017).

Management:

- Application of *Trichoderma viride* (15 g/plant) mixed in well-decomposed FYM at the time of planting should be applied around the root zone of the plants.
- A ring irrigation system should be used to irrigate the crop, so that the water does not come into direct contact with the plant. In the case of new plantings, the disease can be managed by preventing soil water logging.
- Copper oxychloride drench 2–3 L (3 g/L of water) at 15 days intervals.
- During fruit forming the plant should be sprayed with the same solution at the same time. Additionally, it can also be used with Mancozeb (2.5 g/L). The rotted part of the plant should be scrapped, and copper oxychloride or Bordeaux paste should be added in established crops in case of disease attack.
- All solutions should be mixed together and shaken to form a paste. Three liters of copper oxychloride (3 g/L) will drill the plant foundation. During fruit formation, the plant should be drenched twice at the above specified concentrations with copper oxychloride or Mancozeb at an interval of 15 days (Anonymous, 2012).

21.3 *PHYTOPHTHORA* BLIGHT/FRUIT ROT/BASAL STEM ROT/ROOT ROT

The disease is also been called soft foot rot, collar rot, basal stem canker, soft fruit rot, and root rot. In Philippines, Sri Lanka, Santo Domingo, India, Indonesia, Malaysia, Hawaii, Mauritius, Mexico, Australia , Brazil, Spain, Taiwan, and even elsewhere, the disease occurs.

Symptoms:

High soil temperatures and water-saturated soils and low drainage are major favored conditions for causing the disease.

Ripe and unripe fruits are infected because they are still attached to the stem and the lower fruit is first infected, the fruits fall prematurely. Infected fruit is hard and shriveling, coated with whitish mycelium and sporangiophores (Doo, 1963).

There may be lesions at the base of the stem above or below the surface of the soil that cause wilting and yellowing of the leaves, which fall prematurely. Rots develop and become covered in a white growth which contains water mold spores. Fungi and bacteria invade the fruits, increasing the rots, and causing the fruit to shrivel and fall.

Casual Organism: *Phytophthora palmivora* (E. J. Butler)

Pathogen:

Chlamydospores globose, terminal or intercalary on mycelium, are 32–42 μm in diameter. Sporangiophores are sympodial. Mostly sporangia ellipsoid to ovoid, prominently papillate, caduceus, with a short pedicel, 40–60 × 25–35 μm average.

Scientific classification: (Kirk et al., 2008)

Kingdom:	Fungi
Phylum:	Oomycota
Class:	Oomycetes
Order:	Peronosporales
Family:	Peronosporaceae
Genus:	Phytophthora
Species:	*Phytophthora palmivora*

Disease Cycle and Epidemiology:

Hyphae, zoosporangia with zoospores, chlamydospores, and oospores are produced by the organism. For the production of oospores, two separate types of mating (called A1 and A2) are required. Oospores do not play a significant role in the disease cycle because there is low chance that both forms of mating spontaneously occur together. After their release from the zoosporangia, zoospores are motile and infective. In soils, chlamydospores can thrive.

Optimum rising temperature is 30 °C. This pathogen's propagules are distributed through wind-blown rain, splashing rain, and rodents.

Management:

- Plant on land that was not traditionally used for papaya cultivation.
- Make sure the seedlings before planting are disease free.
- Three liters of copper oxychloride solution (3 g/L) should be drilled at the base of the field.
- Virgin soil technique: when planting in infested soil, dig a 12″ long × 4″ deep hole; fill with clean soil creating a 1–2″ high mound; plant healthy seedling on the mound; The plant will no longer be susceptible when roots have spread to infested soil (3 months).

21.4   POWDERY MILDEW

Powdery papaya mildew was described originally from Brazil in 1898. Across several tropical and subtropical regions of the world, including Hawaii, Australia, Bermuda, Florida, Taiwan, India, and New Zealand, it has been recognized since. This common disease usually causes little damage to bearing trees or yield loss where lesions are usually found on senescent leaves. However in areas with moderate rainfall and temperatures, powdery mildew can severely damage young plants.

Symptoms:

The disease occurs as small, slightly blurred areas on leaves and seeds, which later become white powdery pustules. These spots spread and cover the entire area of the leaves. Before dropping heavily infected leaves can become chlorotic and distorted. The fruits affected are of small size and malformed (Anonymous, 2017).

Casual Organism: *Oidium caricae*

Pathogen:

The pathogen is an obligate parasite. Mycelium is hyaline, ectophytic, septate, and creeping. Hyaline, elliptical, and barrel-shaped asexual spore is conidia. The ascospores are the reproductive spores which are globose and gregarious. Each ascocarp has one ascus in it (Anonymous, 2017).

Scientific Classification: (Kirk et al., 2008)

Kingdom: Fungi
Phylum: Ascomycota
Class: Leotiomycetes
Order: Erysiphales
Family: Erysiphaceae
Genus: *Oidium*

Disease Cycle and Epidemiology:

Primary source of inoculums—Dormant mycelia

Secondary source of inoculums—Air-borne conidia.

Spread: Air-borne conidia.

The resting spores in the debris are ascospores formed in cleistothecium. The primary infection can be triggered either by ascospores released from cleistothecia or by conidia carried to papaya by wind from collateral hosts.

Wind blows the conidia away and triggers secondary infections. The mold is purely a mandatory parasite (Anonymous, 2017).

High humidity (80%–85%) encourages the production of powdery mildew in papaya with a temperature of 24 °C –26 °C and a sunshine period of 9.1 h (Raabe, 1964).

Management:

- Spray wettable sulfur 0.25% or dinocap 0.05% or chinomethionate 0.1% or tridemorph 0.1% (Anonymous, 2015).

21.5   ANTHRACNOSE

Papaya anthracnose is more extreme in transit and storage which causes destruction of papaya fruits. Papaya anthracnose caused by *Colletotrichum gloeosporioides* (Penz.) Penz. & Sacc (Bolkan et al., 1976).

Symptoms:

Disease begins with the skin's superficial discoloration, which grows into circular, slightly sunken areas and a diameter of 1 to 3 cm. Little by little the lesions join and development occurs on spot margins. Early contamination of the fruit results in mummification and deformation. It develops necrotic spots on the leaves and stems. Latex from the fruit oozes out in sticky mounds of horns (Anonymous, 2018d).

Casual Organism: *Colletotrichum gloeosporioides*

Pathogen:

Rana (2001) found the mycelium is immersed, branched, and consists of a very short, sparsely septate, hyaline hyphae that becomes slightly darkened with age. Acervuli produced on dark stroma and conidia are subhyalin, pink in mass, in shape variable, usually oblong to cylindrical, smooth, with obtuse ends, often slightly curved, containing 1 or 2 oil globules and measuring 10–25 × 3.5–7.0 µm usually 12–16 × 4–6 µm (Raabe, 1964).

Scientific Classification: (Kirk et al., 2008)

Kingdom: Fungi
Phylum: Ascomycota
Class: Sordariomycetes
Order: Gloemerellales
Family: Gloemerellaceae
Genus: *Colletotrichum*

Disease Cycle and Epidemiology:

The disease is spread through wind-borne conidia and splashes of rain. Excessive humidity is very favorable. Old leaves appear to be more susceptible (Anonymous, 2018d).

Management:

- Spraying with carbendazim (0.05%–0.1%), mancozeb (0.2%–0.25%), and benomyl (0.15%) showed complete inhibition of growth of *Colletotrichum dematium*.
- Spray any one of the following Carbendazim 1 g/L at 45 days interval. Chlorothalonil 2 g/L at 10 to 15 days interval and Mancozeb 2 g/L at 10 days interval (Anonymous, 2018d).

21.6 PAPAYA RING SPOT

Hawaiian papaya production has been severely impacted twice by papaya ringspot virus (PRSV). The virus was introduced to Oahu as early as 1937

(Parris, 1938). Over a number of years, the disease was mild before it either mutated, or a more violent form was introduced around 1950. The quantity of land under papaya production fell 94 per cent within 12 years. Production was then transferred under strict quarantine from Oahu to the Puna region of Hawaii Island (the "Big Island"). PRSV was found in home gardens in 1971, however efforts were made to prevent its spread. The virus originated in commercial farms in 1992, and it was impossible to cultivate in Puna by 1995. Commercial growers moved again to the coast of Hamakua but with only limited success. By the end of the decade Hawaiian papaya production had halved. Transgenic, PRSV-resistant papaya varieties entered production in 1998 and resuscitated the industry (Tripathi et al., 2008).

Symptoms:

In papaya, the leaves produce prominent mosaic and chlorosis on the lamina of the root, and on the petioles and upper part of the stem, water soaked oily streaks. Also known to cause systemic necrosis and wilting along with mosaic and chlorosis is an extreme isolate from Taiwan. Older contaminated plants do grow fruit (Anonymous, 2018b).

Casual Organism: *Papaya ring spot virus*

Pathogen:

The general properties of PRSV are similar to the vast majority of viruses in the genus *Potyvirus* of the family *Potyviridae*. Aphids spread the virus nonpersistently to the papaya and cucurbits. The particle virus, or virion, is a nucleocapsid. Virus particles usually contain 94.5% protein and 5.5% nucleic acid by weight, and do not have (non-enveloped) outer membranes (Anonymous, 2018b).

Scientific classification:

Group: Group IV ((+)ssRNA)

Order: *Unassigned*

Family: *Potyviridae*

Genus: *Potyvirus*

Disease Cycle and Epidemiology:

It is distributed by aphid vectors nonpersistently, and does not multiply in the stream. The process of the disease will begin with aphids feeding on infected papaya for as little as 15 seconds, and then feeding on a healthy papaya. There is no time for the incubation (Anonymous, 2018b).

Management:

- Seedlings free from plant disease.
- Grow sorghum/maize as a barrier crop prior to papaya planting;
- Rogging out infected plants on finding symptoms immediately.
- Do not cultivate cucurbitaceous plants (Anonymous, 2015).

21.7   LEAF CURL OF PAPAYA

Papaya has important economic significance for Indian agriculture. Thomas and Krishnaswamy (1939) first identified the disease of papaya leaf curls in 1939. This disease is of moderate occurrence, and is prevalent in India. Recent findings of papaya fields in India suggested that the incidence of papaya leaf curl disease (as shown by symptoms) has continued to increase, leading to significant economic losses.

Symptoms:

Curling, crinkling and leaf bending, reduction of leaf lamina, rolling inward and downward of leaf margins, and thickening of veins. Leaves are brittle, leathery and twisted. Plants stunted. An affected plant produces neither flowers nor fruits. Spread by whitefly *Bemisia tabaci*. Those symptoms also affect all of the plant's top leaves. Defoliation occurs at advanced stages of the disease, and growth of plants is halted. PLCV is a member of the Geminiviridae family. It is not physically transmitted. The virus vector is the silver leaf whitefly, *Bemisia tabaci* (Anonymous, 2018c).

Casual Organism: *Papaya leaf curl virus*

Transmission:

The virus was readily transmitted through grafting and white fly (*Bemisia tabaci*).

Management:

- Plants affected with uproot.
- Avoid growing tomato, tobacco around papaya.
- Systematic insecticidal spraying for protection of the vector.

KEYWORDS

- papaya
- diseases
- symptomatology
- epidemiology
- management

REFERENCES

Anonymous. (2012). http://agropedia.iitk.ac.in/content/papaya-diseases-its-control

Anonymous. (2015). http://agritech.tnau.ac.in/crop_protection/papaya_diseases_2.html

Anonymous. (2017). https://ecoursesonline.icar.gov.in/mod/page/view.php?id=9812

Anonymous. (2018a). https://en.wikipedia.org/wiki/Pythium_aphanidermatum

Anonymous(2018b).https://www.apsnet.org/edcenter/intropp/lessons/viruses/Pages/PapayaRingspotvirus.aspx

Anonymous. (2018c). http://vikaspedia.in/agriculture/crop-production/integrated-pest-managment/ipm-for-fruit-crops/ipm-strategies-for-papaya/ginger-diseases-and-symptoms#section-6

Anonymous. (2018d). http://www.ikisan.com/tn-papaya-disease-management.html

Bolkan, H. A., Cupertino, F. P., Dianese, J. C. and Takatsu, A. (1976). Fungi associated with pre and postharvest fruit rots of papaya and their control in central Brazil. *Plant Dis. Reptr.* 60: 605–609.

Doo, S. Y. A. C. (1963). The Influence Of Temperature And Inoculum Potential On Disease Development. In *Carica papaya* Caused by *Phytophthora parasitica.* MS Thesis, University of Hawaii.

Kirk, P. M., Cannon, P. F., Minter, D. W. and Stalpers, J. A. (2008). *Dictionary of the Fungi.* 10th edn. Wallingford: CABI.

Parris, G. K. (1938). A new disease of papaya in Hawaii. *Proc. Am. Soc. Hortic. Sci.* 36: 263–265.

Patel, K. D. and Joshi, K. R. (2002). Efficacy of different fungicides against *Colletotrichum gloeosporioides* Penz. and Sacc. The causing leaf spot of turmeric. *J. Mycol. Pl. Path.* 32: 413–414.

Raabe, R. D. (1964). Fungus diseases. *Hawaii Agr. Exp. Sta. Bienn. Rept.* 1962–64: 91–92.

Rana, S. K. (2001). Fungal diseases of papaya. Diseases of fruit crops, Oxford and IBH Publishing Co. Pvt. Ltd. New Delhi (India). pp. 81–91.

Thomas, K. M. and Krishnaswamy, C. S. (1939). *Curr. Sci.* 8: 316.

Tripathi, S., Suzuki, J. N. Y., Ferreira, S. A. and Gonsalves, D. (2008). Papaya ring spot virus-P: Characteristics, pathogenicity, sequence variability and control. *Mol. Plant Pathol.* 9(3): 269–280.

Wallace, G. B. and Wallace, M. M. (1948). Diseases of papaw and their control. *East African Agr. J.* 13: 244.

CHAPTER 22

IMPORTANT DISEASES OF WALNUT (*JUGLANS REGIA* L.) AND THEIR MANAGEMENT

SABA BANDAY[1,*], EFATH SHAHNAZ[1], ROMAN NISSAR,[2] and NASREEN FATIMA[3]

[1]*Division of Plant Pathology, (FOA), Sher-E-Kashmir University of Agricultural Sciences and Technology (SKUAST-K) Srinagar, Kashmir, Jammu and Kashmir, India*

[2]*Division of Agronomy, (FOA), Sher-E-Kashmir University of Agricultural Sciences and Technology (SKUAST-K) Srinagar, Kashmir, Jammu and Kashmir, India*

[3]*Krishi Vigyan Kendra, Ladakh, Sher-E-Kashmir University of Agricultural Sciences and Technology (SKUAST-K) Srinagar, Kashmir, Jammu and Kashmir, India*

Corresponding author. E-mail: sababanday@gmail.com

ABSTRACT

Walnut (*Juglans regia* L.) is an important dry fruits species and belongs to the family Juglandaceae. Walnut is mainly grown in China, USA, Iran, and also in India. In India, major walnut growing states are Jammu and Kashmir, Himachal Pradesh, Uttarakhand, and Arunachal Pradesh. In India, Jammu and Kashmir is the principal walnut growing state having a monopoly in the production and export of walnut contributing 98% foreign exchange of worth Rs 231.64 crore. Walnut provides dry fruits. It is also used to make many high-nutritional dishes, bakery products, confectionery, and oils. It is a rich source of protein, omega-3 fatty acids, vitamins, and minerals with excellent flavor. Many biotic and abiotic factors affect the quality and quantity of walnut. Walnut is severely affected by many fungal as well as bacterial diseases, which cause heavy damage to the crop every year. Major diseases of walnut are discussed in this chapter.

22.1   INTRODUCTION

Walnut (*Juglans regia* L.) is an important dry fruits species. This belongs to the family Juglandaceae. Its centre of origin is Iran (Arora, 1985). World walnut production accounts up to 2.14%, and is mainly grown in China, USA, and Iran. As far as India is concerned, it stands seventh in production (Anonymous, 2010) and major walnut growing states are Jammu and Kashmir, Himachal Pradesh, Uttarakhand, and Arunachal Pradesh. Walnut is grown in Jammu & Kashmir in Rajouri, Poonch, Badrawah, Kupwara, Baramulla, Bandipora, Ganderbal, Budgam, Srinagar, Anantnag, and other temperate and hilly areas with an annual production of 20,873 tons (Anonymous, 2012). In India, Jammu and Kashmir is the principal walnut growing state having a monopoly in the production and export of walnut contributing 98% foreign exchange of worth Rs 231.64 crore (Sharma, 2014). Farmers living in Jammu and Kashmir state high-altitude and for-flung areas have very poor economic conditions and are extremely fragile. Thus the cultivation of walnut can play a vital role in elevating their economic condition (Anonymous, 2012a). Walnut provides dry fruits. It is also used to make many high-nutritional dishes, bakery products, confectionery, and oils. Walnut shells are used for cleaning and polishing surfaces in glue and plastics as well as in dusting and solution making (Bal, 2006). Walnut wood is useful in furniture, fuel and veneer industry, dying and pharmaceutical industries (Zamani et al., 2011). It is a rich source of protein, omega-3 fatty acids, vitamins, and minerals with excellent flavor (Rana et al., 2007). This is also known for cardio-protective effects as it increases the lipoprotein cholesterol-to-total cholesterol ratio, reduces inflammation and enhances arterial function (Diousse et al., 2001). The fruit contained melatonin, which helps to control sleep (Reiter et al., 2005). It also decreases cancer incidence, and slows ageing neurodegenerative diseases.   Many biotic and abiotic factors affect the quality and quantity of walnut. Among the major biotic factors, the important fungal diseases of walnut are anthracnose (*Marssonina juglandis* (Lib.) Magnus), root and crown rot [*Phytopthora cactorum* (Lebert and Cohn) Schrot], armillaria root disease (*Armillaria mellea*), downy leaf spot [*Microstroma juglandis* (Berenger) sacc.)], powdery mildew (*Mycosphaerella juglandis*) besides it some another fungal diseases have also been reported in walnut such as stem canker and die-back diseases caused by fungi like *Cytosperma leucosperma* (Pers. ex Fr.) Fr., *Nectria galligena* (Bres.), and *Fusarium solani* (Mart. Sacc), etc. Important viral disease in walnut is Black line caused by cherry leaf roll virus (Anonymous, 2013). The following diseases are discuss under this chapter.

22.2   ANTHRACNOSE

Introduction and Economic Importance:

Anthracnose or black spot/leaf blotch is the most serious fungal disease of black walnut (*J. nigra* L.) and Persian or English walnut (*J. regia* L.) in various parts of the world like North and South America, Europe, Iran, China, and other Asian countries (Salahi and Jamshidi, 2009).

Kaul (1962) reports the occurrence of walnut anthracnose disease from Kashmir Valley in India for the first time. It is endemic to North America (Todhunter and Beineke, 1984), and Bulgaria's most common and dangerous illness (Kalkism, 2012). In Iran, yield losses of 60%–80% decrease walnut quality due to this disease reported by Saremi and Amiri (2010). Walnut anthracnose diseases are productive effects and quality in various ways, as stated by various workers such as reduction in quantitative parameters such as size, mass and actual nut production, failure of metabolic processes in leaves, and change in biochemical indexes. Premature leaf loss leads to poorly filled, poor quality, and darkened kernels (Zamani et al., 2011). Due to anthracnose contamination, walnut yields that ranged from cultivar to cultivar decreased (Pinter et al., 2001; Kalkism, 2012). The disease results in premature defoilation of the leaf; thus reducing tree vigour, quality, and quantity of fruit. Late nut contamination results in premature fall of the fruit (Worste and Beineke, 2001).

Symptoms:

The symptoms occur primarily on leaves, twigs, and fruits but occasionally on shoots during the current year. The disease first appears on the leaves to irregularly circular spots with slightly sunken lesions as a brown to black circular color. The size and shape of the spot area varies from oval to round and is often surrounded by a yellow halo that eventually enlarges and coalesces into large necrotic areas. Similar lesions usually also develop on small leaflets, stems, spurs, shoots, and fruit. However, in late summer, leaf severity is shown, and infected trees have become defoliated. Apart from necrotic spots, the disease also affected fruit and nutmeat as nut from diseased trees showed dark and shriveled meat. It is also observed that all leaf, shoot, and fruit lesions in the middle of the season are filled with solitary, dark acervulli, 0.2 mm in diameter, sometimes in concentric circles, which produce bicellular spindle-shaped conidia. Early infection causes drop in the nut, while late infection causes the nut to shrivel and discolor (Hassan and Ahmad, 2017).

Causal Organism: *Marssonina juglandis* or *Marssonina californica* *(Gnomonia leptostyla)*

Disease Cycle and Epidemiology:

The fungus primarily winters as acervulli on diseased twigs and on fallen diseased leaves which produce conidia to initiate primary infection in early spring. The subsequent generations of conidia from primary infection initiate secondary infection and the disease continues to spread until autumn. High humidity and frequent rainfall with a temperature range of 20–25 °C favors the disease's rapid spread. Wound or injury is not necessary to cause infection. Black walnuts are more vulnerable than the paradoxical trees (Hassan and Ahmad, 2017).

Integrated Disease Management:

- The disease can be managed through the integration of both cultural as well as chemical practices.
- Prune the diseased/cankered twigs and collect and destroy the fallen leaves/debris to reduce the primary inoculums.
- Avoid foliage wetting while irrigating.
- Trees should be spaced to allow good air circulation.
- Application of Nitrogen in spring can delay leaf maturity and reduce lesion development and the number of acervulli.
- Spring fungicide sprays effectively regulate the illness. 3–4 fungicidal sprays starting from the initiation stage of the leaf are effective at 10–12 days intervals.
- The fungicides include Mancozeb 75 WP @ 300 g/100 L of water or Dodine 65 WP@ 60 g/100 L or Thiophanate metyl @ 50 g/100 L or Carbenadazim 50 WP/100 L or Carbendazim + Mancozeb 75 WP @ 250/100 L water (Hassan and Ahmad, 2017).

22.3 POWDERY MILDEW

Introduction and Economic Importance:

Powdery mildew is a minor walnut disease. Powdery mildew disease may be caused by poor growth of the tree, dead buds, reduced area of the leaf, reduced growth of the shoot, and reduced fruit quality. Continuing heavy infection over several seasons may reduce yield.

It may cause enough loss by Powdery mildew but it is rarely a tree killer. Moist conditions are prerequisite for its occurrence. Hence, it is commonly seen in the wetter spring and fall seasons. During dryer weather, the fungus usually retreats and hence the incidence of powdery mildew decreases.

Symptoms:

Powdery mildew fungi attack leaves, fruit, and green shoots. The spots are initially small (3–6 mm) and dusty white but eventually cover the entire surface of infected leaves and fruit. Small, round, black fruiting bodies (cleistothecia) are visible in the white growth of colonies (Braun, 1995).

Causal Organism: *Mycosphaerella juglandis* on English walnut

Disease Cycle and Epidemiology:

Conidia are disseminated quickly by the wind. Powdery mildew fungi may be in the buds of the previous season as cleistothecia or mycelium during the winter. Even through the dispersal of cleistothecia appears during season of high rainfall and humidity in spring primary infection. The disease does little harm to walnut (Gottwald et al, 1984).

Integrated Disease Management:

- Fertilization promotes fungal infection dissemination (Anonymous, 2002) and therefore it is better to avoid fertilization of trees and shrubs if they suffer from powdery mildew.
- Primary infections can be managed by eliminating the primary sources of inoculum (i.e., leaves, flower buds, and shoot buds infected last year).
- Applications of foliar fungicides can control secondary infections and fruit infections.
- Effective control can be achieved by spraying with 7–10 day interval with wettable sulfur 0.2% or Dinocap 0.25% or Carbendazim 0.1%

or or Karathane or Calixin 0.1% at disease development stage with a prophylactic treatment.

22.4   DOWNY LEAF SPOT

Introduction and Economic Importance:

Downy leaf spot has been reported from India, New Zealand, Spain, United States, and Bulgaria. It is of minor importance usually in commercial walnut production but it may be widespread during wet springs. All species of *Juglans* are susceptible (Garcia, et al., 1995)

Symptoms:

On leaves, lesions are a soft to bright yellow on the upper surface and more or less angular, following patterns outlined by the leaf veins and vein lets. The corresponding opposite areas on the undersurface is chalky white where the fungus sporulates. The lesion becomes necrotic as they age, and many coalesce, causing most of the leaf to turn brown and dry. Some leaves fall, but the trees are not defoliated. Occasionally, severe infection causes very long leaves to distort.

On fruit, the early symptom is a small, roughly circular, downy area on the hull, although it is rarely seen. The lesion expands with the growing fruit, to about 3 cm in diameter. The edge of the lesion is covered at first with the soft, downy growth of the fungus and later turns brown to black, creating wide circle around a lighter green center. The margin is slightly sunken, and center of the lesion somewhat raised and rough. The infected side of the fruit may be flattened. Nut size and kernel weight are reduced in infected fruit, but the kernel quality and percent kernel are not affected (Ogawa and English, 1991).

Causal Organism: *Microstroma juglandis* (Berenger) sacc.

Disease Cycle and Epidemiology:

Little is known about the life history of *M. juglandis* or the development of downy mildew spot. The disease is more common in wet areas and in orchards with poor air circulation (Anonymous, 2002)

Integrated Disease Management:

- Downy leaf spot control is very rarely required. Orchards that are regularly treated with copper materials to control walnut blight usually have very little disease incidence (Anonymous, 2002).

22.5 ROOT AND CROWN ROT OF WALNUT

Introduction and Economic Importance:

Root and crown rot of fruit and nut trees in orchards around the world are causing major losses. It is brought about by *Phytophthora* spp. The issue is more serious in areas of clay to clay-loam soils, as these soils have higher capacity to retain water and lower drainage compared to sandier soils (Browne et al,. 2010)

Symptoms:

The *Phytophthora* spp effect on root systems, symptoms are often not noticed or assumed until the parts of the tree above show symptoms. Infected trees show poor tree vigor, loss of scaffolding branches, and gumming or bleeding found around the tree trunk and yellowing and leaves wilting. Crown rot is considered to cause tree loss quickly, in comparison to tree loss in the case of root rot particularly during the spring. In the case of root rot, several trees are in a decreasing state due to a long period of infection over the years by the time symptoms are recognized. Careful and timely observations in the orchard can help to detect the disease and take the remediation before the manifestation of severe orchard losses.

Crown rot symptoms generally include cancers centered at the crown or lower part of the trunk. The affected areas reveal brown or black sapwood that may be slimy when the bark of such trees is scrapped or removed. Cankers are clearly delineated, with very little discoloration found outside of the necrotic areas in the wood. Crown rot damages and kills the tree by damage to the phloem which prevents nutrients from conducted in the tree. In case of *Phytophthora* root rot, cankers found above the soil line are rare. In this case, the disease affects roots of all sizes severely, eliminating the ability of the trees to coduct or pull water and nutrients from the soil. Trees with root rot may have dark, mushy roots that can be seen during uprooting or digging. Rotten roots and soil could smell like a "rotten egg." However, this odor can be attributed to the disease itself, but from the anaerobic conditions caused by excessive moisture in the soil. Tree death occurs only when a significant portion of the root system gets infected (Browne et al., 2010). Wet and cold climates are ideal for the development of diseases because these are pathogenic environments. Wet and saturated soils have higher chances of disease rates than those below the point of saturation. Regardless of the season, cycles of rainfall or water over 24 h will provide enough moisture for infection with *Phytophthora*. Cool temperatures favor *Phytophthora's* development, thereby increasing pathogen reproduction (Anonymous, 2009).

Causal Organism: *Phytopthora cactorum* (Lebert and Cohn) Schrot

Disease Cycle and Epidemiology:

Phytophthora spp. can be disseminated in several ways, including soil movement on nursery stocks, irrigation water, and infected root pieces. Irrigated citrus often suffers from the biggest problems as runoff water can carry the pathogen into canals, streams or rivers. Water from these sources may then contaminate previously uninfected areas. The fungus may be carried in soil on farm equipment (Graham and Menge, 1999).

Integrated Disease Management:

- Prevention is the first step in treating disease.
- Proper sanitation and removal of debris should be taken
- Proper selection of locations for orchard soil that do not hold water
- Plant the trees high on raised burns, soil line above the graft union as the scions tend to be highly susceptible to *Phytophthora*.
- They strongly support and promote the use of resistant rootstocks.
- Irrigation practices in areas at risk for *Phytophthora* and/or limited to more regular, shorter watering periods should be changed and dependent on need.
- Managing water properly. It should be avoided overwatering, especially during periods of low tree water use.
- Paradox rootstock is significantly more resistant to *Phytophthora* by the use of genetic resistant or tolerant rootstocks. To treat *Phytophthora* efficiently it is best to apply as many, if not all, orchard management methods (Browne, 2011).

22.6   *ARMILLARIA* ROOT DISEASE

Introduction and Economic Importance:

Also known as oak root fungus, mushroom root rot is a chronic problem of many woody plants, eventually leading to death. Walnut sustains more damage when attacked. *Armillaria* species are native to forests worldwide, and the disease is common in orchards planted on recently cleared hardwood forests (Baumgartner et al., 2011)

Symptoms:

Trees with *Armillaria* root rot have thin canopies with yellow leaves and twigs, shoot, or limb dieback. Crown and foliar symptoms of *Armillaria* root disease are typical of most root diseases; poor shoot growth, stunted leaves, premature yellowing, defoliation, and branch dieback. Excavation of the root crown and major lateral roots is necessary to confirm the presence of the causal fungus. A white fungal layer (mycelial fan or plaque) is usually present on lateral roots or on the root crowns of infected trees between the bark and the wood. The root surface sometimes features dark brown to black rhizomorphs. Mycelial fans of *A. mellea* are continuous; those of *A. tabescens* often protrude from cracks in infected bark (Baumgartner and Rizzo, 2002)

Armillaria spp. causes a white rot, decaying wood by removing cell wall components. In advanced stages of decay, the wood is light colored and spongy or stingy. Honey colored mushrooms may form at the base of infected trees in autumn (Rizzo et al., 1998)

Causal Organism: *Armillaria mellea* (vahl:Fr.) P. Kumm.,

Once considered as single species, is a complex of at least nine species (Baumgartner and Rizzo, 2002).

The fungus, *Armillaria mellea*, occurs sporadically on root zone of many trees. The most distinctive sign of *Armillaria* infection is the honey-colored mushroom which grows out of plant roots and base. The fungus is particularly prevalent on oak but also affects many different kinds of fruit and nut trees, ornamental plants, and herbaceous plants.

Disease Cycle and Epidemiology:

Armillaria spp. survives for up to 100 years in woody debris in the soil. Primary spread into orchards happens when the nut trees' roots come into contact with the roots of a previous crop or native plants that have been affected. Secondary spread between infected and healthy plants occurs through root-to-root or rhizomorph to root contact, causing infection centers or foci to form, which may be roughly circular. Rhizomorph can grow several meters away from its food source, through the soil. Basidiospores are not an important part of the life cycles of these fungi under most circumstances (Baumgartner and Rizzo, 2002).

Armillaria root disease occurs in many soil types but is most common in light, well drained sandy soils. Trees that are stressed, especially by drought, are more susceptible. But healthy trees may also be attacked where inoculums pressure is high (Rizzo et al., 1998).

Integrated Disease Management:

- There is no truly effective control for *Armillaria* root disease once it has been established in an orchard.
- Dead as well as dying trees should be removed. Before an orchard is replanted, all woody roots 2.5 cm or more in diameter should be excavated, and ground fallowed for at least 1 year.
- Soil fumigation with methyl bromide when the soil is dry is somewhat effective, but the success of the treatment depends greatly on the soil type and moisture. Fumigation of the commercial plantings is useful in certain situations but rarely eradicates the pathogen, because *Armillaria* spp. can survive in large root pieces and in roots away from the fumigation zone (Anonymous, 2002)
- The most effective management of the *Armillaria* root disease is planting resistant rootstocks. The relative resistance of the root stocks with walnut is uncertain. Northern California Black and Paradox hybrid rootstocks are typically superior to English walnut, but there has been considerable variability in the field, and susceptibility can be highly dependent on site conditions. (Baumgartner et al, 2011)
- Cultural practices have received little attention in the management of *Armillaria* root diseases in nut crops. Water management may be an important factor, but there is little information on reducing disease by increasing plant vigor (Baumgartner et al, 2013).

22.7   BLACK LINE OF WALNUT

Introduction and Economic Importance:

Black line is a grafted walnut disease and was first observed on Persian walnuts grafted on Black walnut and Paradox root stock in California. Blackline disease is considered a serious threat limiting English walnut (*J. regia*) production in Italy and worldwide if walnut species other than *J. regia*, for example, "Paradox" hybrid (*J. regia* × *J. hindsii*), French hybrid (*J. regia* × *J. major* or *J. regia* × *J. nigra*) or northern California black walnut (*J. hindsii*), are used as the rootstock (Ferretti et al., 2017)

Symptoms:

Symptoms of blackline in the canopy area mimic any root rot disease, nutrient deficiencies, and incompatibility with scion-rootstock. Initial symptoms are poor terminal growth of shoots; premature leaf yellowing, drooping, and leaf fall. The disease is characterized by the presence of a strip of darkened cambium and phloem tissues at the graft union, below the bark. A narrow corky nonconductive layer extends between graft unions and cuts off the transport of water and minerals. Initially only a small segment of the graft union can be involved, but in subsequent years blackline slowly girdles the tree resulting in decline, often followed by profuse rootstock suckering. Gradually, necrosis encircles the trunk, girdling the tree altogether. The scion is put to death in a couple of years. When the tree blossoms, trees can become infected at any age, but blackline is more common in trees aged 15–40 years. Both walnut cultivars are resistant to the virus, but the rate of viral spread varies according to cultivar (Srecko et al., 1980).

Causal Organism: Cherry leaf roll virus (CLRV)

Transmission:

Blackline is a viral illness caused by a cherry leaf roll virus (CLRV) strain. The virus moves from the scion to the grafting union where it causes change and phloem tissue necrosis. The virus does not survive due to hypersensitive reaction and the rootstock remains free. The virus is transmitted through seed and pollens besides grafting (Srecko et al., 1980, Mircetich and Rowhani, 1985).

Integrated Disease Management:

- Virus free plants or scion wood resistant plants should be selected for grafting. *J. regia* rootstock would usually be chosen for Persian walnut.
- Monitoring of orchards during late summer or fall for blackline symptoms. Observations on rootstocks of dying trees with sucker shoots. To confirm the appearance of the disease, examine the grafted union by making small cuts and removing the bark for the characteristic blackline or canker. Cut around the tree's circumference at around 4-inch (10-cm) intervals, so the blackline might not be continuous. The virus-free content should be checked with ELISA (Belisario et al., 2006).

KEYWORDS

- **walnut**
- **diseases**
- **symptomatology**
- **epidemiology**
- **management**

REFERENCES

Anonymous, (2009). http://thealmonddoctor.com/2009/09/28/phytophthora-root-and -crown-rot-of-walnut-and-almond.

Anonymous, (2012). Foreign exchange earned on important fruit crops of Jammu and Kashmir. Directorate of Horticulture, Marketing and Produce, Govt of Jammu and Kashmir, pp. 1–2.

Anonymous, (2010). Food and Agriculture Organisation Statistics. Food and Agriculture Organisation of United Nations, Rome, Italy. <http://faostat.fao.org/default. aspx?PageID=567#ancor>.

Anonymous. (2012a). Area, production and yield of important fruits of Jammu and Kashmir state. Statistical Section, Department of Horticulture, Government of Jammu and Kashmir, Srinagar, pp. 1–2.

Anonymous. (2013). *Marssonina juglandis*. Fungal Databases Nomenclature and Species Banks. MycoBank. www.mycobank.org/BioloMICS.aspx Link.

Anoymous, (2002). In: Compendium of Nut Crop Diseases in Temperate Zones. (Eds: Beth L. Teviotdale, Themis J. Michailides, and Jay W. Pscheidt). The American Phytopathological Society.

Arora, R. K. (1985). Genetic resource of less known cultivated food plants. *NBPGR Science. Monograph*, New Delhi, p. 1.

Bal, J. S. (2006). Fruit Growing. Kalyani Publications, New Delhi, India, p. 377.

Baumgartner, K. and Rizzo, D. M. (2002). Spread of Armillaria root disease in a California vineyard. *Am. J. Enol. Viticult. 53:197–203*.

Baumgartner, K., Coetzee, M. P. A., Hoffmeister, D. (2011). Secrets of the subterranean pathosystem of Armillaria. *Mol. Plant Pathol. 12:515–534*.

Baumgartner, K., Phillip Fujiyoshiand and Greg T. Browne. (2013). Evaluating paradox walnut rootstocks for resistance to *Armillaria* root disease. *Hort. Sci. 48(1):68–72*.

Baumgartner, K. and Rizzo, D. M. (2002). Spread of *Armillaria* root disease in a California vineyard. *Am. J. Enol. Viticult. 53:197–203*.

Belisario A., Maccaroni M., Vettraino A. M., Valier A., Vannini A. (2006). *Phytophthora* species associated with decline and death of English walnut in Italy and France. *Acta Hortic.* 705: 401–407.

Browne, G. T., Prichard, T. L., Schmidt, L. S., and Krueger, W. H. (2011). Evaluation of phosphonate treatments for control of Phytophthora crown rot of walnut. *Online. Plant Health Progress doi: 10.1094/PHP-2011–0601-01-RS*.

Browne, G. T., Schmidt, L. S., Bhat, R., Leslie, C. A., Hackett, W., Beede, B., Hasey, J. (2010). Etiology and management of crown and root rots of walnut, *Walnut Research Reports*. California Walnut Board. pp. *225–236*.

Diousse, L., Pankow, J. S., Eckfeldt, J. H., Folsom, A. R., Hopkins, P. N., Province, M. A., Hong, Y. and Ellision, R. C. (2001). Relation between dietary linoleic acid and coronary artery disease in the National Heart, Lung and Blood Institute Family Heart study. *Am. J. Nutrit.* 74: 612–619

Ferretti, L., Corsi1, B., Luongo1, L., Dal Cortivo C., and Belisario1, A. (2017). A survey of *cherry leaf roll virus* in intensively managed grafted English (Persian) walnut trees in Italy. *J. Plant Pathol.* 99(2): 423–427

Garcia-jimenez, J., Armengol, J., and Martinez-Ferrer, G. (1995). First report of downy spot of walnuts caused by *Microstroma juglandis* in Spain. *Plant Dis.* 79: 860.

Graham, J. H. and Menge, J. A. (1999). Root diseases. In: "Citrus health management" (Eds. Timmer, L.W. and Duncan, L.W.) American Phytopathological Society, St. Paul, MN, USA, pp. 126–135.

Gottwalld., T. R., Wood., B. W and Bertrand., P. F. (1984). Effect of powdery mildew on net photosynthesis, dark respiration and kernel compostion of pecan. *Plant Dis.* 68: 519–521

Hassan, M and Ahmad, K. (2017). Anthracnose disease of walnut—a review. *Int. J. Environ., Agric. Biotechnol.* 2(5), Sep–Oct- 2017 ISSN: 2456-1878.

Kalkisim, O. (2012). *In vitro* antifungal evaluation of various plant extracts against walnut anthracnose (*Gnomonia leptostyla* (Fr.) Ces et de Not.) *J. Food Agric. Environ.* 10: 309–313.

Kaul, T. N. (1962). Occurrence of *Gnomonia leptostyla* (Fr.) de not on walnut in India. *Current Sci.* 31: 349

Mircetich S. M., Rowhani A., (1984). The relationship of cherry leaf roll virus and blackline disease of English walnut trees. *Phytopathology* 74: 423–428.

Ogawa, J. M., and English, H. (1991). Diseases of temperate zone tree fruit and nut crops. Publ. 3345. University of California, Division of Agricultural and Natural Resources, Oakland.

Pinter, C., Fischl, G., Kadlicsko, S., Danko, J., Gara, M. and Mako, S. (2001). Walnut pathogens in Hungary. *Acta Phytopathol. Entomol. Hungarica* 36: 269–273.

Rana, J. C., Singh, D., Yadav, S. K., Verma, M. K., Kumar, K. and Predheep, K. (2007). Genetic diversity collected and observed in Persian walnut (*Juglas regia* L.) in the western Himalayan region of India. *Plant Genet. Res. News Letter* 51: 68–73.

Reiter, R. J., Manchester, L. C. and Tan, D. X. (2005). Melatonin in walnuts: influences on levels of Melatonin and total antioxidant capacity of blood. *Int. J. Appl. Basic Nutrit. Sci.* 21: 920–924.

Rizzo, D. M., Whiting, E. C., Elkins, R. B. (1998). Spatial distribution of *Armillaria mellea* in pear orchards. *Plant Dis.* 82: 1226–1231.

Rowhani, A., Mircetich, S. M., Shepherd, R. J., Cucuzza, J. D., (1985) Serological detection of Cherry leafroll virus in English walnut trees. *Phytopathology* 75: 48–52.

Salahi, S. and Jamshidi, S. (2009). Reaction of different walnut cultivars to *Gnomonia leptostyla,* causal agent of walnut Anthracnose. *J. New Agric. Sci .*5: 55–61.

Saremi, H. and Amiri, M. E. (2010). Evaluation of resistance to Anthracnose (*Marssonina juglandis*) among diverse Iranian clones of walnut (*Juglans regia* L.). *J. Food, Agric. Environ.* 2: 375–378.

Sharma, M. R., Kour, K., Singh, B., Yadev, S., Kotwal, N., Rana, J. C. and Anand, R. (2014). Selection and characterization of elite walnut (*Juglans regia* L.) clone from seedling origin trees in North Western Himalayan region of India. *African J. Crop Sci.* 8: 257–262.

Srecko M., Mircetich S. M., Sanborn R. R., Ramos D. E., (1980). Natural spread, graft transmission, a possible etiology of walnut blackline disease. *Phytopathology* 70: 962–968.

Srivastava, J. N., Sharma P. K., Mishra A. and Shahid Ahmad (2009). Diseases of walnut and their management. In: *Plant Disease Management for Sustainable Agriculture.* (Ed.) Shahid Ahmad. Daya Publishing House Delhi, pp. 110–120. ISBN: 978-93-830-4859-5

Todhunter, M. N. and Beineke, W. F. (1984). Effect of anthracnose on growth of grafted black walnut. *Plant Dis.* 68: 203–204

Woeste, K. E. and Beineke, W. F. 2001. An efficient method for evalvating black walnut for resistance to walnut anthracnose in field plots and the identification of resistant genotypes. *Plant Breeding* 120: 454–456.

Zamani, A. R., Imami, A., Mirza, M. A and Mohammadi, R. (2011). A study and comparison of control methods of anthracnose disease in walnut trees of Roodbar region. *Int. J. Nuts Rel. Sci.* 2: 75–81.

CHAPTER 23

WATERMELON [*CITRULLUS LANATUS* (THUNB) MANSF.]: KEY DISEASES AND THEIR MANAGEMENT

S. K. MAHESHWARI*, RAMYASHREE DEVI, B. R. CHOUDHARY, P. P. SINGH, and B. R. KHATRI

ICAR-Central Institute for Arid Horticulture, Bikaner, Rajasthan. India

Corresponding author. E-mail: maheshwariskciah@gmail.com

ABSTRACT

Watermelon or *Tarbooja* [*Citrullus lanatus* (Thunb) Mansf.] is a member of Cucurbitaceae family. Watermelon is a popular desert crop across all of the world's tropics and Mediterranean regions. Actually, it is not only a fruit of the summer season but also is a daily fruit like apples and bananas. Watermelon is mainly grown in China, Turkey, USA, Iran, and also in India. Watermelon fruit is a rich source of B-carotene, vitamin B, C, and E, minerals (K, Mg, Ca, and Fe), amino acid citrulline, and phenolics. Among the various abiotic and biotic factors, diseases are major constrains for production as well yield of Watermelon. Major diseases watermelon are discussed in this chapter.

23.1 INTRODUCTION

Because of their short duration, high yield, nutritional richness, economic viability, and the ability to generate on-farm and off-farm employment, vegetables are important constituents of Indian agriculture and nutritional security. Our country is blessed with diverse agro-climates with distinct seasons, enabling a large array of vegetables to grow. India is the world's second-biggest vegetable manufacturer. Development of enhanced vegetable varieties/hybrids/technology through systematic research coupled with adoption by farmers and government development policies culminated in tremendous growth in vegetable area (10.1 million ha) and production (169.06 million tonnes) in India during 2015–2016. Throughout, our daily diet vegetables are essential sources of protein, vitamins, minerals, dietary fibers, and micronutrients. Besides protein, they do have a possible phytochemicals and antioxidants. In India, vegetables constitute valuable biological assets, particularly genetic resources. In plants, fungi, bacteria, viruses and phytoplasms cause the diseases that reduce crop yields (Khoury and Makkouk, 2010). Increasing per capita income, health awareness, urbanization, growing working women, shifting farmers to high-value vegetables as a result of higher income, and annual growth rate for domestic demand are also important ingredients in fueling the country's vegetable development.

Watermelon [*Citrullus lanatus* (Thunb) Mansf.] is a popular desert crop across all of the world's tropics and Mediterranean regions. Actually, it is not only a fruit of the summer season but also is a daily fruit like apples and bananas. China is the leading country in watermelon production led by Turkey, the USA, and Iran (Huh et al., 2008). In India the production of watermelon during 2015–2016 is 2,325,000 metric tons with an area of 95,000 ha. The fruit is a rich source of B-carotene, vitamin B, C, and E, minerals (K, Mg, Ca, and Fe), amino acid citrulline, and phenolics. Watermelon is also known as *Tarbooj* in some parts of India and *Mateera,* particularly in Rajasthan. *Mateera* is a dry, hardy crop of landraces (Anon., 2003). It is an indigenous form of watermelon and is extensively cultivated in Thar Desert with mixed cropping on sand-dunes landscape. Watermelon/*mateera* fruits are due to edible flesh (pulp) that is sweet and refreshing and consumes fresh and has juicy properties. This crop constitutes 6.8% of the world's vegetable-producing area (Goreta et al., 2005). *Mateera* in Rajasthan is an important landrace of *Citrullus lanatus* cultivated by farmers using their own saved seeds. It is capable of withstanding damage from biotic and abiotic factors and has ample variability in productivity and efficiency.

Watermelon/*mateera* is subjected to attacks of fungal, bacterial, and viral diseases that each year inflict heavy damage on the crop. Among the fungal diseases are essential diseases in Rajasthan as well as major watermelon growing states of India, such as *Alternaria* leaf blight, anthracnose, downy mildew, and viral diseases such as mosaic disease and bud necrosis disease. Attempts have been made for the management of watermelon diseases in arid and semiarid regions.

23.2 FUNGAL DISEASES

23.2.1 *ALTERNARIA LEAF BLIGHT*

Introduction/Economic Importance:

Among many fungal diseases, *Alternaria* leaf blight incited by *Alternaria cucumerina* is a significant disease in Rajasthan as well as India's major growing watermelon states. It was observed during August–September in the field and in river beds of various parts of Rajasthan, as well as Uttar Pradesh. *Alternaria* leaf blight disease is a serious bottleneck in watermelon production under Rajasthan's hot arid conditions, as well as in semi-arid regions. In India the disease was first reported on watermelon from Jobner (Rajasthan) by Khandelwal and Prasada (1970), where it occurs almost every year. Occurrence of this disease in watermelon due to *A. cucumerina* was also reported from China by Wie et al. (1991). This disease causes significant losses in the development of watermelon in the Rajasthan region (Prasada et al., 1973). Depending on the form of cucurbitous crop and its vulnerability, yield losses due to leaf necrosis and foliage loss differ, reaching 88% on watermelons in India (Bhargava and Singh, 1985). The frequency of this disease ranges from 5.75% to 21.50% in the district of Bikaner depending on the weather conditions prevailing.

Symptoms:

Infection is first found near crown of plants in mature leaves. Characteristic symptoms first appear on leaves as small, circular, and light brown to reddish brown spots of different sizes that later grow to form large necrotic areas in a concentrate of larger lesions. Leaf drop is going to be a serious one. *Alternaria* leaf blight is not commonly infecting fruit but yield and fruit quality will be reduced due to reduced plant vigor and sunscald due to sun exposure. **Casual Organism:** *Alternaria cucumerina* (Ell. & Ev.) Elliot

Taxonomic Position:

Division—Eumycota
   Subdivision—Deuteromycotina
      Class—Hyphomycetes
         Subclass—Sporomycetidae
           Order—Moniliales
              Family—Dematiaceae
                 Genus—*Alternaria*
                    Species—*cucumerina*

FIGURE 23.1 *Alternaria* leaf blight in watermelon.

Disease Cycle:

The fungus perpetuates as mycelium or as spores in infected plant debris. Perennial weeds or host collateral lead to primary inocula. The initial infection begins from the lower leaves, which develop numerous conidia. These are transmitted by wind, water, rain splash, or insects and cause the disease to spread secondarily. The conidia germinate at temperature approximately 25 °C–30 °C in the presence of moisture, giving rise to germ tubes entering the host tissue.

Epidemiology:

It is considerably influenced by humid weather. The disease ranged in trace to moderate form depending on the environmental factors during rainy season. Availability of high humidity (80%–95%) coupled with a temperature of about 25 °C–28 °C with bright sunlight and dew or rain was conducive for fungal spore germination at which maximum disease development is obtained.

Outbreak of this severe disease can be forecasted based existing temperature (25 °C–28 °C) associated with high relative humidity (above 90%) coupled with good amount of rainfall.

Integrated Disease Management:

Efficient disease management is accomplished solely through integrated strategies. The application of cultural and chemical control methods to integrated disease management approaches can be effective against this disease.

Cultural Control:

- The infected crop debris and weeds should be collected and burnt.
- Use of disease free and certified seeds is mandatory.
- Deep ploughing in summers is effective in destroying the inoculum.
- Soil solarization during May–June in Rajasthan is very effective.

Chemical Control:

- Sprayings of indofil M-45 (0.2%) at regular intervals is very effective against this disease.
- Pushpa et al. (1999) reported the benefits of bavistin to manage the seed-borne infection of *Alternaria* species in pumpkin, cucumber, muskmelon, and watermelon.
- Among the fungicides, Topsin is effective against the leaf spot disease caused by *A. cucumerina* in watermelon (Bharath, et al., 2005).
- Among all the treatments, foliar spray of mancozeb (0.25%) proved the most effective fungicide for the management of this disease in *watermelon/mateera* under field conditions with the minimum disease incidence (10.0%) and disease severity (8.88%) and the maximum disease control of 69.39% (Maheshwari et al., 2015).

Biological Control:

- Application of bio-agent (*Triochoderma viride* and *T. harzianum*) @ 5% was also effective against this disease.

23.2.2 CERCOSPORA LEAF SPOT

Introduction/Economic Importance:

It is also a serious problem in watermelon growing areas and causes yield losses in semiarid areas of Rajasthan and other states of India.

Symptoms:

This leaf spot is usually restricted to the leaves, but can impact the stems in an area that is favorable. Dark spots are usually first seen on older leaves and are circular to irregularly shaped, with darker margins at lighter colored centers. Tissue surrounding the spots is often yellows. Spot centers can fall out, leaving holes in the leaves. As lesions grow they also unite and blight whole leaves. If the disease is severe, lower quality fruit may turn smaller.

Casual Organism: *Cercospora citrullina*

Disease Cycle:

The fungus in the cucurbit family overwinters in crop debris and on weeds. The spores may be borne in splashing water or wind blown.

Epidemiology:

For infection, which is favored by temperatures of 79–90°F, free water on the leaf surfaces is required. At these temperatures, the disease progresses rapidly, and new leaf infections will occur every 7–10 days.

Integrated Disease Management:

Cultural Control:

- Stop overhead irrigation in the early hours of the morning when leaves are still damp with dew.
- Cut and kill old cucurbit vines and spores, as the fungus lives in winter here. Keep the garden well weeded to avoid humidity building up in the canopy.
- Rotate 2–3 years away from the cucurbits (melons, squash, and cucumbers).

Chemical Control:

- Mancozeb (2 g/L) is effective against this disease.
- Chlorothalonil (kavach) @ 2 g/L can be used as a protective application at 7 days interval.
- 2–3 application of difenoconazole (score) @ 0.5 mL/L per season is also effective.

23.2.3   DOWNY MILDEW DISEASE

Introduction/Economic Importance:

Downy cucurbit mildew was first recorded from Cuba in 1868 and this disease is considered as one of the bad problems. It occurs in tropical and subtropical areas of India. It can cause reduction in crop yield due to early infection of downy mildew up to 60% (Gupta et al., 2001).

Symptoms:

Symptoms first appear on the infected topside of older leaves as small green to yellowish spots, or water-soaked lesions. The lesion center eventually turns brunette or brown, and dies. The disease can migrate onto the upper crop canopy during prolonged wet periods. The lesion often develops a downy growth at the underside of the light yellow lesions observed at the top of the leaf under humid conditions. On the underside of the lesions, the downy growth is often dotted with dark purple to black sporangia (spore sacks) that can be detected with a hand lens. A key to diagnosing this disease is the appearance of the downy growth on the underside of the lesion.

Casual Organism: *Pseudoperonospora cubense* (Berk & Curt) Rostow.

Taxonomic position:

Class—Oomycetes
    Order—Peronosporales
        Family—Peronosporaceae
            Genus—*Pseudoperonospora*

Disease Cycle and Epidemiology:

Downy growth is particularly noticeable after a period of wet weather in the evenings, or when conditions promote dew formation. Cold, damp, and humid conditions prefer downy mildew. Over a wide range of temperatures (5 °C–30 °C), the pathogen produces microscopic saclike structures called sporangia. The production of optimum sporangia occurs between 15 °C–20 °C and requires a high humidity of at least 6 h. Sporangia act analogous to spores. They are quickly transferred by the air currents or splashing rain to healthy plant tissue. They germinate after they land on a vulnerable host, and may infect the leaf directly within an hour. The sporangia can also burst open during extended cool wet periods, and release several zoospores. The zoospores swim toward the stomata through the film of water along the surfaces of the leaves. These natural pores are the primary entry point for the pathogen and result in multiple leaf infections.

Integrated Disease Management:

Cultural Control:

- Control the crop to encourage air circulation and to reduce the amount of humidity within the canopy.
- Stop unnecessary irrigation at the overhead. Consider irrigating late in the morning to facilitate fast drying of the leaves.
- Ensure good on-field weed management.
- Ensure that field workers wash their hands before moving from one field to another, and wear freshly laundered clothes every day, if possible. Function at the end of the day in diseased fields where possible.

Chemical Control:

- Mancozeb (2 g/L) was effective against this disease.
- Chlorothalonil (kavach) @ 2 g/L can be used as a protective application at 7 days interval.
- Two sprays of mancozeb (0.1%) and metalaxyl (0.05%) mixture starting with the onset of conducive weather at 10 days interval give good control (Gupta et al., 2001).

23.2.4 FUSARIUM WILT

Symptoms:

Infection with the fungus can occur at any plant age. Damping can occur in young seedlings, leading to rotting in the soil, where the hypocotyls are watery, resulting in soft rot causing the plants to stun. Wilting occurs later in more mature plants and causes the plant to die. A side of the plant will wilt and another side will stay safe. Flaccid, withered, and brown leaves and vascular discoloration are typical symptoms of the disease. The roots of infected plants may be healthy, but the vascular tissue is brown and discolored. .

Casual Organism: *Fusarium oxysporum f. sp. niveum.*
It produces microconidia, macroconidia, and chlamydospores.

Taxonomic Position:

Class—Hyphomycetes
   Order—Tuberculareiales
      Family—Tuberculariaceae

Disease Cycle:

The disease does not spread from plant to plant above ground, but rather from the fungal spores in the soil. The fungus may be introduced in the field by infected crops, by soil, by farming equipment and vehicles, and with drainage water. When the plants are dead, the fungus remains on dead vines or in the soil in the form of white mycelia, macroconidia, and chlamydospores until a new susceptible host is available.

Epidemiology:

Symptoms of the outbreak of the disease are predicted to be most extreme at temperatures 25 °C–27 °C. In fields with light sandy soils with pH 5.5–6.5 and less than 25% soil moisture and high nitrogen content, this disease can be very serious.

Integrated Disease Management:

Cultural Control:

- This disease is very difficult to manage once the soil has been contaminated. Crop rotation in soil for 4–10 years may decrease fungal density and disease incidence with it. For crop rotation, noncucurbit vegetation should be utilized. Avoid seeding watermelon in the same region for at least 5–7 years.
- Use of disease-free transplants and seeds for greenhouse transplantation, the soil used must be pathogen-free or steam-fumigated or disinfected.

Bio Control:

- This disease can be suppressed by the highly fluorescent siderophore producing pseudomonands and involved in the mechanism of competition (Gupta et al., 2001).

Chemical Control:

- Soil drenching of carbendazim (0.1%) is found to be the most effective against this disease.

23.2.5   ANTHRACNOSE DISEASE

Symptoms:

Water-soaked lesions appear on the leaves and the cotyledons. Infected plants with anthracnose rot, fail, and die. Small water-soaked or yellowish spots appear on brown leaves attaining angularity. Those spots later coalesce and cover the entire area of the veins. Infected juveniles can fall. Symptoms occur on mature fruits as mild, round, water soaked lesions.

Causal Organism: *Colletotrichum* sp.

Epidemiology

Humid or rainy weather is essential for infection, dissemination and sporulation of the pathogen. Fungus occurs best between 22 °C and 27 °C.

Integrated Disease Management:

- Fallow the crop rotation,
- Destruction of weeds and infected plant debris.
- Use of healthy seeds to reduce the chances of infection.
- Seed treatment with thiram or carbendazim (2 g/kg seed) is recommended.
- Sprays of zineb/ mancozeb (0.2%) at 7 days interval is effective.

23.3   BACTERIAL DISEASES

23.3.1   BACTERIAL FRUIT BLOTCH

Symptoms:

Characteristic signs are water-soaked, "oily" areas on the underside of the cotyledons, or seedling leaves that frequently overlap the veins with a yellow halo. The polluted areas are drying up and becoming elongated, pointed, necrotic, and black patches. Some seedlings are about to fail and die instantly from bacterial fruit blotch infection. Others can retain the bacterial infection and will not display any symptoms until the fruit has been placed. The lesions appear to be low, dark, and angled on the leaves. Leaf lesions are significant bacterial reservoirs for infection with the fruit. The sign is a dark green mark, or blotch on the growing fruit's upper surface. At first, the blotch can be 0.5 inches in diameter but will grow rapidly within a week to cover the entire fruit surface.

Causal Agent: *Acidovorax avenae subsp. citrulli*

Disease Cycle:

For this pathogen, decayed fruit and infected seeds and cucurbit weeds (e.g., citron) are hosts.

Epidemiology:

The greenhouse is an excellent environment for BFB (bacterial fruit blotch) distribution. High temperatures and humidity are requisites for development of BFB. Many seed production greenhouses spray top water, which helps to splash pathogen dispersal to other plants. Transplant irrigation at the bottom will help monitor BFB. The movement of contaminated plants into the field from the greenhouse can have significant consequences. Once in the field, high temperature, humidity, and wet weather favors spread of the disease.

Integrated Disease Management:

Cultural Control:

- In a seedling grow-out check, buy watermelon seeds that have been tested as negative for bacterial fruit blotches. Watch for signs of transplantations.
- In the greenhouse, begin distinguishing triploid seedless varieties and seed lots from diploid seeded varieties.
- Diminish physical contact with seedlings.
- Extract infected plants and treat them immediately for blotching of bacterial fruits. If transplants confirm getting BFB, kill all transplants from that lot of seeds. Transplants from other varieties should not be planted in the field at the same location.

Chemical Control:

- Copper hydroxide (54%) @ 700 g added immediately after 7- to 10-day planting.

23.4 VIRAL DISEASES

23.4.1 MOSAIC DISEASE

Introduction/Economic Importance:

Mosaic disease is prevalent in cucurbits growing areas of the world and causes severe yield losses. During 1999, epidemic of this disease was prevalent on muskmelon, cucumber, and squash in Punjab and in adjoining states causing huge losses to the crops (Anonymous, 1999). Mosaic disease has become potential threat to the cultivation of watermelon/*mateera* in western Rajasthan and other watermelon growing states of India. It is recorded from 14.28% to 35.0% disease incidence in Bikaner and nearby areas of Rajasthan.

Symptoms:

Infected plants showed characteristic symptoms of dry, foliar yellow mosaic or mottling with chlorotic patches, development of vein clearing, reduction of leaf size and stunting of the plants. Systemic symptoms consist of downward leaf curling or rolling and mosaic. Disease incidence is found low in this crop due to unfavorable environmental conditions.

FIGURE 23.2   Mosaic disease in watermelon.

Causal Organism: Cucumber mosaic virus (CMV)

The cucumber mosaic virus (CMV) causes this illness. In most host plants, CMV creates a systemic infection with the older plant tissues that formed rarely before the virus became infected.

Disease Cycle:

The virus is known to infect various plants, and cultivated crops serve as a virus reservoir for transmission and other crop spread. The virus is spread by aphid vectors in a nonpersistent way. It is mainly transmitted by aphids, but also mechanically and to a lesser degree in seed. CMV is easily transferable in a watermelon field by sap borne on hands, clothing, and weeding or turning of vines. Acquisition of the virus occurs through aphid instars. Once they begin studying an infected plant, the virus is absorbed by aphids within 10 s. Aphids will spread the virus to other plants in less than a minute. Aphids lose the ability to transmit CMV after about 2 min and completely lose the ability after 2 h.

Transmission: Aphid

The incidence of this disease varies depending upon the genotype, prevalence of vectors, virus strains, and prevailing environmental conditions.

Integrated Disease Management:

Cultural Control:

- Field sanitation and clean cultivation: Eliminate contaminated plants as soon as virus symptoms are observed to avoid or mitigate thrip spread.
- Destruction of weed hosts.
- Use of virus free seeds.
- The use of reflective mulch decreases aphid infection in the early season and provides an additional 2–4 weeks of virus-free cucurbit area. The reflective mulch ceases to be effective deterrent once the plants cover the material.

Chemical Control:

- Imidacloprid (0.05%) was the most effective as seed treatment + one foliar spray for mosaic disease reduction due to reducing insect vector-aphid transmission (Anon., 2015).
- Spray acephate seedlings (0.15%) or monocrotophos (0.1%) produce good results.

- Chemical spray followed by neem seed extract (2%) is also effective in insecticidal rotation.

Host Resistance:

- Among 13 watermelon varieties/genotypes, one variety "Asahi Yamato" showed resistant and 10 genotypes/varieties was categorized as moderately resistant under hot arid conditions in western Rajasthan (Maheshwari et al., 2013).

23.4.2   BUD NECROSIS

Introduction/Economic Importance:

It is also a serious problem and causes yield losses in Bikaner district and also in western Rajasthan of watermelon/*mateera* growing areas. It is also a viral disease.

Symptoms:

Symptoms include crouching of the stems, mottling, yellowing, necrotic lines on the vines, reduced internodes, upright roots, necrosis, and bud dieback.

FIGURE 23.3   Bud necrosis in watermelon.

Causal Organism:

It is caused by the watermelon strain of tomato spotted wilt tospovirus (TSWV-W).

Transmission: Thrips

Disease Cycle:

Diseases occur all year round but the occurrence of the disease is typically higher during dry and hot periods when populations of thrips are increasingly growing. The virus is transmitted in a persistent (propagative) manner by thrips species (After several minutes to hours of feeding, the vector can acquire and spread the virus: the virus replicates within the vector) young thrips (larval stage) acquire the virus and adult thrips spread the virus from plant to plant during feeding. The virus is not transmitted by seed.

Integrated Disease Management:

- Using the seedlings free from viruses.
- Remove infected plants as soon as virus symptoms have been detected to prevent / minimize thrip spread of the virus.
- After the crop season eliminate field litter, weeds and other sources of thrips.
- Ploughing and harrowing can help reduce the soil's vector population.
- Using plastic mulch, blue sticky traps to monitor the population of thrips.

Chemical Control:

- Sprinkle monocrotophos/acephate (0.15%) at intervals of fortnight after transplantation before flowering.
- Chemical spray followed by neem seed extract (2%) is also effective in insecticidal rotation.
- Imidacloprid (confidor) spraying at a rate of 0.5 mL/L of water to control insect sucking.

KEYWORDS

- **watermelon**
- **diseases**
- **symptomatology**
- **epidemiology**
- **management**

REFERENCES

Anonymous, 1999. Annual Report. Department of Plant Pathology, Punjab Agricultural University, Ludhiana.

Anonymous, 2003. Annual Report (2002–2003), Central Institute for Arid Horticulture (ICAR), Bikaner, Rajasthan.

Anonymous, 2015. Annual Report (2014–2015), ICAR—Central Institute for Arid Horticulture, Bikaner, Rajasthan.

Bharath, B.G., Lokesh, S., Rai, V. R., Prakash, H. S., Yashovarma, B. and Shetty, H. S. (2005). Role of foliar spray in the infection biology and management of fungal diseases of watermelon [*Citrullus lanatus* (Thunb.) Matsum and Nakai]. *World J. Agric. Sci.* 1 (2): 105–108.

Bhargava, A. K. and Singh, R. D. 1985. Comparative study of Alternaria blight, losses and causal organisms of cucurbits in Rajasthan. *Indian J. Mycol. Plant Pathol.* 15: 150–154.

Goreta, S., Perica, G., Dumicic, G., Bucan, L. and Zanic, K. 2005. Growth and yield of watermelon on polyethylene mulch with different spacings and nitrogen rates. *Am. J. Hort. Sci.* 40: 366–369.

Gupta, A. K., Sharma, R. C. and Sharma, S. 2001. Fungal diseases of cucurbits. In: Diseases of Vegetable Crops (Eds., Gupta, V. K. and Paul, Y. S.), Kalyani Publisher, New Delhi, p. 71–86.

Huh, Y.C., Solmaz, I. and Sari, N. 2008. Morphological characterization of Korean and Turkish watermelon germplasm. 1st Cucurbitaceae 2008, Proceedings of the IXth EUCARPIA Meeting on Genetics and Breeding of Cucurbitaceae (Ed., Pitrat, M.), INRA, Avignon (France), May 21st–24th.

Khoury, W. El. and Makkouk, K. (2010). Integrated plant disease management in developing countries. *J. Plant Pathol.* 92 (4): 35–42.

Maheshwari, S. K., Choudhary, B. R., Singh, D., Sharma, B. D. and Sharma, S. K. (2013). Screening of watermelon genotypes for resistance to mosaic disease under hot arid conditions of Rajasthan. *Indian J. Arid Hortic.* 8: 16–17

Maheshwari, S. K., Choudhary, B. R., Singh, D., Sharma, B. D. and Sharma, S. K. (2015). Management of *Alternaria* leaf blight of *Mateera* (watermelon) by fungicides, bio-agents and botanical. *Indian J. Arid Hortic.* 10: 64–67.

Prasada, R. and Khandelwal, G. L. (1970). Taxonomy, physiology and control of *Alternaria* leaf spot of watermelon (*Cucumis lanatus*). *Indian Phytopathol.* 23: 32–36.

Prasada, R., Khandelwal, G. L. and Jain, J. P. (1973). Epidemiology, forecasting and control of *Alternaria* blight of cucurbits. *Proceedings of the Indian National Science Academy.* 37: 301–308.

Pushpa, K., Borkar, G. M., Patil, D. V. and Kamble, P. (1999). Studies on seed-borne pathogens of pumpkin, cucumber, watermelon and muskmelon. *J. Soils Crops* 9: 234–238.

Wei, S. Q., Zhong, Y., Ma, Z. T. and Jiang, H. (1991). A survey on watermelon diseases in the Northern China. *China Fruits* 1: 36–37.

INDEX

For Product Safety Concerns and Information please contact our EU
representative GPSR@taylorandfrancis.com
Taylor & Francis Verlag GmbH, Kaufingerstraße 24, 80331 München, Germany

www.ingramcontent.com/pod-product-compliance
Lightning Source LLC
Chambersburg PA
CBHW060424220326
41598CB00021BA/2284

9 781774 639429